# HANDBOOK OF CRITICAL INCIDENT ANALYSIS

# HANDBOOK OF CRITICAL INCIDENT ANALYSIS

**Richard W. Schwester, Editor**

Routledge
Taylor & Francis Group
LONDON AND NEW YORK

First published 2012 by M.E. Sharpe

Published 2015 by Routledge
2 Park Square, Milton Park, Abingdon, Oxon OX14 4RN
711 Third Avenue, New York, NY 10017, USA

*Routledge is an imprint of the Taylor & Francis Group, an informa business*



**Library of Congress Cataloging-in-Publication Data**

Handbook of critical incident analysis / edited by Richard W. Schwester.
p. cm.
Includes bibliographical references and index.
ISBN 978-0-7656-2724-7 (hbk. : alk. paper)
1. Emergency management—Social aspects. 2. Crisis management—Social aspects. 3. Disasters—Social aspects. 4. Accidents—Social aspects. I. Schwester, Richard Wilmot, 1977–

HV551.2.H35 2012
363.34—dc23 2011017395

ISBN 13: 9780765627247 (hbk)

# Table of Contents

# Introduction

RICHARD W. SCHWESTER

The *Handbook of Critical Incident Analysis* brings together academics studying extraordinary incidents that dominate the public discourse and challenge our sense of security. These incidents have the potential to define periods of time. The scope of these incidents is vast—including, but not limited to, school and university shootings, public health crises, natural disasters, and acts of terror and mass violence. Through this handbook, we strive for a better understanding of these incidents. This handbook will be foundational to the field of critical incident analysis—a field that is grounded in several disciplines, including public administration, emergency management, planning, psychology, and social psychology.

Part I of this handbook examines the evolution of critical incident analysis as a field of study, how researchers and practitioners might analyze critical incidents, and includes discussions of what might constitute a "critical incident." More specifically, in Chapter 1, Ochberg takes the reader through a personal journey of critical incident involvement, discussing specific incidents, his role as an intervener, and interactions with key people who have shaped this field of study. In Chapter 2, Kirby introduces a conceptual model designed for critical incident analysis. This model is intended to enhance the ability to understand, mitigate, and negotiate critical incidents. The model consists of three essential elements: the event, the political arena, and the authorized interveners, each of which is influenced by the media and the context of the critical incident. An overview of model components is discussed, which helps establish a common lens for rich and meaningful analyses. Finally, in Chapter 3, Schwester, Horning, and Dank attempt to identify the factors that might define a critical incident.

Part II emphasizes theoretical, disciplinary, and policy perspectives that shape critical incident analysis. In Chapter 4, Cinti examines public health incidents and discusses the recent H1N1 pandemic in the context of the model Kirby lays out in Chapter 2. In Chapter 5, Hawdon and Ryan provide a theoretical basis for explaining certain social responses to critical incidents, whereby emphasis is placed on social networks. They also examine the impact of critical incidents on multiple dimensions of communities—namely community space, social structure, and sentiment. In Chapter 6, Till, Durak, and Goodman present an engineering perspective, discussing the value of using simulation modeling—referred to as agent-based modeling—as a means of preparing first responders for a multitude of incidents. Through simulated incidents, first responders can hone their decision-making skills and test reactive and proactive measures designed to deal with a specific incident. The authors provide results from their own simulation dealing with a chlorine gas leak and a hypothetical emergency evacuation plan. In Chapter 7, Groner deals with the prospect of predicting critical incidents, examining the quantitative and qualitative research methods used to generate empirical models. He notes the inherent challenges to predicting when and where a critical incident might occur and how changes in different factors might impact the magnitude of such incidents. In Chapter 8, Sainato discusses how we make sense of massive repositories of data through a technique known as data mining. In short, data mining uses vast computer technologies to sort through, categorize, and analyze massive datasets. In Chapter 9, Stern examines the policy implications of critical incidents. He provides examples of policies enacted in the aftermath of specific incidents for the purpose of preventing future incidents. However, Stern points out that these policy changes are typically "suboptimal"—representing only small, incremental changes to previous policies. This incremental change results from the nature of our political system, according to Stern, which is hampered by interest group pressures and institutional inertia that impedes wholesale policy changes. In Chapter 10, Chandrasekhar describes how planning addresses natural disasters. The first section of this chapter examines the role of planning in the context of disaster recovery. The second section discusses the impact of disasters on human settlements. This chapter concludes with a description of planning activities relevant to post-disaster recovery. In Chapter 11, Corbett examines the National Institute of Standards and Technology and its investigation of the World Trade Center attacks of September 11, 2001.

Part III focuses on critical incident management. In Chapter 12, Henderson discusses how the individual behavior of first responders—or street-level bureaucrats—is shaped by a multitude of factors. Most notably, Henderson discusses the role of occupational culture and client characteristics in shaping first responder discretion. In Chapter 13, Col presents a disaster mitigation success story as she chronicles the mitigation efforts taken by Qinglong County officials (China) before and during the Great Tangshan Earthquake incident of 1976. This success story provides public administrators and emergency managers with lessons that can be applied to a wide range of incidents. In Chapter 14, Oksanen, Räsänen, and Nurmi examine school shootings from a Scandinavian perspective, focusing on local residents' experiences of school shootings in Finland. In Chapter 15, Levy discusses trauma mitigation as it relates to victims of college and university incidents. She analyzes Antonovsky's Sense of Coherence model, which emphasizes a "health" orientation rather than a "disease" orientation to dealing with stressors associated with critical incidents. Levy further discusses the concepts and practices of psychological first aid, and offers an alternative model for trauma intervention—the so-called Open Door model. In Chapter 16, Maurer and Mosser discuss first responder preparedness. They begin by defining the concept and identifying factors of preparedness. They then examine organizational preparedness as it relates to responding to acts of terrorism and natural disasters. The state of Florida is used as the backdrop for identifying and analyzing six factors of incident preparedness, which include communications, equipment, personnel, planning, procedures, and training. In Chapter 17, Colvin examines the Washington, DC Metropolitan Police Department (MPD) and its efforts to address the needs of a community often overlooked by police organizations, such as the lesbian, gay, bisexual, and transgender (LGBT) community. Colvin looks at the evolution of the MPD as an eventual ally of the LGBT community stemming from the aftermath of three critical incidents. These incidents led to the creation of the Gay and Lesbian Liaison Unit (GLLU) within the MDP. Colvin examines measurable outcomes that resulted from the formation of the GLLU and discusses the broader applicability of this microlevel case study. In Chapter 18, Gorelick examines the roles of "experts" in the management of critical incidents, distinguishing among seven types of media experts. He further touches upon the challenges of managing critical incidents in the digital age. The book concludes with Benton's Chapter 19, which explores the ethical dilemmas most inherent to critical incidents. The author explores four specific types of ethical dilemmas that emerge during critical incidents: disclosure and transparency, planning and preparation, prioritizing rescue and treatment, and the exceptional circumstances rationale for exceptional interventions. The chapter concludes with a discussion of how an "ordinary" incident could be made "critical" based on the perceived ethics of certain decisions at the time of a crisis.

# PART I

# UNDERSTANDING THE FIELD OF CRITICAL INCIDENT ANALYSIS

CHAPTER 1

# The Emergence of Critical Incident Analysis as a Field of Study

FRANK OCHBERG

This *Handbook of Critical Incident Analysis* gathers scholars from remarkably different disciplines who share a common quest. We recognize that certain newsworthy events explode into public awareness, dominate discourse, challenge our sense of equanimity, and have the potential to live on as icons of an era. We seek a better understanding of these episodes so that we can define their elements, recognize their antecedents, anticipate their consequences, and gather evidence for scholars and interveners who will confront future incidents. Our work follows other organized efforts to analyze incidents, but emphasizes the creation of an academic enterprise, rather than the critique of crisis management from the perspective of public officials. Our quest, therefore, is the field of critical incident analysis itself. To create this interdisciplinary discipline we need a compendium of cases; an archive of data; an approach to analysis that includes science, history, and other tools of the academic trades; and a common language. Along the way, we must set boundaries and limits, but understand relevant issues that abut on those perimeter walls. We need to recognize and respect differences among scholars who represent conflicting perspectives. For example, one of our founders, a natural scientist who designed and taught a course in critical incident analysis to honors undergraduates, limits his purview to destructive episodes, whether natural disasters (the San Francisco earthquake), colossal mistakes (the Exxon Valdez oil spill), or intentional acts (the Oklahoma City bombing). This distinguished professor argues forcefully and persuasively for a model of community disruption that reflects, in the aggregate, what post-traumatic stress disorder (PTSD) defines in the individual: a diagnosable disorder of thought, feeling, and function that can be studied and ameliorated. But other scholars, including several chapter authors of this handbook, contend that positive occurrences, such as the first manned mission to the moon (those memorable steps on lunar soil and the view of Earth as a beautiful but vulnerable celestial orb), fit the definition of "critical incident" and should be included in our list of cases for contemplation. Good or bad, these advocates argue, an incident is critical when it attracts widespread attention, changes the way we think or act, and lives on in the collective consciousness.

Since we began this journey in the early 1990s, lively debate has engaged and motivated our participants. My role has varied from instigator to facilitator to avuncular occupant of a seat at the table as others ably lead. I am very, very grateful to all who have contributed, particularly the W.A. Dart family, who saw fit to fund this new field from its inception at MSU; through incarnations at the University of Virginia (where it endures as the Critical Incident Analysis Group) and the National Defense University (where it is called the National Center for Critical Incident Analysis); to our home, the Academy for Critical Incident Analysis at John Jay College of Criminal Justice. Thank you. And no more commercials.

My plan for this chapter is to walk through the past two decades of direct involvement in critical incident activity, focusing on the issues rather than on the actors or the institutions that received Dart Foundation grants. Much has emerged, but the shape of this field remains inchoate. This handbook is the first attempt to define the academic science of Critical Incident Analysis. Chapter 1 initiates discussion, but does not resolve the following fundamental questions:

- What is a critical incident?
- How do we best analyze such incidents?
- What do we hope to achieve through critical incident analysis?

Let us return to 1991, where the journey began.

## SADDAM HUSSEIN CALLS A MEETING

Remember the term "human shields"? It was during the George H.W. Bush administration that Iraq invaded Kuwait and we prepared to retaliate with Desert Storm. To make us think twice about an air attack, Saddam captured Americans and other westerners who worked in the oil industry and held them at various locations so that we would not bomb those locations. Many of these hostages were married to Islamic women who were natives of Kuwait and Iraq. The women evacuated to the United States, congregating on the East Coast. I received a call from a journalist friend who described the hardship of these women, who were filled with fear, alien, and alone, and in need of psychological and social support.

I also heard from a mental health colleague who wanted to mobilize PTSD experts to provide pro bono therapy for these wives of human shields. We formed a unit of volunteers, most of whom were founders of the International Society for Traumatic Stress Studies, and did just that. We called ourselves USA-Give. This activity caused us to convene in weekly phone calls to manage the new charity.

The work went well, but then an extraordinary event occurred. One of the wives decided, on her own, to seek an audience with Saddam Hussein to implore him to release her husband. Another wife learned of this and asked to join. A movement began, with prominent Americans (Andrew Young, John Connelly), interested in coleading the mission. I found an advisor in the Bush administration who agreed to provide anonymous guidance to me and the group about the timing of the visit. She was of high rank in the U.S. State Department, and she said, "If my husband were a hostage I would do this. But we cannot appear to be involved." The very first donation from Dart that I secured as foundation representative supported this mission, once Saddam Hussein agreed to meet with the wives of his human shields. I made sure that the journalist who alerted me to the plight of the wives had a place on the plane to Baghdad.

It worked.

The husbands of these wives were released as a gesture of humanity, and the other human shields were granted freedom several weeks later during the Christmas holiday season.

This "incident" was never critical as a diplomatic, political, or historic occurrence. Other events dominated the public agenda and shaped the flow of history. But it mattered to a small circle of therapists, journalists, and unofficial interveners. We helped Saddam Hussein call a meeting, and we celebrated a significant victory.

## WACO BURNS—AND SO DO THOSE WHO EXPECTED BETTER FROM THE FBI

Fast-forward a few years to the spring of 1993 and the Branch Davidian standoff at Waco, Texas. A cult leader named David Koresh created a bizarre community of followers—men, women, and children—who revered him as a prophet, permitted him sexual access to married women and underage girls, and lived as a collective in an armed compound. Because their weapons were illegal, a field office team from the U.S. Bureau of Alcohol, Tobacco, and Firearms attempted an assault and arrest. In retrospect, this was a badly botched job. Agents were killed, the compound became a barricaded fortress, and the FBI was deployed to intervene. Like the 444-day American embassy siege in Tehran 1979, this 51-day televised incident became popular theater, started a national conversation, and ended in a deadly encounter.

I had a minor role. An ABC *Nightline* producer asked me to participate in a Ted Koppel interview on the question of whether the FBI was trying to influence David Koresh through its press conferences. To better prepare for that interview, I called the former chief of the behavioral science unit of the FBI, who put me in touch with the incident commander, the deputy assistant director of the FBI. I was told that the FBI did not use media to influence Koresh, but I was asked to mention a few things in my *Nightline* interview that might advance a theme in the current negotiation. Interesting paradox! We do not send signals through television interviews, but we will try it this time.

The interview was taped, but it never ran. The FBI attacked the compound that very night and Waco burned. The losses were terrible. Insiders knew about conflict between advocates for prolonged negotiation and advocates for an aggressive end to a national spectacle. The attorney general was new on the job and, in the opinion of all those I knew and trusted as insiders, was misled into believing that Koresh posed an immediate threat to his followers when actually he was injured and was focused on his idiosyncratic analysis of the book of Revelation, which was central to his terms for surrender to authorities.

Possibly because of this limited personal involvement, possibly because of my optimism from the mission to free human shields, I was delighted to receive separate requests from a retired FBI leader and an investigative journalist to help each analyze the tragedy at Waco. One wanted to prepare a white paper for the FBI director, constructively criticizing the erroneous use of force. The other was tasked by a prestigious journalism review to criticize the reporting of Waco. I decided to invite both to MSU, to ask the Dart Foundation to donate a small amount to cover costs, and to invite an eclectic group of faculty to attend a private discussion of emerging ideas.

Both individuals eventually wrote their critiques. Our faculty group never made a public report, but we did conduct a conversation at a television studio that resulted in a training tape for the FBI Academy. And we generated several ideas for seminars, research papers, and college courses.

Waco became a major national incident, with a national commission charged to investigate and report to the government and the people. A dissent was authored by a Harvard University professor who is a psychiatrist and a lawyer. He came to MSU to address a large audience. The sponsor of this event was the new Critical Incident Analysis Group, and the host was the dean of the College of Natural Science. The audience was an interesting mix of students, faculty, and members of the Michigan Militia.

Waco was a critical incident because children were burned to death as a large television audience watched in horror. Waco was critical because the FBI was the crisis management arm of the new administration, and the attorney general and the president were clearly responsible for approving the "terms of engagement." But Waco was also critical because the nation was experiencing a strange phenomenon called "the patriot movement" or the "citizen militia." This was not entirely new, but it took shape in that era with military uniforms, insignia of rank, training camps in rural counties, and leaders who were preachers and gun store owners and frequent guests on *Nightline.* Michigan had one of the largest militias in the nation. They were incensed by the slaughter at Waco, and they wanted government agents held accountable.

I found myself more concerned about the rise of the militia than about radical cults like the Davidians. Would the movement grow into a powerful political force? Could it undermine legitimate law enforcement as a rural vigilante operation? And would the FBI find itself drawn into a series of battles with this new adversary, further wounding its agents and its reputation?

The losses at Waco were profound, but something was growing out of the disaster that could be even more disastrous: a breach of trust between government and the governed.

## DOMESTIC DIPLOMACY

To address this dire consequence of a bungled federal operation, my ex-FBI colleague and I decided to search for allies in the militia movement. We reasoned that men who cared about the U.S. Constitution and who claimed to abhor violence would appreciate respect and recognition from an odd couple of retired federal employees.

I asked a rural Michigan newspaper editor about approachable militia leaders, and he arranged a meeting in his office. The man who agreed to talk with me was, from 9 to 5 on workdays, a sanitary engineer with a county health department. After hours, he trained with the Wolverine Militia. We eyed each other warily, but gradually found common ground, deploring the tactics at Waco and regretting the shooting of Vicki Weaver at Ruby Ridge in Idaho in a previous incident involving a family of white separatists and an FBI marksman who shot and killed a woman during a tense negotiation. Ultimately we agreed that no harm could come from arranging a larger pow-wow. The next meeting would involve my friend from the FBI, a professor from MSU, and leaders of the militia.

We had that summit at a back room in a lodge near Lake Michigan. All we did was introduce ourselves in long, personal biographies. We talked about parents and values and ways we felt called upon to help others. We ate a meal. And we recognized that we all wanted the FBI to succeed without bloodshed when future incidents arose.

At the same time that this link was forming, the FBI reorganized, in part due to its analysis of Waco. They established a unit based at the Quantico training academy called the Critical Incident Response Group (CIRG). No longer would a regional office run an operation that had national implications, a national audience, and a complicated set of diplomatic and tactical options. The CIRG would take charge.

Our MSU Critical Incident Analysis Group testified in favor of the creation of the CIRG. The newly appointed director of the CIRG visited our classes at MSU. But more significantly, he agreed to establish a "secret hotline" with the militia leaders we identified up north.

This hotline was a form of so-called domestic diplomacy and a direct outgrowth of critical incident analysis. We kept the hotline "warm" rather than "hot" through phone calls and face-to-face meetings as various opportunities arose. For example, a militia leader's wife was arrested at a demonstration, and an MSU representative was able to ensure appropriate attention to her medical needs.

Soon afterward there were two dangerous episodes requiring FBI action. One was the Freemen standoff in Montana and the other involved a radical separatist group advocating a Republic of Texas. In both instances, the hotline was activated and the militia leaders helped with mediation. No lives were lost. The FBI operational commander told me, "That was the best thing that happened on my watch," referring to the relationship we designed, thanks to critical incident analysis after Waco.

## A MODEL EMERGES

During our academic conversations and our operational interventions, we came to realize that three very different aspects of any critical incident required attention. The first was the place and people directly involved in a dire or deadly event. We drew a circle and called this the event (Figure 1.1).

Figure 1.1 **The Event**

The event was bound by time and space and constituted a theater of action; lives were threatened, demands were made, destructive forces were unleashed, oil was spilled, wildlife was destroyed, and buildings collapsed. At some point, the damage was done and an aftermath of suffering began. We argued about the time that should be assigned to this circle, since the arena of action in which critical incidents occur can feel prolonged to those at the eye of the storm. A hurricane hits and a city is devastated. The incident doesn't end when the wind stops blowing or when the floodwater recedes. So when does an incident end, and how far does the boundary of an incident extend? Due to the unclear perimeter in time and place, we eventually changed this circle to an amoeba shape. This blob symbolizes the event in our current model of critical incident analysis (Figure 1.2).

Within this arena there are often perpetrators and victims. In an armed cult event such as Waco, the line between perpetrators and victims may be difficult to draw. The twenty children who died were victims in every sense of the word. But what about the men who fired at law enforcement officers and died as martyrs for their radical beliefs? They were not as clearly culpable as the Columbine killers who left a record of their contempt for fellow classmates. The fact that Waco was a long siege and received national television coverage contributed to the impact of the event—beyond Waco, beyond Texas, beyond America. And because those responsible for resolving the crisis needed to consider and possibly employ diplomacy, SWAT techniques, fire safety, cult awareness, and public relations, the event was a complicated incident, indeed (Figure 1.3).

As we, the self-appointed critical incident analysts, thought about such complicated events, we realized that there was a second concept to contemplate and illustrate. This concept referred to the interveners who were responsible for crisis containment, crisis management, and many related functions. In the case of Waco, this group included the ATF, the FBI, local authorities, political leaders at several levels of government, intelligence gatherers, and technical experts—an array of groups and individuals who were not necessarily organized into a coherent whole. We labeled this the bureaucracy and adhocracy. Eventually, we turned this into a triangle, symbolizing a hierarchy with supervisors and subordinates who were deployed according to policy and procedure (Figure 1.4).

This model is a vast oversimplification. There is no coordinated structure to manage a critical incident according to plan. In some cases, an incident is resolved in a way that meets with relative approval from the citizenry. More often, there is a blame game fueled by litigants, pundits, and media. The public perception of competence on the part of those

Figure 1.2 **The Blob**

Figure 1.3 **The Waco Incident as Represented in the Blob**

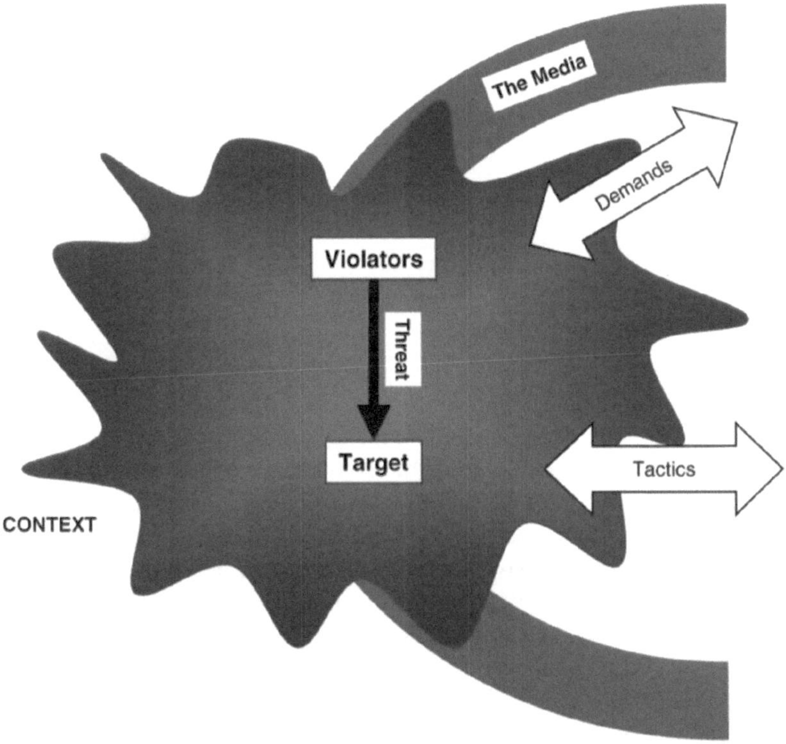


in the triangle mitigates the larger impact of the tragedy in the blob. This emerged as a very important conclusion and is a fundamental feature of the model.

To analyze a critical incident, we must consider not only the forces unleashed at ground zero (the blob), but also the behavior of those we hold responsible for safety before, during, and after the event (the triangle; Figure 1.5). The critical incident is critical when the interveners bungle the job, and it is even more critical when those interveners are agents of the nation rather than of a county or state.

Figure 1.4 **The Bureaucracy and Adhocracy**

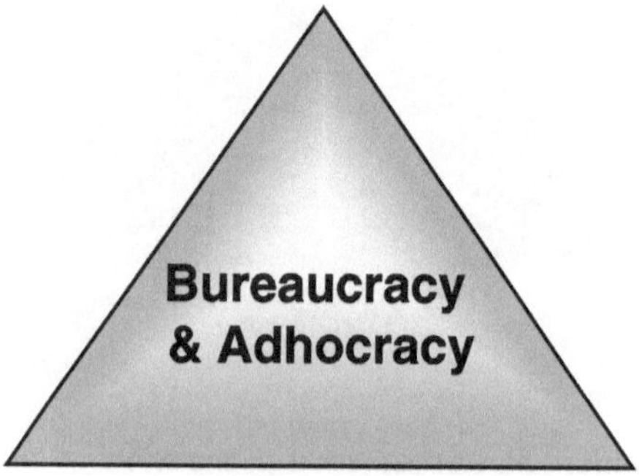


Figure 1.5 **Critical Incident Analysis Model, Part 1**

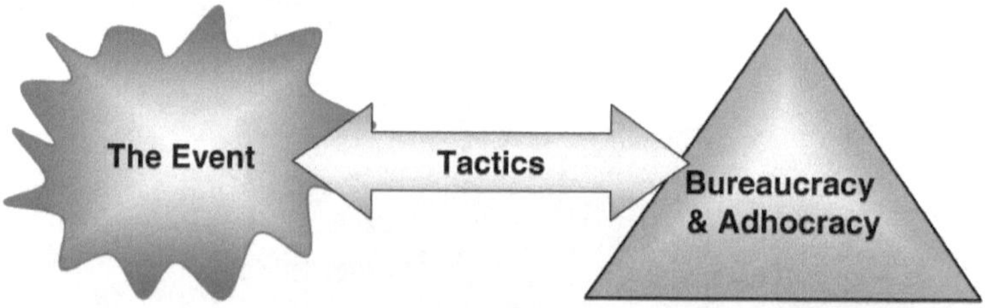


There is a third arena, and it is the most important of all. We eventually made this a box and called it the political arena (Figure 1.6). This box represents the government and the governed. It also contains the larger community that resonates to the events, identifies with the victims, and cares deeply about the competence of the responsible authorities. The political arena is rife with preexisting social fault lines. These are the lines that divide race, class, partisan party members, and other factions of the populace.

Some communities are cohesive and withstand trauma and tragedy well. Others are brittle and vulnerable; easily moved to recrimination, regression, riot, or, at the extreme, genocide, by a combination of combustible conditions instigated by an incident.

My greatest fear is that a mismanaged event will alienate government and drive open the fault lines that divide factions. Terrorists plot ways to accomplish this destruction of democracy. Demagogues smell such opportunities and rouse the rabble. Certain media outlets prefer to add fuel to these fires, while the best professional journalists (a diminishing breed) offer fact and fair discussion as a remedy for bloviation. The media is usually part of the incident—informing, connecting, and influencing the political arena and the other arenas of action (Figure 1.7).

The model is elaborated in several chapters of this handbook and is still evolving. Models should simplify complicated concepts, not add to complexity. But our model is currently a bit too cluttered, awaiting the Einstein among us who will discover a formula like $E = MC^2$. Perhaps the Intensity (I) of the blob times the Efficacy (E) of the triangle

Figure 1.6 **The Political Arena**

Figure 1.7 **Critical Incident Analysis Model, Part 2: The Best Model We Have So Far**

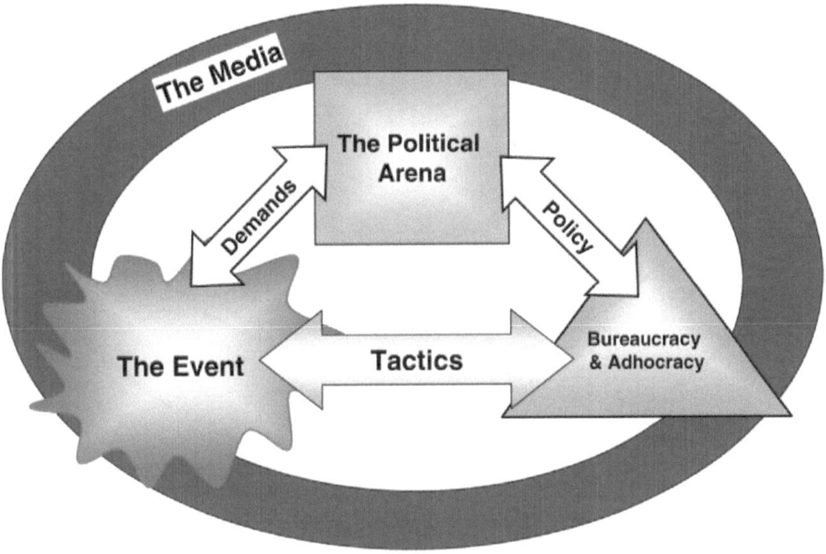


equals the Quality (Q) of the polity in the box. An interdisciplinary science of incident analysis may, eventually, find appropriate metrics for elusive concepts.

An incident, then, is all of the above. It is the time and place in which a profound event occurs, including those individuals directly impacted. It is the people and the agencies that rush to rescue, including those who are responsible for preventing and managing such matters. It is the body politic that is affected, for better or worse, as this event unfolds. Other models portray the aftermath of an incident and speak to such concepts as impact, consequence, and long-term meaning (Figure 1.8).

We are still struggling with these concepts. One of our number, a journalist and author, sees little use in predicting the historic significance of unfolding events. Larger events, such as world war, are significant, he says, but a local incident such as the Kent State massacre makes no lasting difference in the larger theater of human events. Others see the shooting

Figure 1.8 **Critical Incident: The Aftermath**

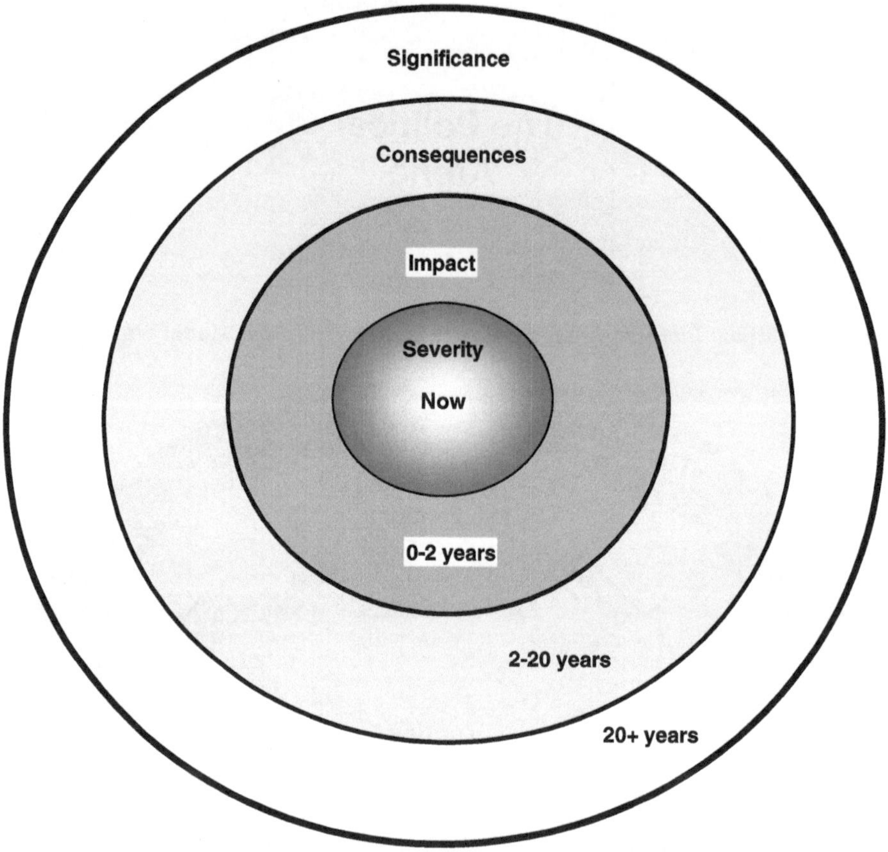


of four students by the U.S. National Guard at Kent State as an enduring icon of an era, emblematic of American conflict about more than war policy in the Age of Aquarius. It was, they assert, about hawks and doves and innocence and arrogance. Had there been no press photographer to capture the grief and horror of a campus coed, I would not remember Kent State, viscerally, the way I do. I'm not the only one of my generation who feels this way. Critical incidents have significance when they elicit common intense recollection.

So we struggle to define and describe how these incidents endure through time and what they mean, long after the dust of destruction has settled.

## A COLLEGE COURSE IN CRITICAL INCIDENT ANALYSIS

Three professors agreed to teach a seminar for two dozen honor students, all undergraduates. One was a paleobotanist, a Renaissance man who could teach any science subject as well as humanities. One was a journalist and a lawyer. I was the third. Our scientist led the team, organized the syllabus, and set a high standard, using every session to demand critical thinking and conclusion based on evidence.

Studies began with earthquakes, floods, and fires. Students learned fundamental facts about tectonic plates, hydraulics, and combustion. They read historic cases and came to understand why some communities were far more vulnerable than others to natural disaster. They compared the journalism of early America to modern times and saw similarities and differences in the style of telling trauma stories. They knew the difference between traumatic stress and grief. Later they learned the concept of human error and the meaning of tort and criminal negligence. Chernobyl, the *Challenger,* and Exxon *Valdez* were contrasted in terms of destructive force, numbers affected, and meaning as markers along a trail of technological advance, marked with pitfalls of carelessness and hubris. During a later phase of study, when incidents of deliberate human design were examined, the Oklahoma City bombing occurred. There were reckless media reports that Middle-Eastern terrorists were responsible, but this class of college students knew better. First, they knew not to rush to rash conclusions. They also knew from previous case analysis that a truck bomb made of fertilizer was far more likely the work of a homegrown dissident. As facts emerged, the students followed the story closely and found that they were knowledgeable participants in a national conversation.

This class and subsequent classes are described in detail in other chapters. We all learned as we literally charted a new course. We learned that the case study method engages a college class. We learned that students eagerly compare current events to headline episodes from a century ago. We found that organizing incidents into three categories—natural disaster, human error, and deliberate design—works well for contrast and comparison. The teachers enjoyed the work as much as the students, but it was hard work for the lead professor. He thought about making it a survey course for hundreds of students, but decided that the intimate interaction in a seminar room was ideal. He loved the chance to think out loud with gifted students.

## OKLAHOMA CITY AND COLUMBINE

The bombing of the Alfred P. Murrah Federal Building by Timothy McVeigh and Terry Nichols was a notable and terrible tragedy. Hundreds were killed, more were maimed, and thousands were directly and deeply affected as witnesses and next-of-kin. The building stood for days with its face ripped off, staring at us from the television screen and the covers of news magazines. If "they" could strike at a public building in the heartland, "they" could strike anywhere. It took a while to learn who "they" were, and longer to sort out mentality and motives. A criminal trial helped with critical incident analysis, and a death penalty closed the legal case.

Our group found itself involved again. That same reporter who traveled to Baghdad with wives of human shields called me from Oklahoma. He was assigned to the story, and he said, "This is different, Frank. I've seen troubled reporters before, but not like this. Maybe it's all the dead children." So I found funds to send our journalism professor down to live with the media on the scene and report back to the class. This stimulated some other initiatives related to incident analysis: training reporters about PTSD; helping reporters maintain their own mental health when covering prolonged, intense tragedy; teaching news managers the importance of anticipating and treating stress in the workforce. The local newspaper, *The Oklahoman,* won the Dart Award for Excellence in Reporting on Victims of Violence in 1996 for sensitive reporting on the victims of the bombing, and the editor became a member of our critical incident analysis group.

Because the trials of the bombers, McVeigh and Nichols, occurred in Denver, there were many reporters who followed the story from Oklahoma to Colorado. The massacre at Columbine occurred less than two years after the trials in Denver concluded. Soon Columbine and Oklahoma were linked in people's minds, and we "incident analysts" found ourselves comparing the two events.

Eric Harris and Dylan Klebold killed a much smaller number at Columbine than McVeigh and Nichols killed in Oklahoma. There were many more children murdered and maimed in Oklahoma. But Columbine seemed to occupy at least as large a space in the nation's attention, and it left a lingering stain and strain in the suburban community where it occurred. Some thought that the criminal trial of the perpetrators allowed better resolution for the Oklahoma incident. Some thought the venue of a high school horrified every parent, while a federal building was somehow less sacred—more akin to a military target. No scientific study has addressed this question in a definitive way. Critical incident analysis, as a new field of research, may give us tools that will help us predict and prepare for the divided community; the lingering, unrequited outrage in some communities; and the methods that work best to ameliorate problems in the "box" of public repercussion.

## BOMBING THE PUBLIC BUILDING

It struck me that both Columbine high school and the Murrah building were public buildings and, in a way, symbols of democracy. Was there something to be learned about the deliberate destruction of such structures? I thought with horror about an attack on the Capitol and what it would do to the morale of the nation. And I was not the only member of our growing Critical Incident Analysis Group who had this prophetic fear. It was in 2000, after Columbine and before 9/11, that we gathered a group of scholars and public officials on the campus of the University of Virginia to tackle the topic. We included the sergeant-at-arms of the House of Representatives; the widow of a Secret Service agent killed in the Oklahoma City bombing; administrators and investigators from Littleton, Colorado; and a wonderful architecture professor. This professor gave us a history lesson starting in classical Greece and moving through modern times, including the forbidding fortress architecture of totalitarian states and the dignified columns of democratic legislatures. "It is all about openness," declared the Columbine supervisor, as she realized, perhaps for the first time, how important it was to preserve the spirit of an open society while ensuring the latest and best physical security for precious occupants. The sergeant-at-arms agreed.

We didn't belabor the model during that colloquium, but we understood that an informed conversation about an assault on the public building had as much to do with public trust as with target hardening.

## CRITICAL INCIDENTS IN CLASSICAL LITERATURE

As a diversion from the grim work of analyzing current catastrophes, a group of us met one evening to read passages from Shakespeare aloud and to consider history, historic fiction, fact, mythology, and the application of Elizabethan literature to modern traumatic events. We included a Shakespeare professor, the deans of the departments of natural science and of communication arts, a prominent political consultant, and an actress who could elevate our amateur attempts to audition for the Royal Theatre. We read passages of *Richard II* and the assassination scene in *Julius Caesar.* And we read a few pages of *Plutarch* and an entry from the *Encyclopedia Britannica.*

The deans fell asleep, not from boredom, but from overwork. The rest of us carried on for a few hours, enjoying our self-entertainment and noting how the Bard distilled historic episode into heroic action; accentuated elements of character, noble and evil; and left us both exhilarated and aggrieved.

A few of us wondered whether modern incidents have particular resonance when they appear to fit the form of ancient drama. Perhaps we, the news-reading audience, search for evidence of tragic flaws, of falls from grace, and of confirmation of natural order. When we transform the news into Greek or Elizabethan tragedy, we feel not only entertained, but satisfied. So we modify reality into myth. The myth may not be totally false. It is remembered in collective memory because it fits a mold that is shaped to the contours of our culture. This line of thought is not just of interest to the Joseph Campbell collector of universal myths or to the literature professor who keeps the classics alive, but also to the political analyst who recognizes Iago in modern dress—the instigator of evil who knows how to manipulate the dark side of flawed character.

## BIOLOGICAL INCIDENTS AND THE THREAT OF MASS DESTRUCTION

As our cadre and our funding moved to the University of Virginia and then to the National Defense University, we gathered experts and interest in bioterrorism. We had sessions on West Nile virus; SARS; avian flu; an episode of deliberate salmonella contamination; the sarin gas attack by the cult Aum Shinrikyo in the Tokyo subway (not a biological agent, but a related scenario). We heard an amazing lecture by a Soviet defector who ran a bioweapons program. He showed maps of existing storage sites in Russia with the capacity to kill a billion people. These were real, not mythical, weapons of mass destruction.

We learned how microbial agents that kill too quickly, such as the Ebola virus, are not as destructive as agents that live longer in the host and therefore spread further (like the H5N1 [avian] influenza virus). We could see how the three-star air force general who convened our meetings at the National Defense University relished our interest in managing a bioterrorism attack, but had less concern about a pandemic instigated by mother nature. The triangle is better funded and better organized when the threat is of human design; when the American military is mobilized, resources follow.

Because one of our leaders was a biological scientist and a military advisor, the critical incident analysis enterprise found itself heavily invested in bioterrorism preparedness. This enriched the discussion through inclusion of current and former leaders of military medicine, such as the U.S. Army's surgeon general.

Our model of incident analysis needed some adjustment when the threat was a germ and the location of impact was global. The triangle of authorized and potential interveners included doctors, hospital staff, vaccine company executives, and administrators of schools where kids spread infection. We realized that avian flu was more likely and more lethal than biowarfare, and we debated quarantine policy and voluntary self-shielding at home and ways to keep vital services afloat while a plague or pandemic raged.

It was very tempting for incident analysts to become public sector preparedness consultants. In fact, much good work was done by these members of our group who facilitated meetings of state and federal officials to hammer out policies for addressing a biological catastrophe. The model did serve a purpose, reminding crisis managers that a crisis in confidence would probably accompany and augment a public health disaster.

## SEPTEMBER 11

The infamous events of September 11, 2001, live on. Not since the assassination of President Kennedy had there been an event with such national and international impact. Events of this magnitude cause something called flashbulb memory. We remember not only the profound action and imagery, but the mundane: where we were; to whom we spoke; and, in some instances, what we ate. Even now, people swap stories about the personal impact of 9/11. We older folks tell the same sort of stories about November 22. Because of flashbulb memory we have idiosyncratic and universal recollection. These extreme incidents are undoubtedly historic: they last, they affect many, and they serve as milestones for individual and national lifelines.

Because so much attention was paid to the World Trade Center destruction, our group, then at the University of Virginia, focused attention on the trauma and tragedy at the Pentagon. We reviewed eyewitness accounts and heard from a former secretary of the army and a congressman about contingency plans for continuity of government.

Our previous exploration of attacks on public buildings helped us understand the impulse to isolate our leaders during a homeland threat, but also to value the need for openness and closeness when the citizenry is anxious.

During one of the unstructured evenings (unstructured time is vital for these gatherings when issues are intense and individuals are surfeited with formal presentations), the former governor of Virginia told me of a recent visit to New York City. It was early summer, a half year after the al-Qaeda attack: "Those people were still shaken. I've never seen a city so scared," he told me. And he shook his head with sadness and concern. I thought to myself that a politician may know how to read a public mood better than a psychiatrist, and took his observation to heart. He had been tasked by the president to lead a commission and recommend remedies. Our conversation went to the psychological impact of those closest to the carnage, whether or not they lost loved ones. Eventually, we and many others digested data that measured mental illness and trauma syndromes among various populations affected by the tragedy. My field, traumatic stress studies, noted how some individuals experienced post-traumatic growth and others experienced vicarious traumatization. These terms are now reduced to acronyms (PTG and VT, respectively), and there are experts who have written books on each.

We still lack consensus on measures of collective impact. For example, we know how many New Yorkers became depressed or met the markers for PTSD and we know where those with emotional damage were likely to have lived and worked. But we don't know how to validate the governor's perception that the people as a whole were frightened and, if so, how that collective anxiety affected collective behavior.

The nation became patriotic. Flags flew from homes that had never raised a flag before. Presidential popularity rose to near unanimous support. By any measure of cohesion or solidarity in the political arena, we were a nation united. But from a perspective of critical incident analysis, we now have to question whether such cohesion is healthy. In a state of unusual solidarity, we made collective miscalculations which most, if not all, regret. This is not meant as a rebuke of political decision making after 9/11, but rather as a challenge to the model builders in our emerging field: How do we map and measure the essential factors in the box so that we recognize when solidarity is a threat to critical thinking?

September 11 was the quintessential critical incident. It occurred on one day. It killed 3,000. It was a shot heard 'round the world. It was a *casus belli.* There were and are many agents, agencies, and disciplines responsible for prevention, response, and aftermath management. Lessons are still being learned. All contributors and readers of this handbook were affected, deeply affected, by 9/11. We have a mammoth archive of data and will mine that data for years to come. We have personal stories of suffering and heroism. It makes good sense for the Academy of Critical Incident Analysis to exist in the city that experienced this critical incident and has the most to gain from a dispassionate and a passionate analysis of its continuing aftermath.

## JOHN JAY ACADEMY STRATEGY AND VISION

The Academy for Critical Incident Analysis (ACIA) at John Jay College was created because the college president recognized a need, a fit, and a potential for scholarship that would make a difference in our lifetime, and because the faculty had the skills, experience, enthusiasm, and diversity to make a hope into a reality.

The John Jay faculty senate approved a structure and a plan. Dozens of faculty members participated in launching the academy, and more have found roles in teaching and research. At the outset, there was strenuous work to create an archive so that data about critical incidents could be assembled, organized, and accessed by students and scholars. The archive of data is the foundation of the academy. Until now, critical incident analysts have assembled sources for individual use but have never built a common repository. The ACIA vision stands on a library of source material.

This handbook is a fundamental feature of the ACIA strategy. It assembles in one place a compendium of chapters that illustrate the scope and style of various participants. It is meant to inform, explain, stimulate, and teach. It is meant to start discussion in an emerging field, not to end debate with definitive conclusions.

John Jay's *Journal of Critical Incident Analysis,* which preceded publication of the handbook, affords scholars a dynamic forum for exchange of ideas. Controversy initiated by chapters in this volume can find expression in the journal and will inform future editions of the handbook.

There is an online undergraduate course in Critical Incident Analysis that is well attended by enthusiastic students. As students learn from professors, professors will learn from students. A strategy of ACIA is to encourage diverse opinion and "let many flowers bloom." Succeeding generations will absorb incidents in different ways and redefine the meaning of critical incident analysis.

Case conferences that digest specific incidents have occurred since we gathered to critique Waco in 1993. Occasionally, a conference has compared and contrasted several related incidents. Large incidents are too complicated for a brief conference, so under-researched aspects have been chosen for study, such as "Analyzing the Aftermath of the Virginia Tech Shootings" (July 2009 at the campus of Virginia Tech). Despite strong interest in the problems of prevention, in understanding the motives of the killer, the conferees limited attention to the aftermath.

Reports of case conferences have been written in several formats, usually in journalistic style with links to prepared papers and supporting evidence. There is no standard for an ACIA case study. Perhaps a template will be designed so that we can have a shelf of case reports, more or less uniform in style and topics covered. Perhaps there will never be a formula for critical incident analysis. There are books and articles that we have identified among ourselves that illustrate excellence in incident analysis. Examples include Kai Erickson's *Everything in Its Path,* treating the Buffalo Creek disaster, and Dave Cullen's *Columbine,* a *New York Times* best seller. ACIA will attend to the model mentioned here, and, eventually will apply the science of computer modeling to incident analysis. We may never reach a formula that relates the quantities and qualities of blobs, triangles, and boxes. But we will learn more about each and more about the dynamics of critical incidents—incidents that have the power to divide us as a people or to bring us closer in ways that preserve freedom and enhance democracy.

CHAPTER 2

# A Conceptual Model for Critical Incident Analysis

ELIZABETH A. KIRBY

The critical incident analysis model introduced in this chapter is a pictorial representation of the thinking that emerged over the last fifteen years among individuals affiliated with various groups, including the Critical Incident Analysis Group (CIAG) at the University of Virginia and the National Center for Critical Incident Analysis (NCCIA) at the National Defense University (Figure 2.1). In 2004, Dr. Frank Ochberg—a psychiatrist; representative of the Dart Foundation; and founder of the Dart Center for Journalism and Trauma, CIAG; NCCIA; and the Academy for Critical Incident Analysis (ACIA) at John Jay College of Criminal Justice (CUNY)—convened a focus group to explore a collective interest in the complex and often-cruel nature of critical incidents. The group embarked on a mission to create a conceptual model that would guide analysis and meet a growing need to understand, manage, and mitigate such incidents.

A model emerged from this effort that provides a conceptual framework of the essential elements of a critical incident and is illustrative of the dynamic interrelationships that exist among them. The resulting value of the model will ultimately be determined within the context of its use. The capacity of the model to enhance critical thinking processes and to propel the consumer toward new learning and meaning making, along with the importance of clearly defining and categorizing critical incidents, are explored in this chapter.

## DEFINING CRITICAL INCIDENTS

A first step in understanding critical incident analysis is definition. Across disciplines, there are many variants as to what constitutes a critical incident, and definitions of critical incident analysis are many and varied. Multiple definitions and interpretations can be confusing, so an effective approach to defining a critical incident may be one that first includes personal experience, reflection, and transformation of knowledge and meaning (Parker, Webb, and

Figure 2.1 **Model for Critical Incident Analysis**

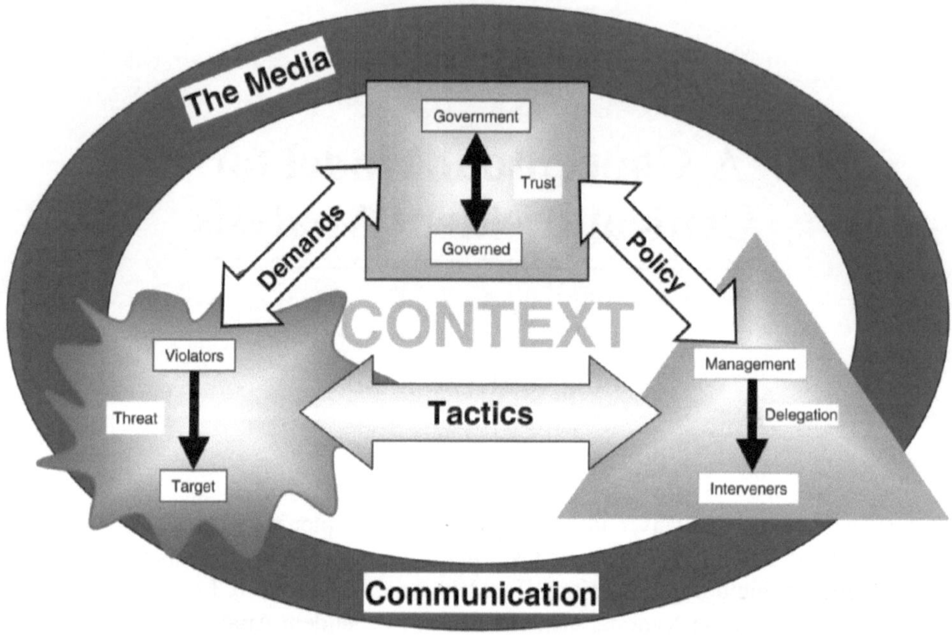


*Source:* Ochberg et al., 2007.

D'Souza 1995). Individuals tend to define a critical incident as a significant event in their lives, such as graduation, marriage, job change, or a traumatic experience. In this way critical incidents are viewed as life-changing events that influence a person's life for a long period of time. Through reflection on the impact of the event, the study of critical incident analysis is first situated within personal experience. This starting point is developmental in nature and marks an incremental step in helping an individual construct a deeper understanding of critical incident analysis. They do this by "making connections, building mental schemata, and developing new concepts from previous understanding" (Roehler and Cantlon 1997).

Unlike individual incidents, critical incidents in a social context are large-scale events that shock communities or nations and may draw global attention as they unfold. They are usually traumatic and tragic, causing death or destruction. Because they are unanticipated and may involve new and unique method or targets, they shake individuals' assumptions and can generate fear or wonder. These incidents "generate extraordinary social concern" and "have the potential to change public opinion, policies, and even social values . . ." (Carter, Kang and Taggart 1999, 6). Through rich analysis and inquiry into these incidents, individuals move from an introspective view of critical incidents toward a new and broader understanding that is collective, or communal, in nature. Feelings anchored by first examining critical incidents in relation to personal experience help to open the door to empathetic understanding of the societal impact and consequences of a major critical incident.

Providing a working definition of critical incident establishes boundaries and removes ambiguity. It narrows and sharpens the focus of the study by intentional grounding without limiting the scope of future development and exploration of the concept. The following definition, attributed to Frank Ochberg, has been adopted for this purpose:

> A critical incident is a relatively brief occurrence involving injury, loss, conflict, discovery or change of significant proportion, usually unscripted and unanticipated, with the potential to alter existing societal norms. Critical incidents are usually traumatic, threatening the bonds of trust that bind communities, but may be positive, initiating historic consequences. (Schwester, Dank, and Horning 2008, 4)

## TYPES OF CRITICAL INCIDENTS

Other ways to reduce ambiguity occur when individuals begin to sort, categorize, and label critical incidents. Carter, Kang, and Taggart (1999) identified three categories of critical incidents: natural disasters, human error or accidents, and purposeful human acts. Natural disasters are those that arise from natural phenomena. Throughout history, these disasters have appeared as earthquakes, tsunamis, hurricanes, and wildfires. Recent examples of natural, uncontrollable events that carried significant consequence include the tsunami in Indonesia (2004); the earthquake in Sichuan, China (2008); and Hurricane Katrina in New Orleans (2005).

Unlike natural disasters, incidents that arise from human complicity, error, and negligence are those that could have been avoided. Examples in this category include aviation, industrial, and environmental crises, such as the nuclear reactor meltdown in Chernobyl (1986); the Bhopal industrial disaster in India (1984); and the crash of the *Hindenburg* (1937).

Incidents that are purposeful and arise from intentional acts create the third category. These acts are deliberate and may endanger one set of values or another. They can be extremely complicated and often involve fervent issues related to religion, politics, and civil liberties. Terrorists attacks, such as those in Oklahoma City in 1995, at the World Trade Center on September 11, 2001, and a series of suicide and car bombs throughout the Middle East in 2008, provide examples of the destructiveness and disregard for human life characteristic of this kind of incident.

Critical incidents do not always have a negative connotation. The first moon landing and the removal of the Berlin Wall are examples of purposeful incidents that are positive. Further, there are multiple examples throughout history of positive consequences and favorable outcomes that emerged after the occurrence of a critical incident.

## CONCEPTUAL MODELS

Critical incidents are multidimensional, and through retrospective analysis we can improve our understanding and management of such events. Given the complexity of these incidents, schematic frameworks can help process information by creating meaningful linkages. Conceptual models and other types of mapping systems "serve as a way for individuals to think explicitly about these connections" (Carnot and Stewart 2006, 2). Important linkages are established that can then be transferred to a broader understanding of a critical incident in a social context. Conceptual models are research-based instructional tools that present frameworks to assist in developing cognitive structures as learners assimilate new concepts. Specifically, a conceptual model utilized for the study and analysis of critical incidents creates a common lens through which a variety of critical incidents can be viewed. By applying a common lens, the similarities and differences among incidents become more evident. These comparisons help narrow the focus and clarify the intent and ramifications related to critical incidents. The model also serves as a powerful means to organize thoughts, generate ideas, and examine multiple perspectives. Critical incident analysis provides a rich context for discourse, and individuals are able to present information and articulate complex ideas more effectively when supported by a conceptual model or framework. Critical thinking processes and the ability to retain complex information are also enhanced.

## Early Model

Critical incident analysis techniques emerged from studies conducted by the military in World War II in an effort to evaluate and classify psychological dimensions of human behavior. The applications of the technique were used to make inferences in related areas, such as performance criteria, training, motivation, and aptitude (Flanagan 1954). Building on the foundation of this seminal work, a conceptual model to analyze critical incidents was first introduced by Ochberg (1978) in the form of a flow chart (Figure 2.2). The early model, called the Hostage Incident Flow Chart, was narrow in scope and designed for analysis of a specific type of incident. It was created during an era of political hostage taking. Hostage-taking incidents, generally used by agents to extort concessions from governments, are highly calculated and publicized events. Throughout history, humankind has generally sought two ways to bring about change: through rational discussion and negotiation or by dramatic and often violent actions. Driven by moral, psychological, and cultural ideologies, people tend to lose patience with the slow speed of change and government bureaucracy. Hostage takers depend on the publicity to raise awareness of their cause and to stimulate a response.

The flow chart was used to analyze hostage-taking incidents and to emphasize the fact that siege managers, those delegated to handling the situation, had to calculate effects beyond the usual goals of public safety officials: saving lives, apprehending offenders, and preserving evidence. With no good options, officials often faced public criticism and, in some instances, outrage. This was the era of the Munich Olympics and the massacre of Israeli athletes (1972) and Carlos the Jackel (1975). The volatility of hostage-taking events, especially terrorist hostage takers, demonstrates the complicated challenges negotiators and tactical response teams face as they attempt to resolve hostage/crisis events. The tragedies of failed negotiations and tactical rescue attempts at Ruby Ridge (1992) and Waco (1993) highlight the difficulty of peacefully resolving hostage/crisis situations.

Figure 2.2 **The Hostage Incident Flow Chart**

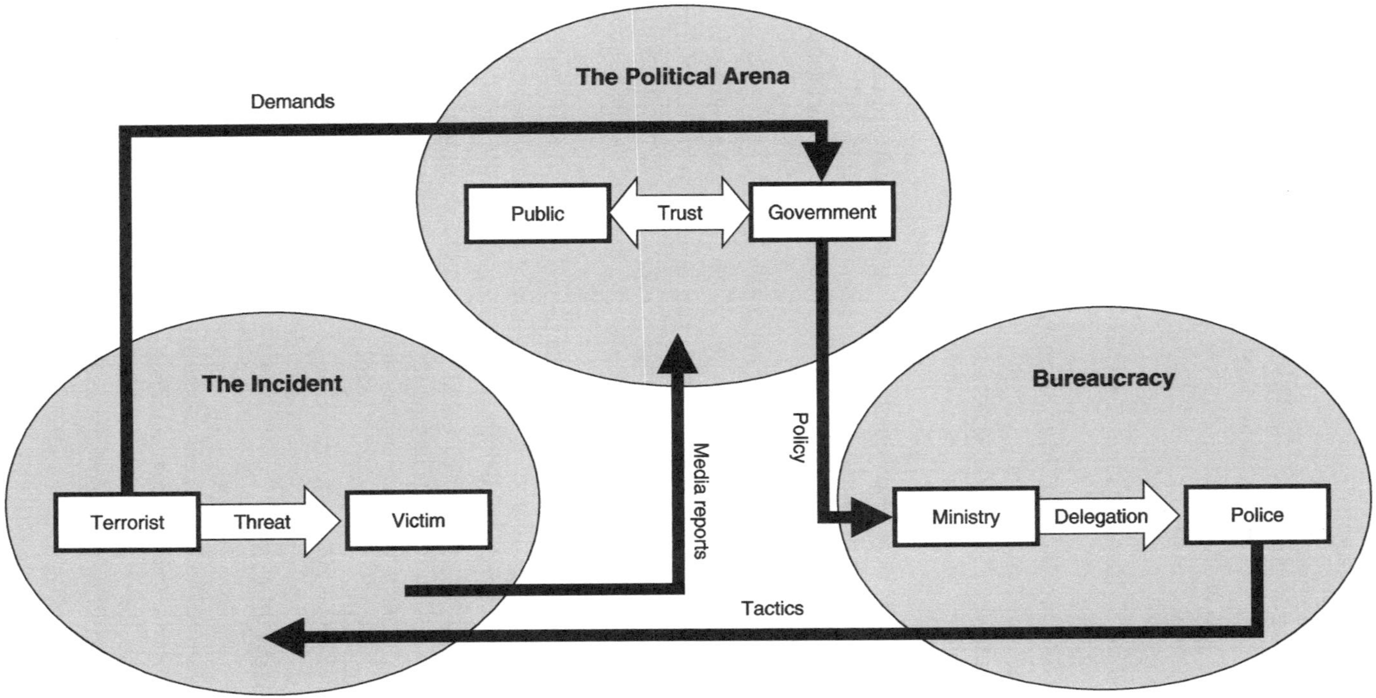


*Source:* Ochberg, 1978.

*Note:* An illustration of how demands from a hostage-taking incident cycle through the political arena and bureaucratic structures. These demands result in interventions and tactics used to negotiate and manage the incident.

Understanding human behavior in response to critical incidents will not necessarily provide solutions; but designing new ways to conceptualize and analyze data provides valuable insight into the problem and contributes to the development of more effective protocols and standards of practice. The Hostage Incident Flow Chart represents another dimension of critical incident analysis. It establishes the early origins of the current model and represents an important step in the journey of critical incident analysis as an interdisciplinary field of study.

**Current Model for Critical Incident Analysis**

The purpose of the current critical incident analysis model (Figure 2.1) is to capture the interconnectedness of the different processes, players, and dependencies that converge during a critical incident. The model is not predictive, nor does it simulate what happens in all critical incidents. In summary, the conceptual model

- presents a framework to help organize reporting and analysis of critical incidents,
- accentuates the similarities and differences among incidents, and
- provides a useful tool for training individuals to analyze critical incidents.

The current critical incident analysis model (Ochberg et al. 2007) was designed as a framework for analysis that could be used to enhance the collective understanding of significant events that challenge societal norms. The model serves as a useful tool in training individuals to think critically as they analyze and critique incidents of this nature. It provides a structure for the analysis and helps visualize a critical incident through discrete, yet interrelated components.

The three primary components in the model represent the event (threat); the political arena (trust); and authorized interveners (bureaucracy and adhocracy) of a critical incident. Through labeling and then defining these components, the variability of possible meanings for the decoder decreases. However, each of the components is complex in itself and open to a wide range of interpretations. These interpretations will be unique to individuals and subject to change as the context changes. Tension created by different interpretations among individuals can result in greater insight into the incident (Safayeni, Derbentseva, and Cañas 2003). An overview of each of the components follows.

## MODEL COMPONENTS

The basic structure of the conceptual model is cyclical in nature and intentionally designed to demonstrate the interconnectedness of three key components. Similar to a cyclic concept map, the relationships between components are dynamic. Changes in or a disturbance to one component contribute to changes or reactions from another (Safayeni et al. 2003). The model represents a dynamic system that is constantly subjected to pressures from a wide range of variables inherent to the incident and resonating within the internal and external environment. Beginning on the left and continuing in a clockwise flow, the event or critical incident (blob figure) creates a demand eventually received in the political arena (box figure). Whether or not a political leader is involved, through existing or ad hoc rules, policy and a delegation of authority occurs from this zone to the next. The triangle figure represents bureaucracy and adhocracy and symbolizes the tools and the tactics people deploy to manage an event. Directional arrows between the three components demonstrate the ebb and flow of the model. For example, the demands of participants and targets in the event zone flow back and forth from those in top levels of governance to those engulfed at the scene. The interveners in the management zone, the bodies and agencies through which elected government official work, subsequently have their own demands generally associated with policy needs. The tactics utilized by the interveners are also dynamic in nature. Tactics deployed to the event zone cause a reaction and, based on that reaction, new tactics may be applied to modulate various aspects of the situation.

### The Event

Each of the components in the model is characterized by specific colors and shapes used to symbolize very different zones of action. The blob figure is intended to represent a dramatic event with a pulsating shape that changes continuously as the event unfolds (Figure 2.3). Within this zone, violators may threaten and act upon others, or forces of nature and accidents may disrupt and destroy lives and property. The populations exposed to a critical incident experience a ripple effect. Consequences of the incident spread from the immediate point of impact and reverberate. The overall impact and force of the incident will influence the ultimate shape and size of the blob symbol.

Using terrorism as an example, agents methodically plan and carry out the attack. These agents generally have a larger goal in mind beyond the immediate target and may be supported financially, spiritually, or strategically by individuals or organizations united by a common purpose for the incident—an aim. This may be specific and come in the form of demands, but is usually accompanied by broader aims of showing strength, stirring up feelings, provoking a reaction that will aid in the recruitment of more agents, damaging the regime in power, or spreading fear throughout the populace. Agents pose a direct threat to the victim(s) or targets of the attack; targets are deliberate and the intention is to cause harm and disturbance to society.

Targets of terrorist acts may also be identified for their iconic value, such as the World Trade Center or the Statue of Liberty. These icons often have profound symbolic and cultural meaning attached to them and their destruction can evoke strong feelings, similar to that of a personal attack. When humans are the target, the emotional response may be even more traumatic. Part of the intent of the agents is to get the media and other responders, such as the police, involved, to increase the impact of the incident. The greater the response, the greater the impact the incident has on the targeted audience and society as a whole. Ultimately, for an incident to be labeled as *critical* it must have a significant effect on the prevailing culture. The outcomes of a critical incident will result in a change in the collective behavior and norms by which society operates.

Figure 2.3 **Event Component of the Critical Incident Analysis Model**

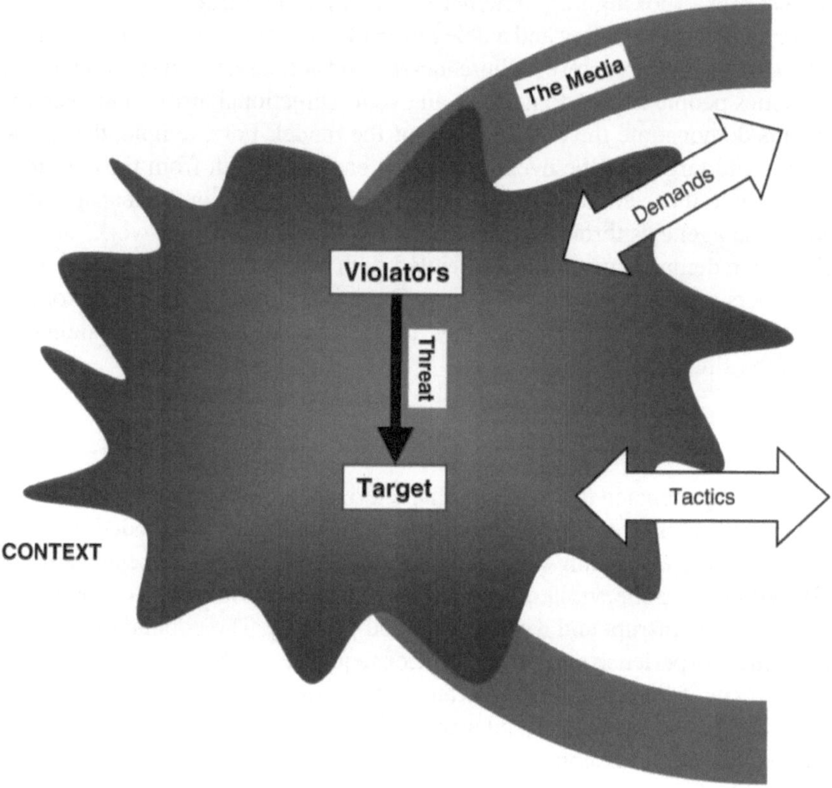


## Political Arena

The box figure represents the political arena; the government and the governed (Figure 2.4). The shape itself is stable and positioned at the top of the model to symbolize the power and significance of all of the forces that exist in the body politic at all levels—local, state, national, and international. The government as portrayed in the model serves as a symbol of ultimate consequence—for communities, the nation, or the world. It is representative of the larger community and includes the various factions, party affiliations, and social groups within society that coexist in a given nation. The individuals and groups within these social clusters include dominant and marginalized classes—those with authorized power or the disenfranchised, those who have access to or are denied resources, and groups that are valued or rendered insignificant in society.

The public and lobby groups can exert great influence on the government. When a triggering event such as a critical incident creates issues related to public health and safety, the government receives urgent demands from the public, which are propelled swiftly to the governmental policy agenda. Policy as defined by Fowler (2009) is "the dynamic and

Figure 2.4 **Political Arena Component of the Critical Incident Analysis Model**

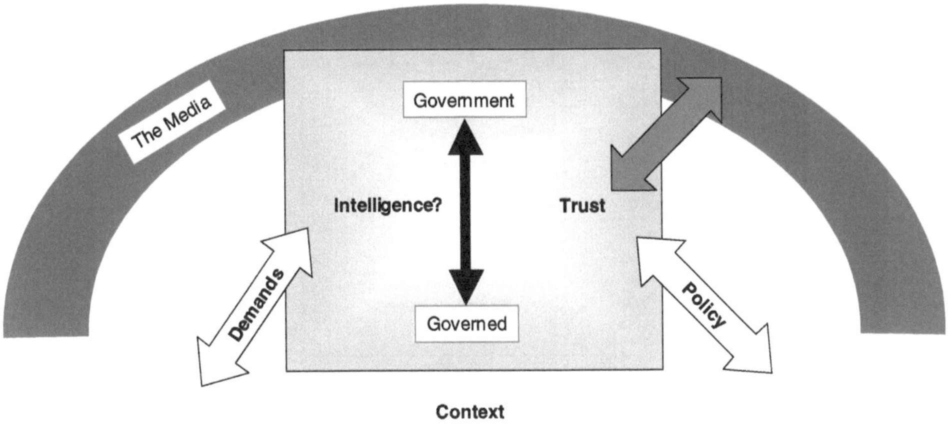


value-laden process through which a political process handles a public problem. It includes a government's expressed intentions and official enactment" (4). During critical incidents, the government receives demands and dictates policy while elected and appointed officials at federal, state, and local levels respond to immediate and future situations. While the government receives general direction from society, it also receives informed opinion and suggestions from the responders in the triangle zone. Strong factions within the body politic often fight and diminish one another because of the event itself, or because of the way the event was managed. An example of this is the critical incident that occurred in Waco, Texas, in 1993. In this incident a standoff between federal law enforcement agents and a sect known as the Branch Davidians resulted in the tragic loss of more than eighty lives. The Waco incident exposed flaws in law enforcement, lowered respect for several political leaders, and intensified the desire for ordinary citizens to form their own army. As a result, the FBI was demoralized and the militia movement was strengthened.

In the face of challenges, vulnerability can undermine trust and shake the foundations of democracy. In critical incidents such as Waco, the manner in which human interactions played out in the blob and triangle zones, the media, and the context, all had an impact on how the box zone (government) was perceived. Trust and confidence between government and the governed are vital outcomes of such incidents and dependent upon this perception.

## Bureaucracy and Adhocracy

The figure, shaped as a classic triangle, represents bureaucracy and adhocracy, terms that bring to mind the notion of a hierarchy and represent the dualistic nature of this zone (Figure 2.5). The points in the triangle suggest direction and purpose, indicative of the many elements and layers of bureaucracy that come into play in this zone, such as training, equipment, coordination, clarity of command, wisdom of response, preparation, and courage. When a critical incident occurs, the government will activate the tools and people

Figure 2.5 **Bureaucracy and Adhocracy Component of the Critical Incident Analysis Model**

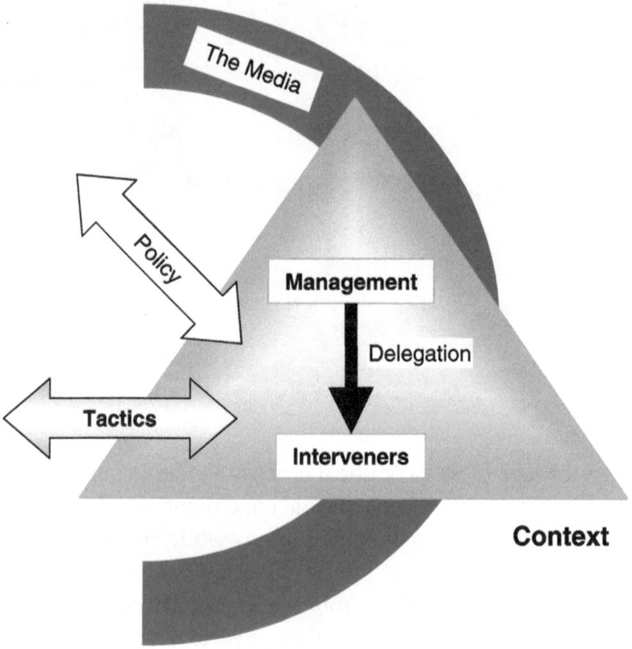


they have in place for handling and recovering from the incident. The people that work in the triangle can be viewed as a subset of the box zone. These individuals will receive direction in the form of policy from the government and deal directly with the agents in the blob zone. As the interveners and various stakeholders and interest groups converge on the scene, the situation heats up. The situation becomes a magnet for both wanted and unwanted attention, presenting a serious challenge to those trying to manage the scene. The response to the incident is highly dependent on the personnel involved and in their ability to inflame or calm the situation.

In some critical incidents, rigid bureaucracy impedes the response. When this occurs, individuals and agencies in the triangle may create ad hoc groups or implement more informal and flexible procedures. The immediate response to the needs of people in New Orleans after Hurricane Katrina provides a classic example of what happens when the interveners become disorganized or hampered by rigid procedures. The agencies in the triangle zone appeared to be poorly managed; their response slow and ineffective. Trust in the government to manage a mega incident was greatly diminished as images and stories of the people affected by the storm came to light in the media. In any case, when the interveners become disorganized, the critical incident becomes even more critical.

The triangle zone is the theater of action during a critical incident and "improvisation is a significant feature of every disaster" (Kendra and Wachtendorf 2007, 1). Preparedness and planning for what *should* be done may vary greatly from what *needs* to be done, given the unique characteristics of critical incidents. Individuals responsible for managing critical situations, responders who have duties to care for the injured or emotionally distraught, and media personnel on the scene of a newsworthy event represent some of the actors on the stage. These actors play dual roles as professionals and participant observers. Their performance will ultimately shape the incident and affect its long-term significance.

Colleges and universities prepare many professionals for operational and analytic roles. By analyzing the actions in the triangle zone, we create a common ground on which to build new, collaborative insights across a range of fields. Firefighters can examine the technology that helps airline officials order passengers out of planes. Red Cross workers can understand psychologists' theory for managing children during emergencies. Military personnel can recognize principles of community policing and compliance techniques applicable to citizen control. Effective incident management requires a professional team with expertise and a wide range of complementary skills.

## Communication and Media

Communication and the media have significant and defining roles in critical incident analysis. The figure representing this component is shaped as a ring to signify the interconnectedness between the event, political arena, and bureaucracy and adhocracy (Figure 2.6). There are infinite of ways in which communication contributes to understanding a critical incident. Richardson and Byers (2007) identified three key areas in which meaning is created during the course of a critical incident: narratives and sense making, compliance gaining and persuasion strategies, and the effects of stressful communication and events on individuals. Sense making involves the construction of meaning as individuals affected directly and indirectly try to make sense of what happened. In most instances, those directly involved find themselves thrust into a situation for which they have no frame of reference. The situation is unfamiliar and unanticipated, which compounds their ability to understand and interpret the incident. The narratives that emerge from a critical incident have great power. Based on the circumstances surrounding the event, they can serve to divide social groups or bind them through a sense of solidarity.

The media is compelling and can influence the outcome of an incident in positive or negative ways. For example, media can be a positive force criticizing the government, exposing corruption, and educating the public. On the other hand, the media can leak information at the wrong time, push the time frame of the incident too fast, sensationalize the incident, or distort reality. An integral part of critical incident analysis is understanding how the media operates and how it impacts an incident. Media consumers are bombarded by calculated and contemporaneous reports, iconic photographs and multiple replays of dramatic images, relentless punditry, and false rumors fanned into forceful themes. Repeated exposure to stressful communication such as this resonates with audiences far beyond the theater of action in ways that are not fully understood. Further, media projected for the public and agents is becoming harder to control due to the immediacy of information available through embedded journalists and other mediums, such as text messaging, blogs, and social networking sites. Information overload and noise generated by technological advancements have

Figure 2.6 **Communication Component of Critical Incident Analysis Model**

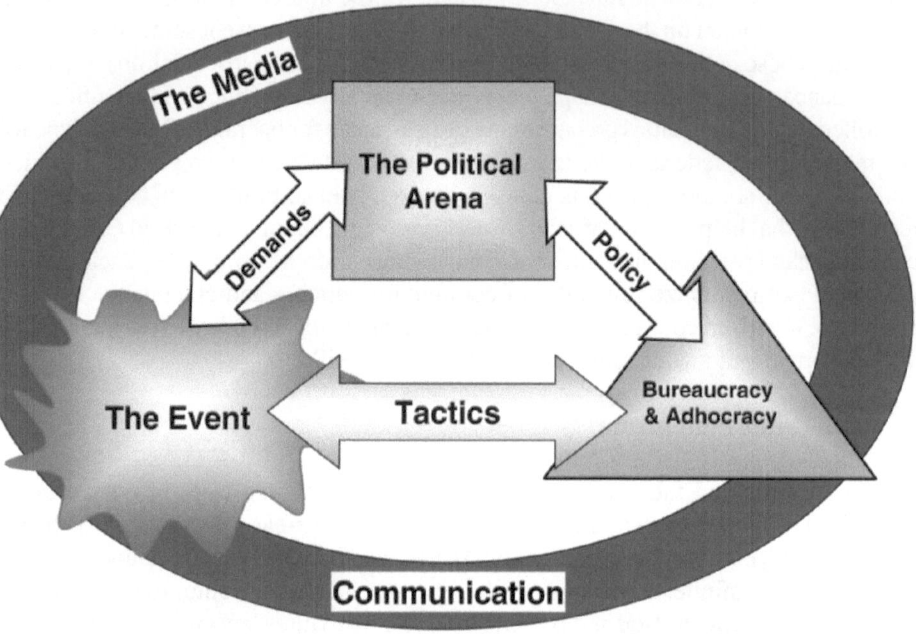


resulted in communication systems that bypass traditional outlets, increasing the ability to sensationalize, generate rumors, and perpetuate myths.

Each of the zones represented in the model will depend on communication and media for different outcomes. The blob zone is used to symbolize the event. Agents, or perpetrators of the incident want to gain attention and therefore increase the impact of the incident. For example, agents welcome extensive media exposure to heighten the importance of their agenda. The triangle zone represents bureaucracy and adhocracy. Agencies and individuals working in the triangle arena want to appear competent and be viewed in a good light. They want to reduce fear and panic in the public and limit or control the information received by the agents in the circle. They may have ulterior motives as well. A positive reflection of their ability to respond effectively may lead to increased recruitment or funding. The box zone, which represents the political arena, wants to maintain public confidence and trust. Those operating in the box hope to be perceived as attentive and helpful in the situation.

### Context

The final component in the model is context, which is represented by the background in the model for critical incident analysis (Figure 2.1). Context refers to such things as the physical situation (the time, place, people; the who, what, where, and when) or the historical circumstances leading up to the incident, which may provide insight into understanding the "why" of the event. Cultural consideration must also be given to the social norms that shape and influence individual and group perceptions of and responses to incidents. Factors such as ethnicity, gender, and social class all contribute to expressing grief and experiencing recovery in different ways. The severity of the incident is another factor to consider. Severity can be measured by mortality and morbidity, but this figure is rendered meaningless when taken out of context. Single deaths, such as the American president, or symbolic loss, such as the first teacher in space, are critical incidents in which the loss of life is minimal but the impact is enormous. Conversely, some severe traumas, such as a cyclone in Bangladesh with more than 100,000 deaths are, unfortunately, expected and tolerated without meaningful change.

The context and immediate effect of an incident is difficult to discern when comparing incidents, as long-term consequences unfold over time. Some events epitomize an era. An example would be the assassination of President John F. Kennedy, followed by Martin Luther King Jr. and Robert F. Kennedy. Each would have been a terrible blow by itself, but the ripple effect of three assassinations in a relatively short time frame was profound. The historical context was significant, and the prevailing mood of the nation as a result of these incidents would ultimately shape and define future generations.

## CONCLUSION

As the frequency in critical incidents increases, citizens around the globe are learning new ways to interpret and predict human responses, both individually and collectively. Critical incident analysis after a disruptive event raises awareness and helps citizens learn how to cope in an imperfect world. The conceptual model and graphic representations in this chapter are devices intended to establish a common ground, enhance dialogue, and prompt new ideas, concepts, and models. Within every critical incident, there will be lessons for future generations—lessons that serve to broaden our understanding and guide our way. Critical incident analysis provides a fertile landscape and conceptual means for generating creative and effective initiatives to successfully mitigate and manage future incidents of this nature.

The purpose of critical incident analysis is to help us to understand, predict, and negotiate the episodes that have deep significance, and to do the task with appreciation of the roles and methods of various academic disciplines and occupational experts. Not every critical incident will matter to historians, epitomize our culture, or change societal norms. Yet all of us are eventually affected by severe critical incidents. By definition, these are the events that resonate well beyond the theater of original action, make their mark on society, and leave a lasting impression.

## REFERENCES

Carnot, M. J., and D. Stewart. 2006. "Using Concept Maps in College Level Psychology and Social Work Classes." Paper presented at the Second International Conference on Concept Mapping, San Jose, Costa Rica, http://cmc.ihmc.us/cmc2006Papers/cmc2006-p232.pdf (accessed May 13, 2009).

Carter, S., M. Kang, and R. Taggart. 1999. "An Interdisciplinary Approach to a Critical Incident Course. *Journalism & Mass Communication Educator* 54 (2): 4–14.

Flanagan, J. C. 1954. "The Critical Incident Technique." *Psychological Bulletin* 51 (4): 327–358.

Fowler, F. C. 2009. *Policy Studies for Educational Leaders* (3d ed.). Boston: Allyn & Bacon.

Kendra, J., and T. Wachtendorf. 2007. "Improvisation, Creativity, and the Art of Emergency Management." In *Understanding and Responding to Terrorism,* ed. H. Durmaz, B. Sevinc, A. S. Yala, and S. Ekici, 324–335. Amsterdam: IOS Press.

Ochberg, F. M. 1978. "The Victim of Terrorism." *Practitioner* 220 (1316): 293–302.

Ochberg, F. M., S. Cinti, E. Goodman, E. Kirby, H. Melia, S. Prior, and R. Taggart. 2007. "The Critical Incident Model." Presentation at the meeting for Critical Incident Analysis at John Jay College of Criminal Justice (CUNY), January 4.

Parker, D. L., J. Webb, and B. D'Souza. 1995. "The Value of Critical Incident Analysis as an Educational Tool and Its Relationship to Experiential Learning." *Nurse Education Today* 15: 111–116.

Richardson, B. K., and L. Byers. 2007. "Communication Studies and Emergency Management: Common Ground, Contributions, and Future Research Opportunities for Two Emerging Disciplines." In *Disciplines, Disasters and Emergency Management: The Convergence and Divergence of Concepts, Issues and Trends from the Research Literature,* ed. D. McEntire, 272–283. Springfield, IL: CC Thomas.

Roehler, L. R., and D. J. Cantlon. 1997. "Scaffolding: A Powerful Tool in Social Constructivist Classrooms." In *Scaffolding Student Learning: Instructional Approaches and Issues,* ed. K. Hogan and M. Pressley, 6–42. Cambridge, MA: Brookline, 1997.

Safayeni, F., N. Derbentseva, and A. J. Cañas. 2003. "Concept Maps: A Theoretical Note on Concepts and the Need for Cyclic Concept Maps," http://cmap.ihmc.us/Publications/ResearchPapers/Cyclic%20Concept%20Maps.pdf (accessed May 14, 2009).

Schwester, R. W., M. Dank, and A. M. Horning. 2008. "Conceptualizing, Defining and Operationalizing Critical Incidents." Paper presented at the Symposium on Critical Incidents, Academy for Critical Incident Analysis (ACIA), John Jay College of Criminal Justice (CUNY), October 2.

CHAPTER 3

# Conceptualizing and Defining Critical Incidents

RICHARD W. SCHWESTER, AMBER M. HORNING, AND MEREDITH DANK

New academic fields frequently endure identity crises; that is, with new or relatively young fields of study there are periods of uncertainty as to foundation and future direction. The field of critical incident analysis is not immune to these challenges, as there is a fundamental yet vital question that needs to be addressed; specifically, what are critical incidents? The purpose of this chapter is to analyze divergent conceptions of what comprises a critical incident. We identify multiple definitions and establish a set of attributes that *may* characterize critical incidents. In doing so, we hope to provide a theoretical and pedagogical foundation to this interdisciplinary field by conceptualizing, defining, and operationalizing critical incident attributes. If this field is to burgeon from research and teaching perspectives, then it becomes paramount to establish a common language of critical incident analysis.

## CRITICAL INCIDENT DEFINITIONS

There are varying and possibly competing conceptions as to what constitutes a critical incident. We conducted a literature review to develop a representative list of critical incident definitions and identified fourteen distinct yet interrelated definitions. Four definitions represent Dart Foundation–sponsored initiatives, including the Critical Incident Analysis Group at the University of Virginia, the National Center for Critical Incident Analysis at the National Defense University, the Academy for Critical Incident Analysis at John Jay College, and the Dart Center for Journalism and Trauma. Four definitions represent academic or university perspectives, namely the universities of Michigan, Richmond, and Virginia, and the Université Catholique de Louvain in Belgium. Two definitions represent international perspectives on critical incidents—the Australia National Code and a joint initiative of the

United Nations and the European Union. U.S. perspectives are represented by several layers of government, including the U.S. Department of Justice, the Santa Cruz County Department of Public Safety (Arizona), and the San Antonio Office of Emergency Management. Finally, one definition incorporates the perspective of a private sector crisis management services firm. The critical incident definitions and sources are presented in Table 3.1.

## CRITICAL INCIDENT ATTRIBUTES

We conducted a content analysis of the fourteen critical incident definitions. In doing this, we simply identified concepts—or smaller coding units such as phrases and sentences. From these identified concepts, we synthesized fifteen critical incident attributes. Figure 3.1 shows examples of coding units pulled from critical incident definitions D and H from Table 3.1.

This process was completed for each of the fourteen definitions (see Table 3.2). The definitional attribute identified most is "significant injury or loss of life." Fatalities and injuries are most associated with critical incidents, as this attribute is noted within ten of the fourteen critical incident definitions. The concept of *social trauma* is referenced in seven of fourteen definitions. Six definitions reference *intergovernmental* or *international coordination* and *changes in societal norms* as attributes. Five definitions cite brevity as a precondition for classifying an event as critical. Critical incidents arguably are *relatively brief occurrences.* They are not chronic conditions. An additional five definitions reference *property and infrastructure damage* and *fear* as preconditions for classification as well. *Undermined public trust* and the *unexpectedness* were attributes cited in four definitions. Less frequently cited attributes of critical incidents include: *declaration of a state of emergency* (three definitions); *limited in scope* (two definitions); *impacted practice of democracy* (two definitions); *positive event* (one definition); *emotional effect on trained people* (one definition); and *attracts significant media coverage* (one definition).

## DEFINING CRITICAL INCIDENT ATTRIBUTES

### 1. Caused Social Trauma

In recent decades, the construct of social trauma has been defined through a number of disciplines such as history, psychology, anthropology, and sociology. Some historical events are so intense that their cultural and personal impact is akin to psychological trauma. The effects of such trauma are a compulsive need to relive and reexperience the event coupled with a pervasive sense of uneasiness. Individuals who partake in or are exposed to violence can suffer from psychological distress. Some of the effects are sleep, learning, and physical disorders, and flashbacks. There are existing clinical definitions of individual trauma and healing; however, the concept of trauma is debatable (Barsalou 2005).

From a sociological perspective, Erikson (1994) contends that a shared trauma can act as a source for creating and impairing a community. The experience of manmade collective trauma can create a loss of confidence in the self, as well as in the institutions of family, community, and government. This may ultimately engender a loss of confidence in society and nature.

Table 3.1

**Critical Incident Definitions and Sources**

| # | Source | Definition |
|---|---|---|
| A | Critical Incident Analysis Group at the University of Virginia | An event that has the potential for causing social trauma and undermining social trust, creating fear that may have impact on community life and even on the practice of democracy. |
| B | National Center for Critical Incident Analysis at the National Defense University | A relatively brief occurrence involving injury, loss, or conflict of significant proportion, with the potential to change existing societal norms. Critical incidents are usually traumatic, threatening the bonds of trust that bind democracies. |
| C | Academy for Critical Incident Analysis at John Jay College | The event is unexpected, at least by those who are not perpetrators or initiators. There is a consequential impact on many at the time of occurrence. The event and its immediate impact are limited in time and space, making it an incident or episode rather than a condition—such as war or poverty or pandemic. There is potential for much larger gain or loss or change, depending upon the event itself, the actions of those accountable for managing such incidents, and other important variables, such as the stability or fragility of the community in which the event occurs. |
| D | Ochberg, Dart Center for Journalism and Trauma | A relatively brief occurrence involving injury, loss, conflict, discovery or change of significant proportion, usually unscripted and unanticipated, with the potential to alter existing societal norms. Critical incidents are usually traumatic, threatening the bonds of trust that bind communities, but may be positive, initiating historic consequents. |
| E | University of Michigan | Any manmade or natural event or situation that threatens people, property, business, or the community and occurs outside the normal scope of routine business operations. Typically requires coordination of numerous resources. |
| F | Australian National Code | A traumatic event, or the threat of such [within or outside Australia], which causes extreme stress, fear, or injury. |
| G | Santa Cruz County (Arizona) Department of Public Safety | A turning point event. Often called a crisis event. A critical incident is any event that has a stressful impact sufficient enough to overwhelm the usually effective coping skills of either an individual or a group. Critical incidents are typically sudden, powerful events which are outside of the range of ordinary human experiences. Because they are so sudden and unusual, they can have a strong emotional effect even on well-trained, experienced people. |
| H | U.S. Department of Justice | Include acts of terrorism, hostage situations, and natural disasters. Typically, these events involve one or more of the following factors (although the presence of one factor by itself does not automatically mean that an incident is critical). Involves threats or acts of violence against government or social institutions; involves significant loss of life, significant injuries, or significant damage to property; demands use of substantial resources; requires coordination among federal law enforcement agencies (more so than usual), state or local law enforcement agencies, local or state prosecutors, emergency relief services, or emergency response services; attracts close public scrutiny through the media. |

*(continued)*

Table 3.1 *(continued)*

| # | Source | Definition |
|---|---|---|
| I | San Antonio Office of Emergency Management | An extraordinary event or series of events that is sudden, overwhelming, and often dangerous to oneself or to a significant other. |
| J | Cohen Associates, P.A. | An extraordinary event that shocks our systems, challenges our usual ways of coping, and thrusts us into a period in which we search for ways to adjust to the harsh reality that has touched our lives. |
| K | University of Virginia | Situations that have the potential to cause injury or loss of life . . . major disruptions of regular activities, or property or environmental damage or loss, or that can threaten the financial standing or public image of the University. |
| L | Centre for Research on the Epidemiology of Disasters, Emergency Events Database (EM-DAT) | A situation or event, that overwhelms local capacity, necessitating a request to national or international level for external assistance; an unforeseen and often sudden event that causes great damage, destruction, and human suffering. Though often caused by nature, disasters can have human origins. Wars and civil disturbances that destroy homelands and displace people are included among the causes of disasters. Other causes can be: building collapse, blizzard, drought, epidemic, earthquake, explosion, fire, flood, hazardous material or transportation incident (such as chemical spill), hurricane, nuclear incident, tornado, or volcano. For a disaster to be entered into the database at least one of the following criteria must be met: Ten or more people reported killed; one hundred people reported being affected; declaration of a state of emergency; call for international assistance. |
| M | The Global Disaster Alert and Coordination System | A major natural, technological, and environmental disaster, which overwhelms the affected country's response capacity and requires international assistance. |
| N | University of Richmond | Events are considered disasters . . . if (1) they represented a threat to life, property, or the environment, (2) they would have required the use of emergency procedures for the limitation and resolution of their impact, (3) they reasonably could have caused a responsible jurisdiction, agency, or organization to invoke or declare the existence of an emergency situation or to mobilize its resources in response, and (4) some significant degree of community or organizational impact was present. These criteria cast a wide net, but the intent is not to include minor, routine emergencies typically handled in today's environment by a single public safety service. |

From a psychological perspective, trauma is a response to events or circumstances that are traumatizing and can manifest in the physical, affective, cognitive, or spiritual realm. Traumatized people feel the world to be dangerous because their experience of the world is unsafe. Their previous schemas of a safe world have been edited out (Burstow 2003). Trauma is centered on "the fact of survival in the face of death." The repetitious dreams that occur after trauma have less to do with trauma itself; having more to do with surviving the trauma. Escaping death is central to trauma (Caruth 1996). According to the *DSM-IV,* post-traumatic stress disorder (PTSD) occurs when

Figure 3.1 **Examples of Coding Units**

**Definition D**

A relatively [1] brief occurrence involving [2] injury, [3] loss, [4] conflict, discovery or [5] change of significant proportion, usually [6] unscripted and unanticipated, with the potential to [7] alter existing societal norms. Critical incidents are usually [8] traumatic, [9] threatening the bonds of trust that bind communities, but [10] may be positive, initiating historic consequents.

**Definition H**

Include acts of terrorism, hostage situations, and natural disasters. Typically, these events involve one or more of the following factors. Involves [1] threats or [2] acts of violence against government or social institutions; involves [3] significant loss of life, [4] significant injuries, or [5] significant damage to property; demands [6] use of substantial resources; [7] requires coordination among federal law enforcement agencies (more so than usual), state or local law enforcement agencies, local or state prosecutors, emergency relief services, or emergency response services; attracts [8] close public scrutiny through the media.

the person experiences, witnesses, or was confronted with death or serious injury, the response to which is intense fear and a sense of helplessness. Some of the symptoms are: reoccurring thoughts or dreams of the events, behaving in a manner as if the event was reoccurring, and avoiding stimuli related to the trauma (American Psychiatric Association 2000). Caruth (1996) described post-traumatic stress not simply as pathology or illness but as "a symptom of history." Those who are traumatized carry this history with them.

The psychoanalytic perspective argues that trauma occurs when communities and institutions fail to serve as "containers" for individuals. There are many failures that can be classified in a multitude of categories. First, the social structure fails to care for and protect individuals. Some examples of this are "the lack of assistance and compassion towards the victims of poverty, disease, natural catastrophe, social turmoil, economic crisis, violence, or war" (Hernandez de Tubert 2006). The second type is when there is an overt attack by hegemonic groups on marginalized populations; for example, social repression,

Table 3.2

**Definition Attribute Map**

| Definition Attributes | A | B | C | D | E | F | G | H | I | J | K | L | M | N |
|---|---|---|---|---|---|---|---|---|---|---|---|---|---|---|
| 1. Caused Social Trauma | X | X | | X | | X | X | | | X | | | | X |
| 2. Caused Fear | X | | | | X | X | X | | X | | | | | |
| 3. Emotional Effect on Trained People | | | | | | | X | | | | | | | |
| 4. Changed Societal Norms | X | X | | X | | | X | | | X | X | | | |
| 5. Undermined Public Trust | X | X | | X | | | | | | | X | | | |
| 6. Impacted Practice of Democracy | X | X | | | | | | | | | | | | |
| 7. Relatively Brief Occurrence | | X | X | X | | | X | | | | | X | | |
| 8. Significant Injury or Loss of Life | | X | X | X | X | X | | X | X | | X | X | | X |
| 9. Significant Property/Infrastructure Damage | | | | | X | | | X | | | X | X | | X |
| 10. State of Emergency Declared | | | | | | | | | | | | X | X | X |
| 11. Unexpected | | | X | X | | | X | | | | | X | | |
| 12. Limited in Scope | | | X | | | | | | | | | X | | |
| 13. Intergovernmental/International Coordination | | | X | | X | | | X | | | | X | X | X |
| 14. Positive Event | | | | X | | | | | | | | | | |
| 15. Attracts Significant Media Coverage | | | | | | | | X | | | | | | |

war, mass killing, or persecution. The third is when the social system champions current social mores and laws while simultaneously undermining them.

## 2. Caused Fear

Fear is a reaction to an external threat and danger, and is often connected to pain. Anxiety is distinct from fear in that it can occur without any external threat. Fear is associated with a need to escape or evade. Anxiety is associated with perceived and unavoidable threats. Fear is a survival tool and often occurs due to negative stimuli (Ohman 2000). Fear is integral to human nature. People are molded by fear. Moreover, fantasy creates imaginary dangers of all kinds. The human condition therefore is characterized by a permanent dialogue with fear. Kemper states that emotions are a manifestation of two fundamental dimensions of social relationships: power and status. Sometimes a social fear may be "born" of a "barbarization" of social life and of a progressive loss of freedom (Cattarinussi 2001).

Erikson's (1976) work regarding disaster victims suggests that individuals by and large experienced "a sense of vulnerability, a feeling that one has lost a certain natural immunity to misfortune, a growing conviction, even, that the world is no longer a safe place to be" (234). Perceived vulnerability, rage, and frustration often stem from fear (DuBow, McCabe, and Kaplan 1979). Under certain circumstances, fear is a reasonable reaction (Sarnoff and Zimbardo 1961). Fear is affected by assessments of risk, but also influences one's judgments (Ferraro and LaGrange 1987).

## 3. Emotional Effect on Trained People

According to Gergen (1995), despite psychology's extensive history of theorizing about emotion, a consensus as to what emotion is has yet to be reached. Most theorists agree that emotions are "palpable states" within individuals. Primarily, emotions are understood as "intrapersonal realities" that are experienced internally (Parkinson 1995). Emotional effect on trained people refers to clinically diagnosed trauma in professionals, such as firefighters, emergency medical technicians, military personnel, and law enforcement officers. Due to the nature of law enforcement, officers are habitually "exposed to dangerous, unpleasant, and horrific situations" (Stratton 1981); that is, being confronted with violent deaths (e.g., homicides and road fatalities) and threats from dangerous people who are often unpredictable, or situations to which most human beings could be expected to elicit an emotional response (de Sousa 1987; Howard, Tuffin, and Stephens 2000).

## 4. Changed Societal Norms

Social norms have been defined as "the rules that a group uses for appropriate and inappropriate values, beliefs, attitudes and behaviors. These rules may be explicit or implicit. Failure to stick to the rules can result in severe punishments, the most feared of which is exclusion from the group." They have also been described as the "customary rules of behavior that coordinate our interactions with others" (Deutch and Gerard 1955).

There are three different ways to conceptualize norms. The first involves a shared "coordination motive," such as using dollars instead of gold to purchase items. Some have to do with shared expectations and do not require social enforcement. Second, norms are maintained by social condemnation or punishment for a violation of norms (Sugden 1986; Coleman 1990). If standing in line is the norm, one will be reprimanded for pushing to the front. The third arises through "the internalization of norms of proper conduct." If the norm is not to litter, one will avoid littering even when out of sight.

### 5. Undermined Public Trust

Trust in political and governmental institutions is significantly undermined when governments deceive the public. Central to public trust are: ensuring accountability and transparency and instilling confidence in the integrity of process (Reissman 2007).

### 6. Impacted the Practice of Democracy

Arguably, there are two concepts central to a democratic system—equal access to power and protected freedoms and liberties (Dahl, Shapiro, and Cheibub 2003). Fair and competitive elections are central to democracies. Any incident that impacts democracy would impede elections and any aspect of political expression, such as freedoms of speech, press, and assembly (Barak 2006).

### 7. Relatively Brief Occurrence

According to Hatry, Abravanel, and Rossman (2006), the incident itself and the immediate aftermath are typically measured in hours or days. The postincident "near-term period" is typically measured in months. Depending on the severity of the incident, the "longer term period" is typically measured in months or years (Hatry et al. 2006).

### 8. Significant Injury or Loss of Life

This refers to the number of fatalities and injuries directly associated with a specific incident. The Centre for Research on the Epidemiology of Disasters (CRED) at the Université Catholique de Louvain (Belgium) maintains an emergency events database. For an emergency event to be included in this database, the event must have caused at least ten fatalities.

### 9. Significant Property/Infrastructure Damage

This refers to extent and monetary damage to personal and real property, in addition to infrastructure. The Federal Emergency Management Agency (FEMA) defines infrastructure as roads, government buildings, wastewater treatment plants, and bridges. Additionally, the Department of Emergency Management in Oklahoma includes water control facilities, public utilities, recreational areas or parks, and debris removal. The Wisconsin

Department of Military Affairs includes schools, nursing homes, hospitals, sewers, and fire/emergency medical service facilities. The United Nations includes power supply facilities as infrastructure as well.

### 10. State of Emergency Declared

This refers to a formal executive decree in response to a crisis or an anticipated crisis. As a result of such decree, normal government functions may be suspended or altered, citizens are informed as to necessary precautions and changes in normal behavior, and government agencies are ordered to implement emergency preparedness plans. States of emergency generally represent formal requests for outside resources from other layers of government. FEMA's Disaster Assistance Policy states that an emergency declaration is a criterion for disaster relief assistance (FEMA 2008). The Emergency Service Act for the State of California defines a state of emergency as:

> Other duly proclaimed existence of conditions of disaster or of extreme peril or the safety of persons and property within the State caused by such conditions as air pollution, fire, flood, storm, epidemic, riot, drought, sudden and severe energy shortage, plant or animal infection or disease, the Governor's warning of an earthquake or volcanic prediction, or an earthquake, or other conditions, other than conditions resulting from a labor controversy or conditions causing a "state of war emergency," which conditions, by reason of their magnitude are or are likely to be beyond the control of the services, personnel, equipment, and facilities of any single county, city and county, or city, and require the combined forces of a mutual aid region or regions to combat or with respect to regulated energy utilities, a sudden and severe energy shortage requires extraordinary measures beyond the authority vested in the California Public Utilities Commission.

### 11. Unexpected

According to Kreps (2001), disasters are atypical events. Lehman and Taylor (1987) explored the effects of a California earthquake—or a so-called "natural" disaster—on a group of students and deemed an earthquake a disaster whose occurrence is highly likely but whose timing is unknown. Palermo and Ross (1999) conducted a study on mass murderers and defined mass murder as "the sudden, explosive killing of a group of people." This type of disaster is referred to as an "unnatural" disaster. The unnatural as compared to the natural disaster may not be considered highly likely. Also, in many instances, the timing as to when a natural disaster will strike can be estimated (e.g., hurricane forecast). This is typically not the case with unnatural disasters.

### 12. Limited in Scope

Scope implies geographic limitations. Depending on the type of incident, the geographic boundaries of incidents vary. For instance, the Virginia Tech massacre was limited to a college campus and the attacks of September 11, 2001, involved two cities and a rural section of Pennsylvania. Hurricane Katrina, however, encompassed most of the Gulf Coast region.

### 13. Requires Intergovernmental/International Coordination

An incident is critical when assistance is received from the emergency and relief services industries at the international, federal, state, or local government levels or nonprofit aid organizations. Under the definition of the North American Industry Classification System (NAICS), the emergency and relief services industries are "primarily engaged in providing food, shelter, clothing, medical relief, resettlement, and counseling to victims of domestic or international disasters or conflicts."

### 14. Positive Event

The very nature of critical incidents implies that its proximal and distal consequences are negative. However, critical incidents might be viewed as positive in nature—that is, events that contribute to the betterment of society.

### 15. Attracts Significant Media Coverage

Charles Wright (1959) notes two classic functions of the mass media. The first is "surveillance of the environment," which simply refers to the dissemination of information. Surveillance further requires that the public's consumption results in learning and understanding (Massey 2002). The second is entertainment. In the context of critical incidents, disseminating information to those affected most reduces exposure to incident-related hazards, as is seen most often with preparation for a hurricane. The media's role as entertainer is telling in terms of judging the magnitude of an incident. Incidents that are covered extensively may signify high magnitude.

We have identified a cross-section of critical incident definitions. From these definitions, we have identified a set of attributes that may characterize critical incidents. The remaining step is to measure these attributes. The least difficult attributes to measure are those that entail "counting" or "documenting." Significant injury and loss of life and significant property/infrastructure damage are measured simply by counting the number of individuals killed or the number of buildings destroyed and the costs of that destruction. A declared state of emergency and whether intergovernmental/international coordination was required are simply documented. Either a state of emergency was declared or it was not. Either multiple layers of government worked together or they did not. These types of measures that deal with counting and documenting are likely to be available through existing data sources. Similarly, the degree of social trauma and emotional effect on well-trained people can be measured by the number new cases of PTSD or other related disorders as documented by existing data or scholarly publications. "Relatively brief occurrence" implies a time frame—when the incident started and when it finished—which is measured in the number of hours or days. "Limited in scope" assumes geographic limitations and specificity, measured in terms of square miles or mile radius. There are questions relating to how brief an incident must be and also how geographically limited it is. Again, the Virginia Tech shootings can be measured in hours and were limited to a college campus and Hurricane Katrina can be measured in days, but its impact was felt across the entire Gulf Coast region.

There is somewhat greater difficulty in measuring attributes that rely less on counting or noting and rely more on perception. In other words, while we can count the number of fatalities associated with an incident, and we can note whether a formal state of emergency was declared, we have less objective bases for determining the extent to which an incident caused fear or changes in societal norms for that matter. Fear may be measured in terms of the level of anger and outrage resulting from an incident, as the literature holds that these emotions are proxies for fear. For example, the anger and outrage following 9/11 were well documented through personal accounts captured by the mass media, but also through people's behavior—namely a small but significant number of abuses reported against Arab-Americans immediately following the attacks. In measuring changes in societal norms, we again draw on 9/11. Changes in societal norms seem relevant to 9/11 as it pertains to profiling as an antiterror tool. Throughout several U.S. law enforcement communities, the issue of racial profiling—or using race as proxy for determining who is more likely to commit a crime—has been exposed. Law enforcement organizations, most infamously the New Jersey State Police, have undertaken steps to reduce the instances of profiling given the unconscionable nature of this practice. However, among some, profiling as it relates to antiterror tactics has gained some measure of acceptance, in spite of documented abuses resulting from this practice. Such instances may signal a change in societal norms.

Regarding undermined public trust, the failed relief effort associated with Hurricane Katrina in 2005 evokes memories of governmental failure. And for many throughout the Gulf Coast region devastated by Katrina, the failure of the federal, state, and local authorities to provide relief and coordinate their efforts served to undermine the trust and confidence in government. This shaken trust and confidence was documented through media accounts and public opinion surveys of the governments' handling of Katrina. "Impacted the practice of democracy" implies that some part of the electoral process was disrupted. For instance, hypothetically speaking, if a terror attack leads to the postponement of a formal election, then one might reasonably consider that the practice of democracy has been affected. The Florida voting technology failures of the 2000 presidential election arguably fit this criterion of a critical incident. For many, the outcome of the 2000 presidential election is still in doubt, and many voting throughout Florida felt as though they were disenfranchised as result of the infamous punch card system and its inordinate number of pregnant, dimpled, and hanging chads.

The unexpectedness of an incident poses measurement challenges as well. If we use "likelihood" as a proxy for expectedness, then one might conclude that 9/11, Virginia Tech, and Columbine were unexpected given the unlikely nature of these incidents.

Natural events, such as hurricanes, occur with regularity, and thus are expected if not predictable. This being said, are we not to include certain natural disasters because they are predictable? In terms of attracting significant media coverage, if we believe that the media function not only to provide information but also to capture our attention as news consumers, then it stands to reason that an extensively covered incident is a more severe incident. Therefore, in measuring this attribute, an incident would have to be covered by major domestic and international news outlets. The problem here is determining for how long an incident must remain at the media forefront. Finally, regarding a positive event, key to measuring this is determining whether the incident contributed to society rather than taking something away from it. Dr. Jonas Salk's polio vaccine serves as a possible example in this case.

## CONCLUSION

The notion of critical incidents is sufficiently complex, as is the field of critical incident analysis given its relative youth and interdisciplinary nature. As such, we have sought a measure of clarity by analyzing multiple definitions of critical incidents, and from these definitions we have synthesized a number of attributes that may characterize such incidents. We do not mean to imply that for an incident to be deemed "critical" it must embody all of these attributes. To use a medical analogy, a patient need not possess all symptoms to be diagnosed with a syndrome. This holds true for identifying critical incidents as well. From here, we have defined and discussed ways of operationalizing these attributes. This chapter should be viewed as a starting point. Future work should critique and expand on the critical incident attributes presented here, their definitions, and ways of objectively measuring them.

## REFERENCES

American Psychiatric Association. 2000. *Diagnostic and Statistical Manual of Mental Disorders,* 4th ed. Washington, DC: American Psychiatric Society.

Barsalou, J. 2005. "Trauma and Transitional Justice in Divided Societies." *Special Report* 135, The United States Institute of Peace, Washington, DC.

Barak, A. 2006. *The Judge in a Democracy.* Princeton, NJ: Princeton University Press.

Burstow, B. 2003. "Toward a Radical Understanding of Trauma and Trauma Work." *Violence against Women* 9 (1): 1293–1317.

Caruth, C. 1996. *Unclaimed Experience: Trauma, Narrative, and History.* Baltimore: Johns Hopkins University Press.

Cattarinussi, B. 2001. "Sentiments." In *Encyclopedia of Sociology,* 2d ed., vol. 1, ed. E. F. Borgotta, 2518–2529. New York: Macmillan Reference.

Coleman, J. S. 1990. *Foundations of Social Theory.* Cambridge MA: Harvard University Press.

Dahl, R. A., I. Shapiro, and J. A. Cheibub. 2003. *The Democracy Sourcebook.* Cambridge, MA: MIT Press.

de Sousa, R. 1987. *The Rationality of Emotion.* Cambridge, MA: The MIT Press.

Deutch, M., and H. B. Gerard. 1955. "A Study of Normative and Informational Social Influence upon Judgment." *Journal of Abnormal and Social Psychology* 51: 629–636.

DuBow, F., E. McCabe, and G. Kaplan. 1979. *Reactions to Crime: A Critical Review of the Literature.* Washington, DC: National Institute of Law Enforcement and Criminal Justice, U.S. Government Printing Office.

EM-DAT Emergency Events Database. The Centre for Research on the Epidemiology of Disasters (CRED) at the Université Catholique de Louvain (Belgium).

Erikson, K. T. 1976. *Everything in Its Path.* New York: Simon and Schuster.

———. 1994. *A New Species of Trouble: The Human Experience of Modern Disasters.* New York: W.W. Norton and Co.

FEMA. 2008. 2008 Federal Disaster Declaration. www.fema.gov/news/disasters.fema?year=2008, accessed December 6, 2011.

Ferraro, K. F., and R. LaGrange. 1987. "The Measurement of Fear of Crime." *Sociological Inquiry* 57 (1): 70–101.

Gergen, K. J. 1995. "Social Construction and the Educational Process." In *Constructivism in Education,* ed. L. P. Steffe and J. Gale. Hillsdale, NJ: Lawrence Erlbaum.

Hatry, H., M. Abravanel, and S. Rossman. 2006. "What Happens to Victims: A Research Guide for Disaster-Response Studies." Washington DC: The Urban Institute.

Hernandez de Tubert, Reyna. 2006. "Social Trauma: The Pathogenic Effects of Untoward Social Conditions." *International Forum of Psychoanalysis* 15: 151–156.

Howard, C., K. Tuffin, and C. Stephens. 2000. "Unspeakable Emotion: A Discursive Analysis of Police Talk about Reactions to Trauma." *Journal of Language and Social Psychology* 19: 295–314.

Kreps, G. L. 2001. "Consumer/Provider Communication Research: A Personal Plea to Address the Issue of Ecological Validity, Relational Development, Message Diversity and Situational Constraints." *Journal of Health Psychology* 6: 597–601.

Lehman, D. R., and S. E. Taylor. 1987. "Date with an Earthquake." *Personality and Social Psychology Bulletin* 13 (4): 546–555.

Massey, K. 2002. "Functions of the Media." In *Encyclopedia of Communication and Information,* vol. 1, ed. Jorge Reina Schement, 348–350. New York: Macmillan Reference.

Ohman, A. 2000. "Fear and Anxiety: Evolutionary, Cognitive, and Clinical Perspectives." In *Handbook of Emotions,* ed. M. Lewis and J. M. Haviland-Jones, 573–593. New York: Guilford Press.

Palermo, G. B., and L. E. Ross. 1999. "Mass Murder, Suicide, and Moral Development: Can We Separate the Adults from the Juveniles?" *International Journal of Offender Therapy and Comparative Criminology,* 43 (1), 8–20.

Parkinson, B. 1995. *Ideas and Realities of Emotion.* London: Routledge.

Reissman, D. 2007. "Reacting to the Crisis: From Public Trust to Panic." *Neiman Reports,* 6 (1), 71–72.

Sarnoff I., and P. G. Zimbardo. 1961. "Anxiety, Fear, and Social Affiliation." *Journal of Abnormal Social Psychology* 62: 356–363.

Stratton, J. G. 1981. "Pressures in Law Enforcement Marriages: Some Considerations." In *Stress and Police Personnel,* ed. L. Territo and H. J. Vetter, 233–241. Boston: Allyn & Bacon.

Sugden, R. 1986. *The Economics of Rights, Cooperation and Welfare.* Oxford: Basil Blackwell.

U.S. Census Bureau. 2011. North American Industry Classification System, 2007. http://www.census.gov/econ/industry/def/d624230.htm (accessed September 26, 2011).

Wright, C. 1959. *Mass Communication: A Sociological Perspective.* New York: Random House.

# PART II

# DISCIPLINARY AND POLICY PERSPECTIVES IN CRITICAL INCIDENT ANALYSIS

CHAPTER 4

# Critical Incident Analysis, Biological Events, and the Case of the 2009 H1N1 Influenza A Pandemic

SANDRO CINTI

This chapter focuses on the study of biological events as critical incidents. I first review how biological incidents became part of the field of critical incident analysis, highlighting several biological incidents from West Nile virus to bioterrorism. This review includes a summary of discussions held at critical incident analysis conferences.

This overview is followed by a discussion of the 2009 H1N1 influenza A pandemic. First, I review the characteristics of flu viruses, how pandemics occur, and what makes 2009 H1N1 influenza A virus unique. Next, I discuss the worldwide response to the pandemic in the context of the Critical Incident Analysis model (see Chapter 2): What went well, why did it go well, and did it matter? What went poorly, why did it go poorly, and did it matter? I focus on specific topics including the World Health Organizations (WHO) communications about the severity of 2009 H1N1 influenza A; the Centers for Disease Control and Prevention (CDC) guidelines on personal protective equipment (especially masks); CDC guidelines on school closure; and the U.S. vaccination effort in the fall and winter of 2009–2010. I attempt to support statements and conclusions using the academic literature; however, the use of media and Internet sources is unavoidable and necessary in order to fully understand the implications of a critical incident in today's twenty-four-hour news cycle.

## BIOLOGICAL EVENTS AS CRITICAL INCIDENTS

### Biological Events Are Different from Other Critical Incidents

Biological events, including bioterrorism, pandemics, and infectious outbreaks, present a unique challenge as critical incidents. First, the scope of these events is often very diffuse, with no definable borders (Roffey, Tegnell, and Elgh 2002). While incendiary events (explosions); weather disasters (hurricanes); and, generally, chemical/radiologic events (Sarin gas) can be contained and even cordoned off, biological events defy borders. Pandemics, like influenza, spread quickly throughout the world following human travel patterns (Mathews et al. 2009). Bioterrorism (BT) attacks, while initially localized, can spread quickly to adjacent areas depending on wind patterns, humidity, radiant light, ventilation systems, and contagiousness of the organism (smallpox) (Waterer and Robertson 2009). As the anthrax attacks of 2001 demonstrated, BT events can also be sporadic, affecting people or areas that were unexpected (Inglesby et al. 2002). Zoonotic diseases, like West Nile virus, can suddenly appear where they have never been before and affect humans and animals simultaneously (Fine and Layton 2001). The diffuse nature of biological events makes it difficult to efficiently focus resources (staff, medications, vaccine) where they are needed most.

Biological events also have a high fear/panic value. Historic accounts of plague outbreaks and pandemics (Barry 2004; Camus 1947), and current descriptions of Ebola outbreaks in Africa (LaFraniere 2005) and flesh-eating bacteria (Altman 2008) are often vivid and terrifying. Furthermore, these accounts and movie renditions of biological disasters (e.g., *Outbreak* [1995]) tend to reinforce the perception that other human beings, even one's own family members, may be the ministers of death and disease. During an outbreak of plague in Surat, India, in 1994, hundreds of thousands of inhabitants fled the city following reports of pneumonic plague spreading from person to person (Ramalingaswami 2001). During the 2001 anthrax attacks in the United States, more than ten thousand people were given antibiotics, even though only twenty-two became infected (Inglesby et al. 2002). The implications of such fear are that responders may refuse to care for victims of biological events and communities may shun infected individuals (Barry 2004; Camus 1947; Inglesby et al. 2002; LaFraniere 2005; Altman 2008; Ramalingaswami 2001). This certainly adds an extra layer of complexity to preparedness efforts.

Finally, biological events are primarily managed by the public health system. Traditionally, emergency management is the lead responder to large-scale critical incidents (Koenig 2009). At the federal level this task falls to the Federal Emergency Management Agency under the leadership of the Department of Homeland Security. Most emergency management personnel do not have sufficient medical or public health training to comfortably manage infected victims. On the other hand, public health responders are unaccustomed to the rapid real-time response required for most critical incidents. Historically, the role of public health departments has been limited to monitoring reportable diseases, promoting vaccination against vaccine-preventable diseases, and preventing spread of infectious diseases on a smaller scale (Cutler and Miller 2005). The rapid response and large-scale coordination of resources required for a diffuse biological event are often outside the scope of public health's capabilities. This has to do both with the type of training public health workers receive and the level of funding given to federal, state, and local public health departments. Recently, the federal government has required that public health responders be certified through the National Incident Management System (Annelli 2006).

## Examples of Biological Events that Were Critical Incidents

### *West Nile Virus Outbreak, 1999*

In the summer of 1999, there was a sudden outbreak of encephalitis among several patients in New York City. This particular form of encephalitis was unusual in that patients also developed severe weakness more consistent with a polio-like syndrome and the mortality was quite high among elderly and immune-compromised people. After a link was established between bird die-offs at a local zoo and human infections, the possibility of a mosquito-borne infection was considered. Soon after this association was made, virologists isolated West Nile virus (WNV) from several of the victims. Before this outbreak, WNV had never been isolated in the Western Hemisphere, although it was relatively common in the Middle East, Africa, and Asia. Even more troubling was the fact that WNV in the United States seemed to cause more severe disease than had been described in other countries (Fine and Layton 2001; Hayes and Gubler 2006). Over the next several years WNV traveled west across the United States and eventually became endemic in almost every state (Hayes and Gubler 2006).

There are several aspects of the WNV outbreak that qualify it as a critical incident. First, this was a sudden, unexpected event with the emergence of a virus never previously seen in the Western Hemisphere (Hayes and Gubler 2006). There were some initial concerns, quickly discounted, that this was a bioterrorism attack perpetrated by groups in the Middle East (Crupi et al. 2003). Second, initial descriptions of deaths and illness were of concern to the public, particularly in the absence of adequate treatment or vaccine (Fine and Layton 2001; Hayes and Gubler 2006). Highly publicized cases of encephalitis were front-page news in New York City (Altman 1999), and, other than avoiding mosquitoes, there was not much citizens could do to protect themselves. Third, the potential for spread of WNV throughout the rest of New York State and throughout the United States had public health officials concerned. The absence of vaccine to control spread, an incomplete understanding of the epidemiology of the virus, and an apparent increased virulence of this WNV compared to its predecessor in Israel challenged the public health infrastructure (Altman 1999; Crupi et al. 2003; Fine and Layton 2001; Hayes and Gubler 2006).

The impact of the WNV outbreak was significant immediately and for several summers after 1999, however this waned over the years and the overall impact after one decade was minimal. As mentioned earlier, there was significant concern that WNV would quickly surge across the United States with movement of the bird and mosquito populations. Initial control efforts included spraying areas with high mosquito populations with DDT and decreasing the number of mosquito breeding grounds. While the latter effort (e.g., removing water-filled tires, emptying birdbaths) was difficult and time-consuming, New Yorkers complied. However, as spraying began there was loud opposition from those who felt that exposure to DDT would be more harmful than WNV infection itself (Lopez 2002; Stone 2006). These initial missteps may have eroded some of the confidence in the New York City Health Department.

Overall, however, public health and the medical community handled this outbreak well. The initial identification of WNV as the cause of the new illness occurred quickly as a result of the smooth collaboration between the medical community, public health officials, and the veterinary community (Crupi et al. 2003; Fine and Layton 2001; Hayes and Gubler 2006). Public health and city officials were immediately able to focus resources on preventing a mosquito-borne illness, and the public was intelligently informed. Over the next few summers, while WNV did spread over the United States and continued to cause morbidity and mortality, it maintained a predictable pattern that had been discerned by epidemiologists at the CDC. Therefore, very little panic occurred in newly infected states as the virus traveled west. A decade later (in 2009), WNV remained endemic in the United States. The overall impact was lessened with improved testing for WNV and continued efforts to find treatments. Also, public health organizations fared very well in this critical incident, which proves that astute observations, epidemiological investigation, and honest communication with the public can stem panic and maintain confidence in government institutions (Crupi et al. 2003; Fine and Layton 2001; Hayes and Gubler 2006; Lopez 2002).

*Bioterrorism: The Anthrax Attacks, 2001*

In October 2001, in the wake of the September 11 World Trade Center attacks, the first case of anthrax occurred in a media worker in Florida (Inglesby et al. 2002; Jernigan, Stephens, Ashford et al. 2001). The sixty-three-year-old man developed a pneumonia and meningitis that was quickly diagnosed as anthrax by an astute infectious diseases specialist (Inglesby et al. 2002; Jernigan et al. 2001). The disease presented as either a skin infection that was easily treated with antibiotics or a devastating inhalational form (Inglesby et al. 2002; Jernigan et al. 2001). Over the next several weeks, twenty-one more people would become infected with anthrax and five would die (Inglesby et al. 2002). It was soon discovered that this outbreak was a biological attack perpetrated through the postal system using a powdered form of anthrax that had been created as a biological agent in the United States (the Ames strain) (Inglesby et al. 2002). Several prominent media figures (Peter Jennings, Tom Brokaw) and two U.S. senators (Tom Daschle, Patrick Leahy) received letters with anthrax spores in them (Inglesby et al. 2002). As a result of the twenty-one anthrax cases, more than ten thousand people received antibiotics as prophylaxis against the disease (Blank, Moskin, and Zucker 2003). In February 2010 the FBI closed its case on the 2001 anthrax attacks naming Dr. Bruce Ivins as the sole perpetrator of the attack (Shane 2010).

The anthrax attacks were clearly a critical incident in the United States and, most would argue, a worldwide critical incident. This critical incident shares some characteristics with the WNV outbreak. The first cases appeared suddenly, with no clear cause, and the infection had rarely been seen in the United States. Although anthrax is a well-characterized disease, the infection is mainly seen in third world countries. The most common form of infection, cutaneous anthrax, involves the skin and is easily diagnosed and treated. Inhalational anthrax and anthrax meningitis are less common and are associated with inhalation of spores, usually from animal fur. This form of disease, called woolsorter's disease, is difficult to diagnose and carries a high mortality rate. Like the WNV outbreak, the public was very concerned, given news accounts that this might be a biological attack. Furthermore, even though treatments were available, a delay in treatment could be catastrophic and antibiotic prophylaxis might be necessary to prevent disease (Inglesby et al. 2002; Blank et al. 2003). Finally, similar to the WNV outbreak, public health officials could not predict where the next cases would occur or how extensive the contamination had been.

There were additional aspects of the anthrax attacks, however, that differentiated this critical incident from the WNV outbreak. First, and most obvious, the anthrax attacks were cases of bioterrorism. The U.S. government had been concerned for several years that certain terrorist groups, with or without state support, might perpetrate a biological attack, and anthrax was identified as one of the most likely agents (category A) of attack (Darling et al. 2002; Inglesby et al. 2002). The al-Qaeda attacks on the World Trade Center had recently occurred, and there was speculation that the anthrax release might be an extension of that event (Roberge 2006). As a result there was instant worldwide attention to the anthrax outbreak even though very few cases occurred. Public fear and trepidation were widespread and the U.S. government had difficulty providing quick answers that might calm the panic. People went to extreme lengths to protect themselves, and some of these methods were propagated over the Internet. For example, after one expert suggested that ironing mail would kill anthrax spores, some people began ironing their mail and wearing gloves before opening envelopes (Roberge 2006). Several websites recommended using duct tape to anthrax-proof a room or house (Stark 2003). People presented to their physicians with requests for ciprofloxacin, an antibiotic used to treat anthrax, and there was concern that supplies of this drug were becoming scarce (Allegra, Cochrane, and Allegra 2004).

Second, the spread and extent of the anthrax outbreak was difficult for public health officials to characterize. The epidemiological pattern of WNV quickly became clear, and public health officials were able to predict subsequent spread to a fairly accurate extent (Fine and Layton 2001; Hayes and Gubler 2006). This was not true for the anthrax outbreak. There had been very little experience with bioterrorist attacks, and, even though some bioterrorism preparedness had occurred, the method of dissemination of anthrax through the mail was unexpected. Not knowing who had perpetrated the attacks, the CDC and FBI could not predict where other letters may have been mailed. In order to protect postal workers and Capitol Hill employees, extensive decontamination was performed and a large amount of antibiotic prophylaxis administered (Blank et al. 2003; Sharp and Roberts 2006).

The impact of the anthrax attacks was tremendous. Unlike the WNV outbreak, the anthrax attacks have had a lasting effect. This event was the first real test of the U.S. government's response to a biological attack. Early on there were missteps when an announcement was made by the Secretary of the Department of Health and Human Services (HHS) that an American photojournalist infected in Florida had contracted the disease from drinking stream water in North Carolina (Maillard et al. 2002). The secretary made these remarks without consulting medical experts, and, once it was revealed that the anthrax had been sent through the mail, the reputation of HHS was compromised (Lemonick 2001; Stolberg and Miller 2001). Furthermore, the delay in acknowledging that the anthrax outbreak might be a biological attack and the failure to immediately suspect the mail as the portal of infection may have "ended up costing the lives of two Postal Service employees who didn't have to die" (Lemonick 2001). Anxiety and resentment among postal workers was high, particularly vis-à-vis the efforts that were under way to protect senators on Capitol Hill. The postal group felt disenfranchised and suspected that they might not be treated as well as the senators (Meckler 2001). Eventually, the federal government mobilized significant resources to decontaminate the postal offices and treat the postal workers with prophylactic antibiotics (Blank et al. 2001; Sharp and Roberts 2006).

The federal government did eventually improve communication with the public and experts in public health, and, specifically, scientists with expertise in anthrax were able to craft accurate messages and propagate useful information. Frequent CDC briefings with the media facilitated by Director Dr. Julie Gerberding ensured that accurate information was being disseminated and that the media had a chance to ask questions (CDC 2001). In addition, the federal government released large amounts of antibiotics (ciprofloxacin) as prophylaxis for potentially exposed people (Blank et al. 2003). This helped calm fears and restore confidence that the government was acting responsibly and quickly.

The long-term impact of the anthrax attacks is difficult to quantify. It does seem, however, that it was a more impactful critical incident than the WNV outbreak. Large amounts of money were spent in cleanup and prophylaxis in the anthrax attacks. A shift in funding at the National Institutes of Health (NIH) made more money available to researchers who were working with category A agents such as anthrax, smallpox, plague, botulinum toxin, tularemia and hemorrhagic fever viruses (Frist 2002). This money was undoubtedly shunted from other valuable NIH-funded projects. International monitoring of biological agents, particularly category A agents, has increased, and scientists are often subject to investigation simply because they work with these agents (Kalish et al. 2009). The career of one prominent scientist was tarnished when he became the focus of an FBI investigation of the anthrax attacks (Apuzzo 2008). Finally, anthrax attacks convinced the public that terrorists were willing to use biological agents as weapons. This has had broad cultural repercussions with representations in literature, movies, and music.

Public health organizations received generally high marks for their response to the anthrax attacks (Harvard School of Public Health 2010). After initial missteps regarding the origin of the first case, the CDC took the lead in communicating to the public. This inspired confidence that the situation was being handled competently and calmed fears of impending doom (CDC 2001). The communication during this critical incident serves as a model for future events.

### *Severe Acute Respiratory Syndrome, 2003*

The severe acute respiratory syndrome (SARS) is thought to have originated in the Guangdong province of China in November 2002. At that time, reports of an unusual flulike illness were circulating in China but not much information was getting out to the World Health Organization (WHO 2003; Sampathkumar et al. 2003). It was not until February 2003 that an outbreak in Vietnam, reported by Dr. Carlo Urbani, brought international attention to this disease (Oransky 2003). At around the same time, a mainland Chinese physician staying at the Metropole hotel in Hong Kong infected sixteen other hotel visitors who then spread disease to Canada, Singapore, Taiwan, and Vietnam (Peiris et al. 2003). On March 12, 2003, the WHO issued a global alert that SARS was spreading to other countries. The CDC soon followed with its own health alert (WHO 2003).

By the conclusion of the epidemic in July 2003, 8,273 people had been infected and 775 had died (case-fatality rate approximately 10 percent). Epidemic outbreaks in Vietnam, Hong Kong, China, Canada, and Taiwan pushed hospitals and health systems to capacity and health care workers to the point of exhaustion (Parashar and Anderson 2004). Canada had two waves of disease that stretched resources. Late in the second wave, a call for physicians from the United States and other countries was made to help fill in for exhausted physicians (Penciner 2003). While no deaths occurred in the United States, twenty-seven people were infected and hospitals were bracing for an epidemic similar to Canada's (Cheng et al. 2007; Peiris et al. 2003).

The SARS epidemic had several qualities that made it a critical incident. As with WNV and the anthrax attacks, SARS was sudden and unexpected. Unlike the other two critical incidents, however, SARS was a completely new virus. Although the virus fit into the category of coronaviruses, it was much more virulent than others in this category. Most coronaviruses cause mild cold symptoms that resolve without any treatment, and patients rarely seek medical attention. The SARS coronavirus, on the other hand, causes a severe pneumonia that is fatal almost 50 percent of the time (Holmes 2003). An infected person starts with a cough, fever, and muscle aches. These symptoms worsen and eventually give rise to a viral pneumonia and acute respiratory distress. Worse, and maybe most concerning of all, is that SARS spreads efficiently from person to person through respiratory secretions (Peiris et al. 2003). WNV and anthrax are limited events, and there is nothing to fear from an infected person. Unfortunately, SARS spread most efficiently among health care workers who were not adequately protected with gowns, gloves, goggles, and masks. Some patients were considered "superspreaders" who could infect eight to ten others (Cheng et al. 2007; Holmes 2003; Parashar and Anderson 2004; Peiris et al. 2003). So as the epidemic spread, exhausted health care workers were increasingly fearful of becoming infected. The public, having little knowledge of this disease, avoided cities like Toronto for fear of becoming infected, and many of the hardest-hit cities experienced an economic downturn (*Baltimore Sun* 2003).

Finally, no antiviral agent or vaccine existed to treat or prevent SARS. Strict public health and infection control measures had to be instituted in order to stop the disease. The lack of treatment or vaccine made it difficult to retain physicians and nurses, as many were concerned about getting infected or infecting their loved ones (Peiris et al. 2003; Parashar and Anderson 2004; Cheng et al. 2007; Holmes 2003). Many health care workers who cared for SARS patients were quarantined until they were without fever for ten days after their last shift. This created great stress on the workforce that cared for ill patients (Hull 2005).

SARS had a widespread immediate impact throughout the world. Given that this was a completely new virus that was causing significant mortality, media attention to the outbreak was constant (Hull 2005). The spread of this virus within the health care community caused fear and trepidation among doctors, nurses, and others who had contact with patients. The dynamics of spread were unclear initially and the existence of superspreaders made transmission more difficult to predict (Peiris et al. 2003; Parashar and Anderson 2004; Cheng et al. 2007; Holmes 2003). It was not initially evident what patient characteristics identified someone as a superspreader. While it was clear that symptomatic patients needed to be isolated, negative-pressure rooms (rooms with a negative pressure to the outside) required for this level of isolation were scarce (Holmes 2003). As a result, some hospitals created SARS wards, which were isolated from other wards (*Asia Times* 2003; Holmes 2003). Health care workers on these wards were required to wear gloves, gowns, N95 masks (masks that filter 95 to 99 percent of particles), and goggles during their shifts (Holmes 2003; Sampathkumar et al. 2003). This garb can be uncomfortable, and many health care workers had difficulty breathing through the masks (Harber et al. 2009).

Some countries went to extreme lengths to contain the disease. China, which had the most cases, built a one-thousand-bed hospital to house SARS patients (*Asia Times* 2003). Toronto established a priority system that designated certain hospitals as SARS hospitals, while keeping other hospitals free of SARS patients (Svoboda et al. 2004). This high level of vigilance resulted in fatigue, anxiety, depression, and frustration among health care workers (Hull 2005). To make matters worse, at the height of the outbreak the supply of N95 masks was running low in several countries (Srinivasan et al. 2004).

Even though the spread of SARS outside hospitals was relatively low, widespread panic did occur in many communities. Tourism and business travel to Toronto essentially ceased for the duration of the outbreak (*Baltimore Sun* 2003; Holmes 2003; Svoboda et al. 2004). Several countries, particularly China, instituted stringent triage protocols in airports and public settings where screening for fever delayed flights and prolonged disembarkments (Chan et al. 2004). Many hospitals severely restricted access to the public, allowing only one family member to visit a sick patient. Furthermore, hospital staff members were triaged with questions and temperature checks every day (Svoboda et al. 2004).

The public health and scientific response to the SARS outbreak was quite impressive. Lessons learned from the anthrax attacks were applied to the SARS outbreak, and communication was excellent. The WHO had regular updates that relayed current information to hospitals, emergency planners, public health organizations, and media outlets around the world (WHO 2003). The CDC director had regular press briefings to keep the U.S. public current on the rapidly changing situation (CDC 2003). The research community very quickly identified the virus, and this information was shared throughout the world. Eleven virology research labs worked collegially and identified the coronavirus within two months of the WHO global alert (Ksiazek et al. 2003). This was remarkable given the secrecy that can sometimes shroud new scientific discoveries. Public health organizations also contributed significantly with epidemiologic data. Rapid responses to outbreaks allowed accurate data collection. There were some countries that limited access to WHO investigation efforts for fear that these investigations might affect tourism or travel (Parry 2003). For the most part, however, countries cooperated in the interest of controlling the outbreak.

The medium (2 years) and long-term (>5 years) impact of the SARS outbreak is difficult to gauge. Communication methods developed during the anthrax outbreak were clearly reinforced and improved upon during the SARS outbreak. The collaboration among labs around the world to rapidly identify the coronavirus was unprecedented and will likely be needed again in the future (Ksiazek et al. 2003). The success of infection control procedures in controlling SARS outbreaks was a victory for hospital epidemiologists, and these lessons are being applied to other outbreaks, including influenza (Ksiazek et al. 2003; Parashar and Anderson 2004; Svoboda et al. 2004). The concept of cohorting patients in order to control an outbreak gained renewed attention during the SARS outbreak and has since been applied to pandemic influenza outbreaks. Negative outcomes are more difficult to measure as the economies of affected cities rebounded eventually and the outbreak subsided before draconian measures such as forced quarantine were required on a large scale (Svoboda et al. 2004). Remarkably, SARS has essentially disappeared, and the last case was the result of a lab accident in 2004 (Normile 2004). While research on the virus continues, there has not been much incentive to increase scientific funding for work related to this virus.

## THE 2009 H1N1 PANDEMIC AND THE CRITICAL INCIDENT ANALYSIS MODEL

This section focuses on the 2009 H1N1 pandemic and how the Critical Incident Analysis (CIA) model can be used to assess the response to this critical incident's impact (0 to 2 years) and make predictions about its overall significance (>10 years). The first section will be a summary of the first wave of 2009 pandemic H1N1 influenza A (April–August 2009) and the challenges faced as the second wave of the pandemic descended upon the Northern Hemisphere (October 2009). The second section applies the CIA model to this incident, separating its components: the event (blob), the bureaucracy and adhocracy (triangle), the political arena (box), and the media. The third section addresses positive and negative interactions among the CIA model components. Particularly, interactions about mask use, vaccine development and distribution, and dissemination of information on the pandemic severity. Finally, the CIA model is used to make predictions about the impact (0–2 years) and long-term significance (>10 years) of this critical incident.

### 2009 H1N1 Influenza A Pandemic: Overview

On June 11, 2009, for the first time in forty years, the WHO declared an influenza pandemic (Chan 2009). The novel H1N1 influenza A virus, which had originated in Mexico in March 2009, was spreading throughout the world. The first wave through the spring and summer of 2009 was mild; however, a second wave beginning in September stressed resources and caused considerable mortality among children and young adults (CDC 2010a).

Worldwide influenza pandemics occur every thirty to forty years and vary in severity. The most devastating incident in the last hundred years occurred in 1918 during World War I. In this event, forty to fifty million people died worldwide and 500,000 died in the United States (Taubenberger, Reid, and Fanning 2000). By comparison, each year about forty thousand people die in the United States from seasonal influenza (Thompson et al. 2009). Pandemics occur when there is a major change (shift) in the circulating influenza A virus. While influenza A viruses do change year to year, only large changes cause pandemics. These changes occur in the hemagglutinin (H) and neuraminidase (N) portions of the virus (Glezen 1996). The 1918 pandemic resulted from a major shift from an H3N8 virus to an H1N1 virus. This virus was so different that most of the world's population had no immunity to it. The result was severe illness, particularly among young people whose strong immune systems turned against them (Taubenberger et al. 2000).

Influenza typically causes a febrile illness with chills, muscle aches, sore throat, headache, and nasal congestion. While "the flu" can be uncomfortable, most people recover within several days. Each year, however, vulnerable patients (elderly, immunocompromised, young children) die of influenza complications. Vaccines and antiviral medications can prevent influenza deaths and complications (Sullivan et al. 2009). While the symptoms of infection with a pandemic influenza strain are similar to seasonal influenza, the virus is more widespread in the population and infected people are more likely to get severe disease (Sullivan et al. 2009; Taubenberger et al. 2000). Severe disease or death from a pandemic influenza A strain results from a viral pneumonia affecting the deeper portions of the lung or from secondary bacterial infections that take over after the immune system is weakened by the influenza A virus (Sullivan et al. 2009; Taubenberger et al. 2000). Because a pandemic often occurs suddenly, there is little time to develop vaccine quickly enough to stop the spread of disease.

An influenza pandemic is a critical incident because it is a sudden, unexpected event that spreads quickly throughout the world causing considerable mortality and much fear. The 2009 novel H1N1 pandemic certainly qualifies based on these criteria. The initial reports from Mexico in April 2009 raised immediate fears and speculation that a pandemic might occur. Deaths in Mexico were sudden and unexpected and the virus spread quickly, first to the United States and then to the rest of the world (Sullivan et al. 2009). The panic died down as more epidemiologic and scientific data proved that the virus was not as deadly as the 1918 influenza strain (Sullivan et al. 2009). However, in the second wave of the pandemic, many more people become ill and there was increasing concern about mortality among children (Libster et al. 2010). Furthermore, hospital resources including masks, personnel, ventilators, and intensive care unit beds were strained as the virus persisted (Schneck 2009). Vaccine eventually became available but only in limited amounts for certain priority groups, including pregnant women, persons caring for children younger than six months old, and health care workers (Kuehn 2009).

## 2009 H1N1 Influenza A Pandemic: CIA Model Components

### *The Event: The Blob*

A biological event can be characterized by extent, severity, and whether it naturally occurs or is a form of bioterrorism. The event can be further characterized by multiple waves (spring/summer and fall/winter). As described earlier, the parameters of the H1N1 pandemic event changed rapidly. It was clear early on that the event occurred naturally. However, the world had not seen a pandemic in more than forty years (1968 was the most recent), and it was not clear what to expect. The most recent biological events, including the WNV outbreak, the anthrax attacks, SARS, and avian flu, all had potential to spread, but worst-case scenarios were not realized for any of these events. As the news about a new influenza A virus began to surface in Mexico and the United States, speculation and fears of a pandemic immediately hit the airwaves and the Internet. Furthermore, avian influenza A was fresh in the public's mind, making the trepidation more acute. Why, when influenza is well known in the scientific and public health communities, is there so much trepidation and uncertainty about such an event in this case? The answer is the unpredictable nature of the virus. Past pandemics have varied from mild (1968) to very severe (1918) (Taubenberger et al. 2000). The length of time since the previous pandemic (1968) was complicated by an expected pandemic in 1976 that materialized but led to a disastrous vaccination campaign (Fineberg 2009).

So, both the severity and extent of the event were initially unpredictable. As it turns out, the first wave was relatively mild in severity and appeared to have characteristics similar to a seasonal influenza A outbreak. It does differ, however, in that most deaths occurred in young people between the ages of fifteen and forty-five (Miroballi et al. 2010). The extent of the event was worldwide, and the WHO declared it a pandemic (Sullivan et al. 2009).

However, uncertainty remains about the second wave of 2009 pandemic H1N1 influenza A in the 2009–2010 influenza season. Was this a separate event or an extension of the first? Information from the Southern Hemisphere following its 2010 influenza season seemed to support the latter (Kotsimbos et al. 2010). The severity was similar to the earlier incident, with very little mixing between "seasonal" influenza and 2009 pandemic H1N1 influenza A (Kotsimbos et al. 2010). The extent of the event was also unchanged, with continued worldwide activity and little evidence of abating. Past experience led to trepidation, as the mild 1918 spring wave was followed by a fall/winter wave of a much more severe pandemic (Taubenberger et al. 2000).

### *The Bureaucracy and Adhocracy: The Triangle*

The triangle represents the responders and managers of the event. In naturally occurring biological events this group is largely comprised of members of the public health and medical communities. Public health interveners include local, regional, and/or state public health employees who interact with hospitals and the community to disseminate messages, perform surveillance, and implement containment strategies including isolation, quarantine, and community mitigation (e.g., closing schools). During the first wave of the 2009 H1N1 pandemic, public health interveners throughout the world worked feverishly to gather surveillance. These efforts paid off early when data demonstrated that H1N1 influenza A caused a relatively mild illness (Garske et al. 2009). Mitigation efforts varied around the world. Some countries monitored incoming flights (China); some slaughtered pigs (Egypt); and many closed schools where H1N1 cases were found (Audi 2009; Embassy of the PRC 2009; Oliver 2009). Each of these interventions and many others required huge efforts by an already strained public health system. For example, the Michigan Department of Community Health operated in a state with a dismal economic situation (Carmichael 2008). This led to public health jobs being cut and mandatory furloughs instituted (Brayton 2009).

Medical interveners include hospitals; clinicians (e.g., physicians, nurses); paramedics and behavioral health experts (e.g., psychiatry, social work). This contingency is responsible for direct patient care and is therefore at high risk. While many critical incidents pose a risk to interveners, a contagious biological event is especially dangerous. Not only must the medical intervener deal with the stress of caring for sick, scared, and anxious victims, but he or she must also be concerned about getting infected or infecting loved ones. Furthermore, the loss of direct care interveners as a result of illness adds another level of complexity and stress to the event. The mild nature of the first wave of the 2009 H1N1 allowed medical interveners to work without much fear, although many were annoyed at having to wear personal protective equipment (PPE) (Garske et al. 2009; Sullivan et al. 2009).

There were some outbreaks, however, that stressed medical interveners. The University of Michigan Hospital and Health Centers, a tertiary care center, received fifteen very sick patients with 2009 pandemic H1N1 influenza A in the intensive care unit (ICU) (Napolitano et al. 2009). Not only did the staff in the ICU have to manage very sick young patients, some of whom died, but they also had to contend with getting infected themselves. Furthermore, the PPE required for protection is uncomfortable and restricting. Although the situation never reached a tipping point, resources and nerves were strained. Had the outbreak in the ICU lasted longer or become more widespread, a greater crisis would have emerged. Other potential medical interveners include dental services and nursing, medical, and dental students. These groups might have expanded caretaking roles if staff shortages reach 30 to 40 percent in a second wave.

Other members of the triangle for this incident included emergency management personnel; public safety and security personnel; and hospital ancillary personnel (e.g., maintenance workers, food service workers). The first wave of 2009 H1N1 influenza A pandemic did not require activation of these groups. If staff shortages occur during the second wave or if civil unrest becomes a factor, these groups will have more of a role as interveners. Particularly, security and safety personnel may be needed if hospitals and emergency rooms become overrun with the sick and worried well (Kimmerly 2009).

### *The Political Arena: The Box*

The major participants in the political arena during the 2009 H1N1 pandemic included the WHO at the international level, the Department of Health and Human Services and the CDC at the national level, and state and local public health departments. In the United States, the CDC was the repository for all guidance on the 2009 pandemic, including recommendations on treatment, vaccination, infection control, and community mitigation (CDC 2009b). While CDC guidelines mostly agreed with WHO guidelines, there were some differences, particularly on mask use. Investigative and security organizations including the FBI and the Department of Homeland Security had very little involvement in the first wave of the 2009 H1N1 pandemic. This is in contrast to the security and investigative issues that were present after the anthrax attacks of 2001 (Warrick 2010). As a result, the box was much less a factor in the first wave of the 2009 H1N1 pandemic than the triangle (Figure 4.1).

### *The Media: The Continuous Stream*

The media played a very important role in the first wave of the 2009 H1N1 influenza pandemic. At first, media reports from Mexico in April 2009 seemed to indicate that the outbreak was severe with a high hospitalization and mortality rate (CDC 2009a). This caused immediate anxiety. As New York City began to report more cases, news spread about school closings and other public health mitigation efforts (CBS News 2009). As more information surfaced indicating that the 2009 pandemic might actually be a mild disease, pundits and experts wondered out loud why schools needed to be closed. These conflicting messages were confusing to state and local interveners until the CDC finally recommended that schools with 2009 pandemic H1N1 influenza A cases should remain open unless more than 25 percent of the students became ill (Park 2009).

While many localities in the United States were inundated with information about 2009 pandemic influenza A, some counties had a limited ability to disseminate information. For example, in Ann Arbor, Michigan, the *Ann Arbor News* had recently closed (Karoub and White 2009). The *News* had been the main source of local news, and its absence was felt acutely. It became difficult to deliver up-to-date information on the pandemic to the public. Recommendations from public health officials on testing, treatment, infection control, and vaccination were only available on the Internet or broadcast radio. This seemed to be adequate in the mild first wave, but in a more severe incident, the absence of print news could make public health's job more difficult.

## Positive Interaction: Dissemination of Information to Responders in the Triangle

The dissemination of information about novel H1N1 by public health organizations (WHO, CDC) to first responders was excellent in the first wave of the pandemic. Understandably, at the outset information coming from Mexico caused concern and seemed to suggest a severe pandemic with high mortality and morbidity rates (Walsh 2009). For one week in April 2009 novel H1N1 dominated the news. However, rapid genetic analysis of the new virus by WHO and CDC labs demonstrated that this strain did not have characteristics similar to avian influenza (H5N1) or 1918 influenza A

Figure 4.1 **The CIA Model in the First Wave of the H1N1 Pandemic**

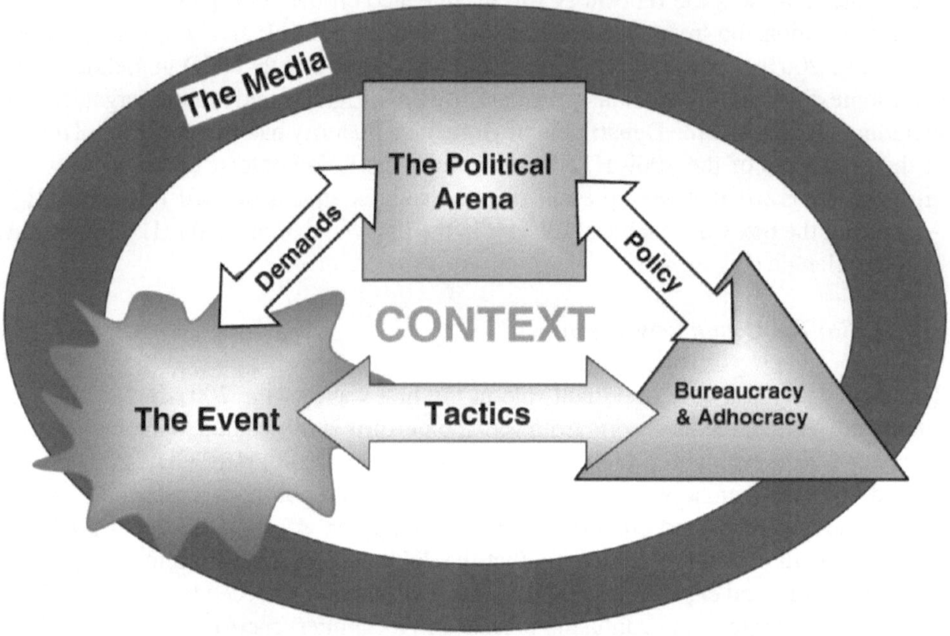


(Saey 2009). Furthermore, epidemiologic studies of the novel H1N1 outbreak in the United States proved that the disease was relatively mild and most similar to seasonal influenza (Sullivan et al. 2009).

As the pandemic spread, more studies and observations helped responders make strategic decisions. Data from the Southern Hemisphere influenza season, which precedes the Northern Hemisphere's flu season by several months, demonstrated that most cities could expect two to three times the volume of influenza cases in clinics, emergency and urgent care facilities, and hospitals (Kotsimbos et al. 2010). This information was rapidly analyzed and guided preparations in the Northern Hemisphere (Kotsimbos et al. 2010; Sullivan et al. 2009). Genetic analysis of the Southern Hemisphere viruses and studies in which ferrets were infected with these strains confirmed that the virus was not changing or becoming more virulent as some experts had feared early in the pandemic (DHHS 2009). Armed with this information, public health officials and hospital preparedness leaders were able to allay fears and develop effective prevention and treatment protocols.

In addition, the CDC and WHO did an effective job updating and disseminating recommendations on treatment, infection control, and vaccine delivery and administration. The CDC website was comprehensive and up to date (CDC 2009a). Important changes in recommendations were widely disseminated to public health organizations, hospitals, and the public in a timely fashion with logical explanations. The one exception to this was the infection control guidelines on mask use (CDC 2009b).

## Negative Interaction: CDC Infection Control Recommendations on Mask Use

During a typical influenza season the CDC recommends that a clinician who is caring for a patient with suspected influenza infection should wear an isolation/surgical mask (droplet precautions) in addition to gloves (CDC 2009b). This mask prevents the clinician from being infected by large droplets made airborne when a flu patient coughs or sneezes. Certain infections, such as tuberculosis and SARS, require a more protective mask called a respirator (N95 mask), which can filter out small airborne particles. These masks are cumbersome to wear, and they are often in short supply. When H1N1 influenza A first presented in April 2009, the CDC recommended that clinicians with close patient contact should wear an N95 mask (CDC 2009b). This seemed appropriate at the time because little was known about the virulence of novel H1N1 and maximum protection seemed prudent.

However, as more data emerged about this virus's similarity to seasonal influenza, infection control experts in the United States felt that infection control practices commensurate with seasonal flu could be reinstituted. The WHO, the Society for Hospital Epidemiologists of America, the Infectious Diseases Society of America, and the Association for Professionals in Infection Control and Epidemiology supported this view (Society for Healthcare Epidemiology of America [SHEA] 2009; WHO 2007). Even the Healthcare Infection Control Practices Advisory Committee, which advises the CDC on infection control issues, recommended going back to using regular isolation/surgical face masks for most patient care and using respirators only for high-risk procedures (e.g., bronchoscopy, intubation; SHEA 2009). Several states and hospital systems followed the new recommendations, because it was clear that respirators were in short supply.

Unfortunately, the CDC, the Institute of Medicine, and the Occupational Safety and Health Association did not agree and continued to recommend that clinicians use fit-tested N95 masks. These groups based their recommendations on several studies that showed that respirators were better than isolator/surgical masks for preventing the spread of influenza (National Academy of Sciences [NAS] 2009). Other studies, however, refuted these findings and demonstrated that both masks were equally protective (Loeb et al. 2009).

It appeared to the first-response community (the triangle) that the governmental bodies, (CDC and OSHA [the box]) were disregarding shortages in N95 masks and overburdening hospitals with just-in-time fit-testing of these masks and cumbersome infection control procedures. Disagreement on the mask issue continued through the second wave in fall 2009. Finally, on October 14, the CDC allowed hospitals with shortages to use isolation/surgical masks in certain situations (CDC 2009a). Unfortunately, the process of prioritizing respirators is very complicated, and many hospitals around the country continue to preferentially use isolator/surgical masks. Recently, a study in China that showed respirators to be more protective than isolator/surgical face masks was retracted after a review of statistical analysis (Roos 2009). On the other hand, a study among nurses in Canada showed no difference between the two different masks (Loeb et al. 2009). As of this writing (March 2010) the mask controversy has not been fully resolved and much anxiety persists about mask availability and the possibility of government sanctions for not following established CDC guidelines.

## The Impact of 2009 Pandemic Influenza A

As of March 2010, there were approximately forty-one thousand hospitalizations and more than two thousand deaths in the United States attributed to the second wave of the 2009 H1N1 pandemic influenza A virus (CDC 2010b). It does not appear that the virus had changed or become more virulent, and most influenza experts believed that the pandemic closely resembled the 1957 pandemic, which resulted in sixty thousand deaths (Henderson et al. 2009). Using the 1957 pandemic as a surrogate to estimate impact, it would follow that, while novel H1N1 is stressed the health care system, its long-term impact would probably be minimal (Henderson et al. 2009). However, there are a few potential long-term consequences worth mentioning. As a result of the mask controversy, lawsuits would be filed by health care workers who felt they were not adequately protected in hospitals that did not follow CDC guidelines (Hennessy-Fiske 2009a). These post-pandemic legal actions might lead to policy changes or actions by OSHA against hospitals that did not comply. Future studies on mask safety might lead to better methods of protecting health care workers against respiratory pathogens.

Another potential area of impact might be influenza vaccine production and distribution. Very few hospitals in the United States mandate that health care workers get seasonal influenza vaccines, and, as a result, only about 60 percent of the health care workforce accepts vaccine (Sullivan et al. 2009). However, in hospitals that tie clinical privileges to getting influenza vaccine, the uptake is greater than 95 percent. The 2009 H1N1 pandemic has placed a spotlight on this issue, and many hospital systems are now considering mandatory influenza vaccination (Sullivan et al. 2009). Vaccine availability was also an issue during the 2009 pandemic. Production was slow (>4 months) and there was not enough vaccine for those who needed it (Kuehn 2009). Long lines of anxious and frustrated people made headlines all over the United States (Hennessy-Fiske 2009a). The 2009 H1N1 pandemic spurred the government to dedicate more resources to improving vaccine production and distribution. Advances that are likely to occur in the future include moving away from egg-based vaccine production, developing vaccine adjuncts that increase the potency of the vaccine and allow the use of smaller doses, and developing a vaccine that protects against all influenza A viruses rather than just the seasonal strains (Horimoto and Kawaoka 2009).

## REFERENCES

Allegra, P. C., D. G. Cochrane, and J. R. Allegra. 2004. "Public Concern for Exposure to Agents of Bioterrorism Leading to Emergency Department Visits in New Jersey, 2001." *Annals of Emergency Medicine* 44 (4; suppl.): S34–S35.

Altman, Lawrence K. 1999. "Scientists Say Virus in Encephalitis Outbreak Is Like an Israeli Strain." the *New York Times,* December 5, http://www.nytimes.com/1999/12/05/nyregion/scientists-say-virus-in-encephalitis-outbreak-is-like-an-israeli-strain.html (accessed July 16, 2011).

———. 2008. "New Bacteria Strain Is Striking Gay Men. The *New York Times,* January 15, http://www.nytimes.com/2008/01/15/health/15infe.html?ref=science (accessed July 16, 2011).

Annelli, J. P. 2006. "The National Incident Management System: A Multi-agency Approach to Emergency Response in the United States of America." *Rev Sci. Tech* 25 (1): 223–231.

Apuzzo, Matt. 2008. "Anthrax Doctor's Suit with Justice Department Settled. *The Huffington Post,* June 27, http://www.huffingtonpost.com/2008/06/27/anthrax-doctors-suit-with_n_109684.html (accessed July 16, 2011).

*Asia Times.* 2003. "Xiaotangshan: Beijing's 'SARS City.'" *Asia Times Online,* May 15, http://www.atimes.com/atimes/China/EE15Ad04.html (accessed July 16, 2011).

Audi, Nadim. 2009. "Culling Pigs in Flu Fight, Egypt Angers Herders and Dismays U.N." *The New York Times,* April 30, http://www.nytimes.com/2009/05/01/health/01egypt.html (accessed July 16, 2011).

*Baltimore Sun.* 2003. "Travelers Told to Avoid Beijing, Toronto." *The Baltimore Sun,* April 24, http://www.baltimoresun.com/bal-te.china24apr24,0,1398836.story (accessed July 16, 2011).

Barry, J. 2004. *The Great Influenza: The Epic Story of the Deadliest Plague in History.* New York: Penguin Viking.

Blank, S., L. C. Moskin, and J. R. Zucker. 2003. "An Ounce of Prevention Is a Ton of Work: Mass Antibiotic Prophylaxis for Anthrax, New York City, 2001." *Emerging Infectious Diseases* 9 (6): 615–622.

Brayton, Ed. 2009. "State Government on Furlough Today." *The Michigan Messenger,* July 6, http://michiganmessenger.com/22187/state-government-on-furlough-today (accessed July 16, 2011).

Camus, A. 1991 [1947]. *The Plague*, trans. Stuart Gilbert. New York: Vintage Books.

Carmichael, Michael. 2008. "Michigan's Economy in 2009: What's Next?" *Corp!* (November), http://www.corpmagazine.com/Departments/CoverStories/tabid/54/itemid/253/pageid/3/Default.aspx (accessed July 16, 2011).

CBS News. 2009. "School Closings over H1N1 Continue in NYC." *CBS News,* May 18, http://www.cbsnews.com/stories/2009/05/18/health/main5022049.shtml (accessed July 16, 2011).

Centers for Disease Control and Prevention (CDC). 2001. "Update on Anthrax Investigations with Dr. Julie Gerberding." CDC Telebriefing Transcript, November 15, http://www.cdc.gov/media/transcripts/t011115.htm.

———. 2003. "Safer, Healthier Summer." CDC Telebriefing Transcript, June 26, http://www.cdc.gov/media/transcripts/t030626.htm.

———. 2009a. "Outbreak of Swine-origin Influenza A (H1N1) Virus Infection—Mexico, March–April 2009." *Morbidity and Mortality Weekly Report* 58: 1–3.

———. 2009b. "Interim Guidance on Infection Control Measures for 2009 H1N1 Influenza in Healthcare Settings, Including Protection of Healthcare Personnel," http://www.cdc.gov/h1n1flu/guidelines_infection_control.htm.

———. 2010a. "2009–2010 Influenza Season Week 7 Ending February 20, 2010," http://www.cdc.gov/flu/weekly/.

———. 2010b. "2009–2010 Influenza Season Week 8 Ending February 27, 2010," http://www.cdc.gov/flu/weekly/.

Chan, L. S., G. T. Cheung, I. J. Lauder, C. R. Kumana, and I. J. Lauder. 2004. "Screening for Fever by Remote-Sensing Infrared Thermographic Camera." *Journal of Travel Medicine* 11 (5): 273–279.

Chan, Margaret. 2009. "World Now at the Start of 2009 Influenza Pandemic," http://www.who.int/mediacentre/news/statements/2009/h1n1_pandemic_phase6_20090611/en/index.html.

Cheng, V. C., S. K. Lau, P. C. Woo, and K. Y. Yuen. 2007. "Severe Acute Respiratory Syndrome Coronavirus as an Agent of Emerging and Reemerging Infection." *Clinical Microbiology Reviews* 20 (4): 660–694.

Crupi, R. S., D. S. Asnis, C. C. Lee, T. Santucci, M. J. Marino, and B. J. Flanz. 2003. "Meeting the Challenge of Bioterrorism: Lessons Learned from West Nile Virus and Anthrax." *The American Journal of Emergency Medicine* 21 (1): 77–79.

Cutler, D. M., and G. Miller. 2005. "The Role of Public Health Improvements in Health Advances: The Twentieth-Century United States." *Demography* 42 (1): 1–22.

Darling, R. G., C. L. Catlett, K. D. Huebner, and D. G. Jarrett. 2002. "Threats in Bioterrorism; I: CDC Category A Agents: *Emergency Medicine Clinics of North America* 20 (2): 273–309.

Embassy of the People's Republic of China (PRC). 2009. "Ministry Orders Stepped-up Flu Monitoring after First Confirmed A/H1N1 Case." Embassy of the People's Republic of China in the United States of America, May 12, http://www.china-embassy.org/eng/zt/t561979.htm.

Fine, A., and M. Layton. 2001. "Lessons from the West Nile Viral Encephalitis Outbreak in New York City, 1999: Implications for Bioterrorism Preparedness." *Clinical Infectious Diseases* 32 (2): 277–282.

Fineberg, H. V. 2009. "Swine Flu of 1976: Lessons from the Past. An Interview with Dr. Harvey V. Fineberg." *Bulletin of the World Health Organization* 87 (6): 414–445.

Frist, B. 2002. "Public Health and National Security: The Critical Role of Increased Federal Support." *Health Affairs (Project Hope)* 21 (6): 117–130.

Garske, T., J. Legrand, C. A. Donnelly, et al. 2009. "Assessing the Severity of the Novel Influenza A/H1N1 Pandemic." *BMJ* 339: b2840.

Glezen, W. P. 1996. "Emerging Infections: Pandemic Influenza." *Epidemiologic Reviews* 18 (1): 64–76.

Harber, P., S. Bansal, S. Santiago, D. Liu, D. Yun, D. Ng, Y. Liu, and S. Wu. 2009. "Multidomain Subjective Response to Respirator Use during Simulated Work." *Journal of Occupational and Environmental Medicine* 51 (1): 38–45.

Harvard School of Public Health. 2010. "National Poll Examines Public Health Response to Anthrax Attack." *The Medical News,* February 20, http://www.news-medical.net/news/20100220/National-poll-examines-public-health-response-to-anthrax-attack.aspx.

Hayes, E. B., and D. J. Gubler. 2006. "West Nile Virus: Epidemiology and Clinical Features of an Emerging Epidemic in the United States." *Annual Review of Medicine* 57: 181–194.

Henderson, D. A., C. Brooke, T. V. Inglesby, E. Toner, and J. B. Nuzzo. 2009. "Public Health and Medical Responses to the 1957–58 Influenza Pandemic." *Biosecurity and Bioterrorism* 7 (3): 1–10.

Hennessy-Fiske, Molly. 2009a. "Nurses Plan Strike over Swine Flu Conditions at Hospitals." *The Los Angeles Times,* October 19, http://latimesblogs.latimes.com/lanow/2009/10/nurses-plan-strike-over-swine-flu-conditions-at-hospitals.html.

Holmes, K. V. 2003. "SARS-associated Coronavirus." *The New England Journal of Medicine* 348 (20): 1948–1951.

Horimoto, T., and Y. Kawaoka. 2009. "Designing Vaccines for Pandemic Influenza." *Current Topics in Microbiology and Immunology* 333: 165–176.

Hull, H. F. 2005. "SARS Control and Psychological Effects of Quarantine, Toronto, Canada." *Emerging Infectious Diseases* 11 (2): 354–355.

Inglesby, T. V., T. O'Toole, D. A. Henderson, J. G. Bartlett, M. S. Ascher, E. Eitzen, et al. 2002. "Anthrax as a Biological Weapon, 2002: Updated Recommendations for Management." *The Journal of the American Medical Association* 287 (17): 2236–2252.

Jernigan, J. A., D. S. Stephens, D. A. Ashford, et al. 2001. "Bioterrorism-related Inhalational Anthrax: The First 10 Cases Reported in the United States." *Emerging Infectious Diseases* 7 (6): 933–944.

Kalish, B. T., C. A. Gaydos, Y. H. Hsieh, et al. 2009. "National Survey of Laboratory Response Network Sentinel Laboratory Preparedness." *Disaster Medicine and Public Health Preparedness* 3 (2; suppl): S17–23.

Karoub, Jeff, and Ed White. 2009. "*Ann Arbor News* to Close in July; Will Be Replaced by Website." *The Huffington Post,* March 23, http://www.huffingtonpost.com/2009/03/23/ann-arbor-news-to-close-i_n_178095.html.

Kimmerly, D. P. 2009. "A Survey of Pandemic Influenza Preparedness and Response Capabilities in Chicago Area Hospital Security Departments." *Journal of Healthcare Protection Management* 25 (2): 16–32.

Koenig, K. L. 2009. "Preparedness for Terrorism: Managing Nuclear, Biological and Chemical Threats." *Annals of the Academy of Medicine, Singapore* 38 (12): 1026–1030.
Kotsimbos, T., G. Waterer, C. Jenkins, et al. 2010. "Influenza A/H1N1_09: Australia and New Zealand's Winter of Discontent." *American Journal of Respiratory and Critical Care Medicine* 181 (4): 300–306.
Ksiazek, T. G., D. Erdman, C. S. Goldsmith, et al. 2003. "A Novel Coronavirus Associated with Severe Acute Respiratory Syndrome." *The New England Journal of Medicine* 348 (20): 1953–1966.
Kuehn, B. M. 2009. "CDC Names H1N1 Vaccine Priority Groups." *Journal of the American Medical Association* 302 (11): 1157–1158.
LaFraniere, Sharon. 2005. "New Outbreak of Deadly Ebola Virus Is Feared in Congo Republic." The *New York Times,* May 14, http://www.nytimes.com/2005/05/14/health/14congo.html.
Lemonick, Michael D. 2001. "Anthrax." *Time,* December 31, http://www.time.com/time/magazine/article/0,9171,1001576,00.html.
Libster, R., J. Bugna, S. Coviello, D. R. Hijano, et al. 2010. "Pediatric Hospitalizations Associated with 2009 Pandemic Influenza A (H1N1) in Argentina." *The New England Journal of Medicine* 362 (1): 45–55.
Loeb, M., N. Dafoe, J. Mahoney, et al. 2009. "Surgical Masks vs N95 Respirator for Preventing Influenza among Health Care Workers. *Journal of the American Medical Association* 302 (17): 1865–1871.
Lopez, W. 2002. "West Nile Virus in New York City." *American Journal of Public Health* 92 (8): 1218–1221.
Maillard, J. M., M. Fischer, K. T. McKee Jr., L. F. Turner, and J. S. Cline. 2002. "First Case of Bioterrorism-related Inhalational Anthrax, Florida, 2001: North Carolina Investigation." *Emerging Infectious Diseases* 8 (10): 1035–1038.
Mathews, J. D., J. M. Chesson, J. M. McCaw, and J. McVernon. 2009. "Understanding Influenza Transmission, Immunity and Pandemic Threats." *Influenza and Other Respiratory Viruses* 3 (4): 143–149.
Meckler, Laura. 2001. "New Precautions Foreseen at Post Offices across the Country as Suspected Exposure to Anthrax Increases." *The Bismarck Tribune,* October 22, http://www.bismarcktribune.com/uncategorized/article_51a5e752-cbce-5e41-b9ee-8fd051a1f12d.html.
Miroballi, Y., J. S. Baird, S. Zackai, et al. 2010. "Novel Influenza A(H1N1) in a Pediatric Health Care Facility in New York City during the First Wave of the 2009 Pandemic." *Archives of Pediatrics & Adolescent Medicine* 164 (1): 24–30.
Napolitano, L. M., P. K. Park, K. C. Sihler, T. Papadimos, et al. 2009. "Intensive-Care Patients with Severe Novel Influenza A (H1N1) Virus Infection—Michigan, June 2009." *Morbidity and Mortality Weekly Report* 58 (27): 749–752.
National Academy of Sciences (NAS). 2009. "IOM Recommends N95 Respirators to Protect Health Care Workers from H1N1 Flu, Cautions against Reliance on a Single Strategy to Control Infection." September 3, http://www8.nationalacademies.org/onpinews/newsitem.aspx?RecordID=12748.
Normile, D. 2004. "Infectious Diseases. Second Lab Accident Fuels Fears about SARS." *Science* 303 (5654): 26.
Oliver, Steve. 2009. "Hundreds of School Closings Due to H1N1 Outbreak." *San Francisco Examiner,* October 29, http://www.examiner.com/h1n1-headlines-in-national/hundreds-of-school-closings-due-to-h1n1-outbreak.
Oransky, Ivan. 2003. "Carlo Urbani" [Obituary]. *The Lancet* 361 (9367): 1481.
Parashar, U. D., and L. J. Anderson. 2004. "Severe Acute Respiratory Syndrome: Review and Lessons of the 2003 Outbreak." *International Journal of Epidemiology* 33 (4): 628–634.
Park, Alice. 2009. "CDC Says H1N1 Outbreak Shouldn't Close Schools." *Time,* August 7, http://www.time.com/time/health/article/0,8599,1915244,00.html.
Parry, J. 2003. "Hong Kong Hampered in Fight against SARS by Lack of Cooperation from Mainland." *BMJ* 327 (7419): 832.

Peiris, J. S., K. Y. Yuen, A. D. Osterhaus, and K. Stöhr. 2003. "The Severe Acute Respiratory Syndrome." *The New England Journal of Medicine* 349 (25): 2431–2441.

Penciner, R. 2003. "I Am a SARS Physician." *Canadian Journal of Emergency Medical Care* 5 (4): 281–282.

Ramalingaswami, V. 2001. "Psychosocial Effects of the 1994 Plague Outbreak in Surat, India." *Military Medicine* 166 (12; suppl.): 29–30.

Roberge, M. R. 2006. "Bacillus Spores in the Mail: 'Ironing' Out the Anthrax Problem." *Journal of Medical Toxicology* 2 (2): 64–67.

Roffey, R., A. Tegnell, and F. Elgh. 2002. "Biological Warfare in a Historical Perspective." *Clinical Microbiology and Infection* 8 (8): 450–454.

Roos, Robert. 2009. "Reanalysis Changes Findings in Respiratory Protection Study." Center for Infectious Disease Research and Policy (CIDRAP) News, November 5, http://www.cidrap.umn.edu/cidrap/content/influenza/general/news/nov0509respirator.html.

Saey, Tina Hesman. 2009. "Genetic Analysis of Swine Flu Virus Reveals Diverse Parts." *Science News* 175 (13), http://www.sciencenews.org/view/generic/id/44068/title/Genetic_analysis_of_swine_flu_virus_reveals_diverse_parts.

Sampathkumar, P., Z. Temesgen, T. F. Smith, and R. L. Thompson. 2003. "SARS: Epidemiology, Clinical Presentation, Management, and Infection Control Measures." *Mayo Clinic Proceedings* 78 (7): 882–890.

Schneck, Bevan. 2009. "A New Pandemic Fear: A Shortage of Surgical Masks." *Time,* May 19, http://www.time.com/time/health/article/0,8599,1899526,00.html.

Shane, Scott. 2010. "F.B.I., Laying out Evidence, Closes Anthrax Case." The *New York Times,* February 19, http://www.nytimes.com/2010/02/20/us/20anthrax.html.

Sharp, R. J., and A. G. Roberts. 2006. "Anthrax: The Challenges for Decontamination." *Journal of Chemical Technology & Biotechnology* 81 (10): 1612–1625.

Society for Healthcare Epidemiology of America (SHEA). 2009. "Interim Guidance on Infection Control Precautions for Novel Swine-Origin Influenza A H1N1 in Healthcare Facilities." June 10, http://www.sgh.org.sa/sf/files/SHEA_Updated_IC_061209_H1N1_Statement.pdf.

Srinivasan, A., D. B. Jernign, L. Liedtke, and L. Strausbaugh. 2004. "Hospital Preparedness for Severe Acute Respiratory Syndrome in the United States: Views from a National Survey of Infectious Diseases Consultants." *Clinical Infectious Diseases* 39 (2): 272–274.

Stark, Judy. 2003. "Duct Tape and Cover." *St. Petersburg Times,* February 22, http://www.sptimes.com/2003/02/22/Homes/Duct_tape_and_cover.shtml.

Stolberg, Sheryl Gay, and Judith Miller. 2001. "A Nation Challenged: The Response; After a Week of Reassurances, Ridge's Anthrax Message Is Grim." The *New York Times,* October 26, http://www.nytimes.com/2001/10/26/national/26RESP.html?pagewanted=1.

Stone, Jess. 2006. "Net Gains." *The New York Times,* September 22, http://www.nytimes.com/2006/09/22/opinion/22stone.html.

Sullivan, S. J., R. Jackson, and G. A. Poland. 2009. "Mandating Influenza Vaccination for Healthcare Workers." *Expert Review of Vaccines* 8 (11): 1469–1474.

Svoboda, T., B. Henry, L. Shulman, et al. 2004. "Public Health Measures to Control the Spread of the Severe Acute Respiratory Syndrome during the Outbreak in Toronto." *The New England Journal of Medicine* 350 (23): 2352–2361.

Taubenberger, J. K., A. H. Reid, and T. G. Fanning. 2000. "The 1918 Influenza Virus: A Killer Comes into View." *Virology* 274 (2): 241–245.

Thompson, W. W., M. R. Moore, E. Weintraub, et al. 2009. "Estimating Influenza-associated Deaths in the United States." *American Journal of Public Health* 99 (2; suppl.): S225–S230.

U.S. Department of Health and Human Services (DHHS). 2009. "Assessment of the 2009 Influenza A (H1N1) Pandemic on Selected Countries in the Southern Hemisphere: Argentina, Australia, Chile, New Zealand and Uruguay." Report, August 26, http://flu.gov/professional/global/final.pdf.

Walsh, Bryan. 2009. "Mexico's Mystery: Why Is Swine Flu Deadlier There?" *Time,* April 29, http://www.time.com/time/health/article/0,8599,1894534,00.html.

Warrick, Joby. 2010. "FBI Investigation of 2001 Anthrax Attacks Concluded; U.S. Releases Details." *The Washington Post,* February 20, http://www.washingtonpost.com/wp-dyn/content/article/2010/02/19/AR2010021902369.html.

Waterer, G. W., and H. Robertson. 2009. "Bioterrorism for the Respiratory Physician." *Respirology* 14 (1): 5–11.

World Health Organization (WHO). 2003. "WHO Issues a Global Alert about Cases of Atypical Pneumonia: Cases of Severe Respiratory Illness May Spread to Hospital Staff." Press release, March 12, http://www.who.int/csr/sarsarchive/2003_03_12/en/.

———. 2007. "Infection Prevention and Control of Epidemic- and Pandemic-prone Acute Respiratory Diseases in Health Care: WHO Interim Guidelines." Report, June, http://www.who.int/csr/resources/publications/swineflu/WHO_CD_EPR_2007_6/en/index.html.

CHAPTER 5

# Space, Sentiment, and Social Structure after Critical Incidents

## Toward a Theory of Recovery

JAMES HAWDON AND JOHN RYAN

Ochberg and his associates (2007) offer a framework for studying critical incidents, known as the critical incident analysis (CIA) model (also see Chapter 2 in this volume). One goal of the *Handbook of Critical Incident Analysis* is to elaborate this framework to enhance its utility and lay a foundation for studying critical incidents. This chapter adds to the CIA model in one of its lesser-studied areas: the larger social context within which critical incidents occur. Kirby (2012, 30) states that "context refers to such things as the physical situation—the time, place, people—the who, what, where, and when" of the incident. The notion of "community" encompasses much of what this definition implies. If we accept that critical incidents are "large-scale events" (Kirby 2012, 20) and not individual incidents, all critical incidents happen to some collective and all collectives can be thought of as a type of community. While the scope of the "community" may vary—from informal groups to formal organizations; from a village or small town to a large city a nation-state or even the world system—all communities, regardless of their scope, have preexisting cultures and social structures. These preexisting cultures will vary in the extent to which they are complex or cohesive, and these preexisting structural arrangements will vary in the extent to which they are differentiated. It is likely that these cultural and structural sources of variation across communities dramatically influence how a critical incident is experienced.

In this chapter, we offer a general theoretic model to explain social responses to critical incidents based on what aspect of community the incident affects and the strength of various social networks operating in the affected community. We begin with a discussion of community. While theorists have conceptualized community in several ways, we rely on Campbell's (2000) conceptual model. We then consider what our model implies for definitions of critical incidents. In the next section, we outline the general responses communities have to critical incidents. Next, we use Hunter's (1985) discussion of the various "social spheres" that comprise a community's social structure and derive specific hypotheses relating these social spheres to the social response a critical incident will likely garner. We then offer evidence to support the theory, relying on observations from several critical incidents. Finally, we conclude by highlighting the implications of this research and offering directions for future research.

## THEORETICAL BACKGROUND ON COMMUNITIES

Researchers and theorists have conceptualized "community" as an analytic category, an empirical phenomenon, and a site of investigation. While these approaches have led to several definitions and theories of community, there are shared elements upon which a general conceptual model can be built. Campbell (2000) argues that the interrelated elements of *space, sentiment,* and *social structure* provide areas of agreement, encompass numerous perspectives of community as conceptualized by sociologists, and offer promising lines of inquiry.

According to Campbell (2000), *space* refers to the geographic location of the community. It includes the physical infrastructure of a community: its public buildings, factories, businesses, homes, parks, roadways, sidewalks, rivers, lakes, bridges, farms, fields, water and waste systems, electrical system, and other human-made and natural elements that make a given location look and function as it does. Space and place are basic components of the world as we experience it and define the scale of life for those who share it (Campbell 2000; Tuan 1987). It roots us to a location and provides stability and predictability to our worlds (Giddens 1990). Space was once limited to geographically bounded territories. But now, space can extend to "virtual space" in "virtual communities." Chat rooms, for example, are the space of the virtual communities that create them.

*Social structure* refers to the social networks that comprise the communal group. These networks, whether based on affection, coercion, or functional interdependency, give the community its social shape. They determine who is and who is not a member, who interacts with whom, who has power over whom, who has access to what resources, how those resources are distributed, and how the community approaches and attempts to solve the collective problems it confronts. A community's networks intersect and overlap; they connect not only community members to each other, but also the community to the larger social system in which the community operates. The linkages among individuals and organizations are based on relationships that create social connections through which information and material resources flow (Campbell 2000; Peacock and Ragsdale 1997). These ecological networks vary in the extent to which they are unified, centralized, and interdependent (Peacock and Ragsdale 1997).

Finally, *sentiment* is the aspect of community that is "felt, experienced, conceived, or communicated" and "exists as much in the hearts and minds of people as in the material components of its makeup" (Campbell 2000, 43). Sentiment is the symbolic and cultural side of community and includes the "things, behaviors, and ways of being that give any particular community its singular character" Campbell (2000, 45). It is the psychological attachment and emotional bond members have with their community and the sense of togetherness, norms of trust and reciprocity, or solidarity that members share with each other. What we mean by "sentiment" has been conceptualized by several similar concepts, including social solidarity (most notably, Durkheim's ([1915] 1966) concept of mechanical solidarity); psychological sense of community (McMillan and Chavis 1986); collective efficacy (Sampson, Raudenbush, and Earls 1997); and social capital (Putnam 1993; 2000).

## WHAT IS A CRITICAL INCIDENT?

Viewing critical incidents as occurring in communities and communities as the intersection of space, social structure, and sentiment allows us to consider what is meant by "critical incidents." If we consider what dimensions of community a critical incident affects and what dimensions are at times unaffected by incidents that many consider critical, a clearer definitional understanding emerges.

A critical incident can affect each of the dimensions of community. Numerous incidents have severely damaged community space (Bolin and Stanford 1998; da Cruz 1993; Reckdahl 2007). Hurricane Katrina, for example, literally transformed the "space" of New Orleans and the Gulf Coast by devastating entire neighborhoods (U.S. Department of Homeland Security 2006; Reckdahl 2007), and Hurricane Andrew destroyed a 30-mile-wide swath of southern Dade County and left an estimated 180,000 residents homeless (e.g., Girard and Peacock 1997). In addition to the devastation it left in North and South Carolina, Hurricane Hugo destroyed the majority of the built environment on St. Croix (Tierney, Bevc, and Kuligowski 2006). Similarly, the 1923 Great Tokyo Earthquake and fire destroyed approximately 381,000 of the approximately 694,000 houses in the Tokyo-Yokohama area (USGS 2009). The December 26, 2004, Indian Ocean tsunami destroyed the coastlines and coastal infrastructure in many Southeast Asia nations, including Indonesia, Sri Lanka, India, Thailand, the Maldives, Somalia, Myanmar, Malaysia, and Seychelles. The damage, estimated at more than $10 billion, included residential and commercial buildings as well as roads, water supply systems, electric power systems, schools, hospitals, and other health care facilities (Hawkins and Rao 2008; Risk Management Solutions 2006).

A critical incident can also alter the social structure of a community. With the loss of individual lives, families and friendship networks are altered. While many facing critical incidents rely on strong family and friendship networks for help (Hurlbert, Beggs, and Haines 2000), these relationships often become stressed after approximately one month (Bolin 1982; 1993; Harden and Vedantam 2005). For example, the vast destruction of the East Marmara Earthquake in Turkey (Kasapoglu and Ecevit 2003), Hurricane Andrew in south Florida (Morrow 1997a), and Hurricanes Rita and Katrina in the gulf region (Harden and Vedantam 2005) forced survivors to rely on their friends and families. While doing so helped the victims tremendously, their reliance on these networks severely strained their familial and neighborly ties. Similarly, critical incidents can alter community organizations and secondary relations. Widespread disasters can bankrupt businesses, which can be lost to the community forever (Dash, Peacock, and Morrow 1997). For example, after the 1985 earthquake in Mexico, 1,200 of the approximately 7,500 small businesses in the most critically affected area were destroyed and closed (da Cruz 1993). Similarly, after the 9/11 terrorist attacks, approximately 40,000 jobs were immediately lost, and an additional 300,000 residents lost their jobs because of street closures in lower Manhattan (Gotham and Greenberg 2008). Critical incidents often strain other community institutions such as the educational system (Kasapoglu and Ecevit 2003) and governmental agencies ranging from local agencies to the federal government (Bates and Peacock 1987; Brinkley 2006; Morrow and Peacock 1997; Rivera and Miller 2007). Yet, critical incidents also provide opportunities for new groups and organizations to form (Bates and Peacock 1987; Gotham and Greenberg 2008; Morrow and Peacock 1997). In either case, as some community groups dissolve and new ones form, the community's social structure changes.

A critical incident can also change residents' perceptions of what their community is; that is, it can change a community's sentiment. A horrific act of violence, for example, can weaken residents' beliefs that their community is safe. The Buffalo Creek flood, which devastated a small West Virginian community, adversely affected the sense of community the town's residents once held (Erikson 1976). Yet, just as critical incidents can both destroy and create new social structures, they can both destroy and create a sense of community. Critical incidents sometimes enhance the collective sense of pride, resolve, and togetherness associated with community; that is, these incidents can act as a source of cohesion or solidarity (Collins 2004; Durkheim 1915; Hawdon and Ryan 2008; Hawdon, Ryan, and Agnich 2010; Portes and Stepick 1993; Ryan and Hawdon 2008). For example, Collins describes a tremendous upsurge in solidarity after the 9/11 terrorist attacks (Collins 2004) and Morrow and Peacock (1997, 226) claim, "[t]he initial responses to (Hurricane) Andrew saw ethnic, racial, and class antagonisms seemingly dissipate, supplanted by community solidarity." There is ample anecdotal evidence of such a collective response after other tragedies and disasters (Barton 1969; Carretero and Angel 2003; Carroll et al. 2005; Collins 2004; Drabek 1986; Dynes 1970; Miki 2002; Pijawka, Cuthbertson, and Olson 1987; Quarantelli and Dynes 1976; Scarisbrick-Hauser and Lewis 1990; Shrum 2007; Siegel, Bourque, and Shoaf 1999; Smelser 2004; Sweet 1998; Turkel 2002). Again, the important point is that critical incidents affect a community's sentiments, either positively or negatively.

Consequently, critical incidents can affect all three dimensions of community. Yet we should recognize that these dimensions of community are highly interrelated. For example, sentiment largely determines how we use, interact in, and interpret a given space, while space and social structure dramatically influence sentiment. Thus, geographically defined community, where interaction occurs in social networks that are rooted in place, is a primary source of trust (see Airriess et al. 2008; Giddens 1990), and sentiments such as trust give shape to the social structure and space that make a community what it is (see Brehm and Rahn 1997; Ferman and Kaylor 2001; Putnam 2000). Thus, when critical events strike communities, they typically affect all three dimensions of community, at least to some degree. Even if the effect is indirect, critical events change how a community looks, behaves, and thinks.

Although violent incidents typically affect all three dimensions of community, examples demonstrate that not all dimensions of community need be affected for an incident to be considered critical. Many incidents that most people would consider "critical" do not affect a community's "space" or social structure. For example, most scholars would undoubtedly consider the assassinations of John F. Kennedy, Robert F. Kennedy, or Martin Luther King Jr. to be critical incidents; yet these murders did not affect the "space" of the American community. Nor did these events radically alter the social structure of the United States. Similarly, the 1986 assassination of Swedish Prime Minister Olof Palme led to mass grieving throughout Sweden and approximately a 21 percent increase in the incidence of live births of infants of very low birth weight (Catalano and Hartig 2001). This event was unquestionably critical to the Swedish community, but it did not alter that community's "space" or change the relationships that existed among Swedes. Thus, a critical incident does not necessarily affect space or social structure.

Conversely, incidents that only affect space or social structure are sometimes not widely defined as critical. Many wildfires, earthquakes, and floods destroy space, but if they do not threaten human life they are not typically considered critical. Even events that cause large-scale loss of life, thereby altering aspects of the community's social structure, are not always defined as critical. For example, in July 1995 a heat wave in Chicago resulted in the deaths of more than five hundred people. Despite the loss of life being twenty times that of Hurricane Andrew, the Chicago government and local press refused to recognize the heat wave as a major emergency and largely blamed the victims' actions for their deaths (Klinenberg 2002). Similarly, airplane crashes may result in hundreds of deaths, but most people consider these simply as tragic accidents. Therefore, events that only affect space or social structure may not be critical incidents. We contend, however, that all critical incidents affect sentiments.

The realization that critical incidents affect community sentiments is consistent with many existing definitions. For example, Carter, Kang, and Taggart (1999, 6) argue that critical incidents "have the potential to change public opinion, policies, and even social values." The Academy for Critical Incident Analysis (ACIA 2009) defines a critical incident as

> A relatively brief occurrence involving injury, loss, conflict, discovery, or change of significant proportion, usually unscripted and unanticipated, with the potential to alter existing societal norms. Critical incidents are usually traumatic, threatening the bonds of trust that bind communities, but may be positive, initiating historic consequents.

"Social values," "trust," and "altering social norm," are aspects of what Campbell (2000) means by a community's sentiments. Thus, critical incidents affect community sentiments. These events can have thousands or millions of victims, one victim, or no victims; they can be natural events or events created by human hands; they can alter a community's space and social structure or they can leave these dimensions unaffected; they can be tragic, but they not need be.

A critical incident occurred when Susan Smith murdered her two young sons by drowning them in a South Carolina lake. This incident was critical despite not influencing a community's space or social structure because, despite her claim that her car was stolen and her sons kidnapped, she had killed her own children. As Rekers (1995) says, a "murdering mother" demands much explaining. Evidence that shock gripped the community included flowers, small gifts, and letters placed on the shores of the lake (Hawdon 2009). Similarly, the Tate-LaBianca murders committed in August 1969 by Charles Manson and his followers was a critical incident, mainly because the murders stunned the nation and not because seven people died. Even in 1969, murder was not unheard of in the greater Los Angeles area; however, the apparent randomness of the murders, the fact they occurred in affluent neighborhoods, and the savagery of the acts terrorized the Los Angeles community (see Bugliosi and Gentry 1974). As Vincent Bugliosi, the lead prosecutor in the Manson trial, stated on the fortieth anniversary of the murders:

> [The Manson Family was] looking and living like typical hippies, but they were mass murderers. That was their religion, their credo. They wanted to kill as many people as they could. *That shocked the nation,* and that hurt the countercultural movement. (quoted in O'Neil 2009, emphasis added)

Finally, the Apollo 11 moon landing in 1969 was a critical incident, despite not being a tragic event, because it affected the sentiments of the American community. The mission's success stimulated a tremendous sense of pride and nationalism among Americans. This pride was in part because of the scientific and technological brilliance that the task required; however, it was also because the United States had unquestionably surpassed the Soviets in the space race. Throughout the 1960s, the Soviets had been winning the space race by launching the first satellite, putting the first animal into space, and putting the first human into space. The United States had trailed in achieving major breakthroughs in space exploration. The moon landing, which fulfilled President Kennedy's 1961 challenge to America, was therefore a critical symbolic victory for the United States in the Cold War (Nelson 2009).

Recognizing that critical incidents affect community sentiments, we now ask how they are affected. More important, if a critical incident shocks and damages a community's sentiments, how are they repaired? By what mechanisms do communities recover from traumatic critical incidents? Which dimension of community is most important for recovery? We now turn to these questions.

## COMMUNITY SENTIMENTS AFTER CRITICAL INCIDENTS

Given our claim that critical incidents necessarily affect a community's sentiments, after such an event, sentiments must change, at least temporarily. Therefore, after a critical incident, sentiments either weaken or strengthen. When they weaken, they expose prior rifts in the community's social fabric and threaten the group's future vitality. When they strengthen, they enhance social solidarity and bring community members closer together.

Some critical incidents weaken the collective sentiments and produce conflict at the community level (Carroll et al. 2006; Fradkin 2005; Quarantelli and Dynes 1976; Ryang 2003; Tierney 2007). Often, this conflict reflects preexisting tension and power differences within the community. As Bates and Peacock (1987, 312) state, "conflicts often emerge in the aftermath of a disaster over the distribution of scarce resources and over the equity principles which should guide the reconstruction effort." For example, the provision and allocation of disaster aid after the 1906 San Francisco earthquake reflected and reinforced the city's racial, ethnic, and class divisions (Fradkin 2005). During the Mississippi River flood of 1927, preexisting racial and class divisions influenced the decision to blow up levees on the Mississippi River to save New Orleans at the expense of poorer, neighboring parishes (Barry 1997). Similarly, after Hurricane Andrew, households in majority-white communities received more aid of all types, including federal assistance, than their equally victimized neighbors in majority-black communities (Dash et al. 1997; Morrow and Peacock 1997). After 9/11 and Hurricane Katrina, tax incentives, enterprise zones, and Community Development Block Grants were the central strategies for rebuilding the community. While these programs were justified as a means of creating jobs that would benefit the entire city, "in practice . . . tax incentives and subsidies are awarded in an inequitable fashion, going disproportionately to large firms and high-income residents, with little to no benefit for low- and moderate-income people" (Gotham and Greenberg 2008, 1055). Thus, in all of these cases, the critical incident experience was influenced by the pre-incident power dynamics embedded in the local social structure, and ultimately, the critical incident exasperated preexisting community rifts.

The conflicts that emerge over the distribution of resources may have serious political implications and result in permanent changes in the relationships between the community and the larger government (Bates and Peacock 1987). Thus, critical incidents can undermine the legitimacy of governments and elites. In the 1985 Mexico City earthquake, for example, the efforts of the victims to provide aid and provisions to and for themselves were much more rapid and effective than the government's response. The government's ineffective response was one factor that led to successful challenges to the Institutional Revolutionary Party (Olson 2000). Similarly, qualitative interviews with residents who suffered in the Big Branch Coal waste impoundment in eastern Kentucky indicate the incident elevated levels of public distrust in corporate, governmental, and regulatory authorities (Scorr et al. 2005). After Hurricane Andrew, the recovery efforts by the Federal Emergency Management Agency (FEMA) were badly mismanaged (see Peacock et al. 1997). Because of the agency's failures and perceived incompetence, many disaster specialists and some legislators thought that it would be better to dissolve the agency than to try to reform it (Franklin 1995); and when the agency mismanaged the Hurricane Katrina crisis, FEMA director Michael Brown was relieved of his duties amid harsh criticism (Baker 2005).

While critical incidents can weaken and threaten collective sentiments, they can also promote a sense of togetherness (Drabek 1986; Dynes and Quarantelli 1971; Quarantelli and Dynes 1976; Tierney et al. 2006; see Tierney 2007 for a discussion of how disasters produce both conflicts and solidarity). Typically, traumatic critical incidents produce responses of collective grieving. For example, shortly after the 1999 Columbine High School shootings, which left thirteen victims and the two gunmen dead, several mass displays of solidarity occurred. Thousands of mourners visited Clement Park and left hundreds of tributes to those who died; thousands visited the crosses placed on Rebel Hill to honor the victims; and thousands attended the formal and informal vigils that were held throughout the week (Cullen 2009). Similarly, a day after the mass murder of ten students in Kauhajoki, Finland, in September 2008, town residents held a candlelight vigil, and memorial services were held on the evening of September 23. On September 28, Finland's President Tarja Halonen, Kauhajoki students, and hundreds of town residents attended another service as the nation watched on live television (BBC News 2008). When a gunman killed thirteen and wounded more than forty people at the Fort Hood army base in November 2009, thousands of soldiers and other mourners gathered five days later to honor the victims (Tedford 2009).

Such mass gatherings not only serve to express grief, they also promote solidarity. As large numbers of persons focus their attention on the event and participate in mass rituals of communal bereavement, a collective mood sweeps over them. Because the affected sentiments are widely held and respected, any threat to them threatens the group's collective identity. The collective, in turn, resists this threat by acting together. As Durkheim (1915, 102) states in a discussion on community response to particularly heinous crimes, "everybody is attacked; consequently, everybody opposes the attack. Not only is the reaction general, but it is collective." The collective action following tragic critical incidents enhances the collective sense of pride, resolve, and togetherness associated with community; thus, the group becomes more united or solidified. Collins explains the process whereby the collective response to a traumatic critical incident stimulates solidarity:

> What creates the solidarity is the sharp rise in ritual intensity of social interaction, as very large numbers of persons focus their attention on the same event, are reminded constantly that other people are focusing their attention by the symbolic signals they give out, and hence are swept up into a collective mood. (Collins 2004, 55)

These displays of solidarity reinforce the sentiment that the group is special and will overcome the collective loss they feel.

The reinforcing of collective sentiments is likely a key factor in a community's recovery. Not only do these sentiments provide emotional support for survivors, they reaffirm that the group, while damaged, remains. For example, based on a longitudinal study of Virginia Tech students following the 2007 shootings there, students who expressed high levels of solidarity within the first five months of the murders had higher levels of emotional and behavioral well-being eight months and one year after the tragedy, even after controlling for initial levels of emotional and behavioral well-being (Hawdon and Ryan 2008). Similarly, after a nineteen-year-old gunman killed eight and wounded five by firing a rifle in a shopping mall in Omaha, Nebraska, residents with higher levels of solidarity expressed higher levels of well-being than did those residents with lower levels of solidarity (Hawdon et al. 2009). Again, this positive relation between solidarity and well-being held even after controlling for initial levels of well-being. The same relationship between solidarity and emotional well-being was observed after an eighteen-year-old student killed eight students, staff, and faculty in Finland's Jokela High School (Hawdon et al. 2009; see Räsänen, Oksanen, Hawdon, and Ryan 2008 for a discussion of how the Jokela incident affected the small town).

In addition to the emotional support the sentiments of solidarity can provide, they also can stimulate collective action (Chavis and Wandersman 1990; Hawdon and Ryan 2009; Hawdon, Ryan, and Mobley 2000; Sampson et al. 1997). For example, after the 1991 riots in Los Angeles, California, residents created *Rebuild LA* to coordinate recovery efforts, and community groups established *We Will Rebuild* after Hurricane Andrew devastated south Florida (Peacock and Ragsdale 1997). Similarly, in response to the Virginia Tech tragedy, the university established the VT-ENGAGE program. In a little more than one academic year, those involved pledged and served more than 350,000 hours of community service (Gilbert 2009). Once formed, groups that emerged in response to a critical incident can continue to provide valuable services to the community (Hawdon et al. 2000; Lochner, Kawachi, and Kennedy 1999). Consequently, critical incidents may damage communities, but they can also strengthen and allow them to recover by stimulating group cohesion. Clearly, for these processes to be effective, parochial relations must be in place before the advent of a critical incident. In other words, communities with strong parochial relations are resilient communities.

## DETERMINANTS OF THE DIRECTION OF SENTIMENT CHANGE

Sentiments of solidarity are central to recovery. But what sustains them? We know, for example, that the amplified sense of community that exists immediately following a crisis decreases with time as things return to normal. After 9/11 Collins (2004) found elevated levels of solidarity two to three months after the attacks; however, these gradually declined toward normalcy in six to nine months. A similar trajectory of elevated levels of solidarity lasting for approximately six months followed by a decline in solidarity was also seen at Virginia Tech after the 2007 murders (Hawdon et al. 2010). Typically, after solidarity returns to pre-incident levels, conflicts within the community reemerge and threaten it. It logically follows that sustaining high levels of solidarity would therefore lessen the probability that conflicts would emerge and slow a community's recovery. What, then, can sustain the heightened sense of community?

We believe the answer to this question lies in the character of the local social structure, because it is through social networks that "things happen." As Peacock and Ragsdale (1997, 25) state, "community recovery [is] a process in which the groups and organizations making up the community attempt to reestablish social networks to carry out the routines of daily life." Indeed, it is through social networks that collective action occurs (Bin and Edwards 2009; Granovetter 1973; Hawdon et al. 2000; Morrow and Peacock 1997; Orr 1999). Moreover, locally based social networks are the basis of a community's social fabric and are critically important for individual and aggregate levels of community attachment, solidarity, and cohesion (Alexander 2006). But what aspect of the social structure is most critical for sustaining community sentiment?

Hunter's (1985) discussion of the spheres of community-level social control can help to answer this question. According to Hunter, a community's social structure consists of networks in the private, parochial, and public spheres. The private sphere includes family and friendship networks. The private sphere is primarily comprised of "strong ties" (Granovetter 1973), although it can include neighbors who interact regularly but are linked by weaker ties than family networks tend to be (Bridge, Forrest, and Holland 2004). The local interpersonal networks and interlocking institutions that serve the community and its members, including the networks of stores, schools, religious institutions, voluntary associations, and clubs, make up the parochial sphere. The public sphere refers to the formal bureaucratic agencies of the state and public works (Hunter 1985). We also include the national media in the public realm. Researchers have found that these various types of networks relate to community functioning in different ways (see Bursik and Grasmick 1995; Hawdon and Ryan 2009), and we predict they are differentially related to the sentiments communities rely on to recover from traumatic critical incidents.

## SUSTAINING SOLIDARITY AS A PATH TO COMMUNITY RECOVERY

Each of the spheres of social structure plays an important role in the process of recovering from critical incidents. However, just as each realm serves a different function during normal times, each realm assumes different roles in the recovery process. We will now consider each sphere's role and specifically detail how the various spheres influence the community's collective sentiments.

### The Private Sphere

The primary relations of friends and family found in the private sphere are critically important for individual mental health (Granello 2001; Hawdon et al. 2009; Ruzek et al. 2007; Ryan, Hughes, and Hawdon 1998; Savage and Russell 2005), and provide valuable resources for those adversely affected by critical incidents (Harden and Vedantam 2005; Klinenberg 2002; Morrow 1997b). Although reliance on kinship and friendship networks diminishes as other forms of assistance become available, it remains important throughout the recovery process (especially see Morrow 1997b). In our research on the Virginia Tech recovery process, one of the primary predictors of emotional well-being six months after the tragedy was the extent to which students were embedded in private-sphere relations (Hawdon and Ryan 2008). We should therefore not underestimate the importance of private-sphere relations for individual well-being following a critical incident.

Yet private-sphere relations cannot sustain community solidarity. While important for individual recovery, private-sphere relations can actually impede the development and continuance of communal sentiments. As Sampson (2001, 95) argues, "social networks are differentially invoked, and dense, tight-knit networks may impede social organization if they are isolated or weakly linked to collective expectations for action." Similarly, Granovetter (1973) argues, individuals involved in a densely knit clump of friends can connect to another densely knit clump of friends through acquaintances; however, these clumps would be connected only by weak ties (Granovetter 1973; 1983). As a result, individuals can be difficult to organize or integrate into larger social movements. The collective movement will not spread beyond the clique and, as a result, "most of the population will be untouched" (Granovetter 1983, 202). While interacting with friends and family may help the individual heal, these interactions are limited in scope and perhaps limited in their effectiveness for community healing. Personal ties are uniquely constructed and negotiated among participants; thus, we expect that *the relations in the private sphere are too unique to the individuals involved to offer much to the community.*

### The Public Sphere

Similarly, relations in the public sphere—those initiated and conducted by official state agencies—are vitally important for a community to return to normal functioning. The public services provided by government are so central to the normal functioning of a community that these must continue to operate—or begin reoperating—if any semblance of normalcy is to return. For example, while the federal government has been criticized for its response to the tragedy of Hurricane Katrina, it provided resources far beyond the capacity of private networks and parochial organizations. As of August 21, 2006, federal spending on Katrina, Rita, and Wilma relief reached $108.9 billion. Of that money, $32.7 billion went to providing temporary housing and building long-term housing. In addition, the federal government allocated $18.2 billion to rebuild the levees, highways, and other aspects of the Gulf Coast's physical infrastructure (U.S. DHS 2006). By comparison, the American Red Cross raised $2 billion for Katrina relief by January 2006, and this $2 billon is six times the amount raised by the Salvation Army, the second-place charity fundraiser for Katrina relief (Reckdahl 2007). Thus, public-sphere resources are needed for recovery efforts, especially when the critical incident causes widespread damage to a community's space. However, relations in the public sphere, while necessary, are unlikely to generate community sentiment.

Public-sphere relations are affectively neutral and formal (Hunter 1985) and therefore offer little in the way of emotional support. In fact, there is evidence that local populations often oppose the rationalized forms of interaction and problem solving used by extra-local organizations attempting to coordinate and implement recovery efforts (Carroll et al. 2006). These ties also are not unique to a specific community. Each locality has governmental agencies involved in its daily operation. This commonality is especially pronounced where a federal government provides most of the services associated with disaster relief. Although an effective handling of the crisis by the government may make you proud of your nation, it is less likely to generate pride in your local community. Instead, conflict often occurs when nonlocal entities are too prominent in a recovery effort because there is a perceived loss of local agency (Carroll et al. 2006). Consequently, we expect that *the relations in the public sphere are not unique enough to offer much to the community's sentiment.*

## The Parochial Sphere

The parochial realm is also vital for community recovery. Often, the organizations within this realm perform important functions as an incident unfolds. Research indicates that the extent of connectedness in the parochial realm influences a population's vulnerability. Residents in communities with strong parochial-realm relations are able to access the resources of these interlocking organizations to find safety. Moreover, the wide-ranging networks of the parochial realm provide a source of surveillance as the community "looks out" for possible victims of a tragic incident. As a result, residents of neighborhoods with strong parochial relations fare better during a critical incident than those in neighborhoods with weak communal ties (Klinenberg 2002; Wisner et al. 2003). Parochial relations are also important in the immediate aftermath of a critical incident. For example, after Hurricane Andrew, in communities where authorities were slow to respond, local churches provided and coordinated relief efforts (Morrow and Peacock 1997). Community realm relations are also the key to sustaining the solidarity we believe is necessary for long-term recovery. Thus, these relations are vitally important during a critical incident, shortly after a critical incident, and for the long-term recovery from a critical incident.

There are several reasons parochial-realm relations are the key to sustaining community solidarity after a critical incident. First, the parochial realm offers relations that are unique to a community. These relations are common enough for community members to feel that they are a meaningful part of them, but limited enough to exclude outsiders. The parochial sphere is comprised of social networks of personal knowledge and trust and is reinforced by interlocking local institutions. These institutions create social boundaries that separate insiders, who are mutually accountable to one another, from outsiders (Hunter 1985). Second, the parochial sphere is based on the values of providing mutual aid and support (Hunter 1985), which may be critical to recovery. Third, the relations in the parochial realm amplify the collective's ability to realize a goal by promoting cooperation (Putnam 1993; 2000). As Granovetter (1973) argued, the residents of the Boston communities of Charlestown were more successful than were those in the West End in opposing urban renewal because they had a rich history of organizational life and its male residents worked in the community.

Thus, those who engage in parochial relations frequently are more likely to form the weak ties that connect tightly knit clumps of individuals to each other, which in turn empower the community. Consequently, we hypothesize that parochial relations will be the most important for producing and sustaining the sentiments that define community cohesion. They will therefore provide the most efficient means by which members of a community can recover from a disaster.

From this discussion, we can state the following theoretical propositions at the individual level:

- Participating in private-realm relations will be positively associated with individual levels of well-being after a critical incident; however, participation in private-realm activities after a critical incident will be inversely related to expressed levels of community solidarity.
- Participating in public-realm activities after a critical incident will positively influence the recovery of a community's "space" (assuming it is damaged); however, participation in public-realm relations will be inversely related to levels of community solidarity and will often be related to heightened conflict in the community.
- Participating in parochial-realm activities after a critical incident will be positively related to expressed levels of community solidarity.

We can also state hypotheses at the community level:

- The greater the percentage of residents who participate in parochial-realm activities, the higher the levels of community solidarity.
- The higher the levels of community solidarity, the lower levels of community conflict after a critical incident.

## SOME EVIDENCE FROM CRITICAL INCIDENTS

Preliminary evidence after the murder of twenty-seven students and five faculty members on Virginia Tech's campus in 2007 supports our propositions. First, solidarity surged after the event. On the day after the murders at Virginia Tech, more than ten thousand people attended a convocation held to honor the victims, and thousands of people gathered at a candlelight vigil that night. On the Saturday after the tragedy, thousands of students and town residents attended a picnic on campus (Hawdon et al. 2010; Ryan and Hawdon 2008). Serendipitously, we had data measuring levels of solidarity among a sample of Virginia Tech students collected prior to the critical incident. These unique data permitted us to compare levels of solidarity among students measured approximately one year prior to the critical incident to levels of solidarity measured approximately five months after the tragedy. Based on analysis of variance models, solidarity increased by approximately 18 percent, and the increased levels of solidarity occurred for all groups of students, regardless of age, race, ethnicity, gender, or academic measure.

Approximately nine months after the shootings, the heightened levels of solidarity were still evident; however, solidarity levels had significantly decreased between five and nine months post-tragedy. Levels of solidarity continued to decrease between nine months and one year (Hawdon et al. 2010). The intense solidarity that emerged after the event slowly decreased, and conflicts began to emerge in the community. For example, the university, police, and town officials faced a lawsuit filed by the families of two victims.

While many observers of prior critical incidents would anticipate the rise and fall of solidarity we documented at Virginia Tech, the influences of the various types of social relations on sustained levels of solidarity need investigating. Based on a repeated measures analysis that predicts levels of solidarity at five months, eight months, and twelve months after the tragedy for 369 respondents for whom all three waves of data were available, participating in parochial relations elevated solidarity among Virginia Tech students, faculty members, and staff (Hawdon and Ryan 2008). Participating in both general parochial activities, such as shopping at local stores and eating at local restaurants, and tragedy-specific parochial activities, such as attending a vigil for the victims, were positively related to solidarity shortly after a tragedy. The effects of parochial relations on solidarity were apparent even after controlling for gender, ethnicity, years at Virginia Tech, age, personal knowledge of a victim of a tragedy, and emotional well-being. Although the positive effects of general parochial-realm relations on solidarity disappeared after eight months, participating in tragedy-specific parochial events within a week of the tragedy significantly predicted solidarity levels after one year (Hawdon and Ryan 2008). In contrast, participation in private-realm activities or public-realm activities was unrelated to levels of solidarity. One public-realm activity, seeing a professional counselor, significantly reduced levels of solidarity, at least in the short term (Hawdon and Ryan 2008).

Similar results were found in Omaha, Nebraska, after the tragic 2007 mall shootings. Using an OLS regression to predict solidarity levels at "time 2" based on participation in private relations, participation in parochial relations, education, age, income, gender, ethnicity, the extent to which the respondent watched television news, and levels of emotional depression at time 1, we find support for our hypotheses. Participating in parochial relations nine months after the tragedy was positively related to levels of solidarity thirteen months after the tragedy (Hawdon et al. 2009). We also analyzed cross-sectional data from Jokela, Finland, collected after the 2008 school shooting. In this analysis, participating in parochial-realm relations was positively related to solidarity at the $p < 0.10$ level (Hawdon et al. 2009).

Based on these analyses from different communities in different cultures, it appears our theoretical assertions are correct. Parochial-realm relations generate the sentiments of solidarity that benefit communities after critical incidents. In these three cases, engaging in community activities helped bring people together. As noted by Collins (2004), participating in activities that are specifically focused on the tragic critical incident appears especially powerful in promoting solidarity.

## CONCLUSION

This chapter adds to the CIA model by elaborating the "context" field. Recognizing that critical incidents affect communities, we argue that, while critical incidents often influence a community's space and social structure, they always influence a community's sentiments. Sentiments may strengthen or weaken, but they always change. Especially after tragic critical incidents, sentiments strengthen as the group is brought together in common grief. As they focus their collective attention on the event, they generate community cohesion and solidarity. However, this solidarity often gives way to conflict as time passes, solidarity wanes, and the recovery process lingers. Yet participating in parochial-realm activities can sustain levels of solidarity.

While participating in private-realm activities can promote individual-level well-being, these activities can also separate the individual from her or his community. And, while the public-realm activities of governments and other extra-local organizations are needed to help with recovery from devastating damage to space, these activities may actually hasten conflict within the community. Thus, it is vital for local businesses, schools, religious organizations, clubs, and neighborhood associations to encourage residents to participate in the activities they offer after a critical incident. The solidarity these activities promote is beneficial to individuals and the overall community.

Our theory and the research that supports it have implications for efforts to help communities recover from tragic critical incidents. Those trying to help should remember that tragedies not only produce grieving individuals, they also produce grieving communities. The failure to distinguish between personal and communal grief may lead to misplaced attempts at healing those not directly affected by the tragedy. For example, many—if not most—of the counselors who arrived immediately after the Virginia Tech tragedy came with the assumption that community grief could be treated using the same techniques used to treat personal grief. The dominant psychological paradigm of grief carries the assumption that because the community was harmed, all individual members were harmed as well. This assumption commits the ecological fallacy. This approach may be especially problematic because interventions that focus solely on treating the individual fail to recognize the healing that can be experienced through engaging in community-level rituals of solidarity and bereavement. Although some psychological interventions suggested by those who came to help at Virginia Tech pointed to the importance of participating in symbolic displays of solidarity with the Virginia Tech community as a way to deal with the tragedy, most recommended seeing professional counselors, going home to families, or "going to the place where you will get the best hugs." While these approaches may be effective for people suffering individual grief, they may not be the best approach for dealing with communal grief. To view grief solely through a psychological paradigm ignores the inherent healing power of communal grief.

Although the effectiveness of grief counseling has been questioned (Center for the Advancement of Health 2003), we are not suggesting that the therapeutic community does not have an important role to play for those individuals who were directly affected by tragedy and for people who do not have the social networks that have historically helped individuals process grief. We in no way mean to downplay the intense personal grief felt by some community members, but communal grief does not need to be treated using a therapeutic approach because, as Durkheim noted, communal grief can itself be healing. When counselors and therapists enter a community experiencing communal bereavement, they can strip the community of its inherent ability to heal and undermine the solidarity-producing effects of having masses of people gather in collective grief and collective support.

Thus, those helping communities recover from tragedies should recognize the difference between individual and community grief. While those individuals who lack social support networks probably need intensive counseling, the efforts to provide that counseling should not interfere with the community's offering of parochial-realm activities. Canceling community events, which often occurs after tragedies out of respect for the victims, may not be the most beneficial approach. While one may find it difficult to continue with everyday mundane activities, these activities bring people together, promote solidarity, and enhance emotional and behavioral well-being. Local businesses, restaurants, religious establishments, voluntary organizations, and social clubs should be encouraged to proceed with meetings, even if only to give members an opportunity to talk.

Yet we caution that community solidarity is not a panacea or cure-all. Intense solidarity, by confining the collective to only those in their tightly bonded clique, can decrease ties between the community and the outside world (Granovetter 1983). Without ties to the larger society around them, the community can have a difficult time securing the resources needed to recover (Dash et al. 1997). In addition, in intensely solidified communities, parochial-realm relations can overwhelm and intrude on private relations. As noted earlier, each of these spheres of social networks performs its own function, and private-realm relations are important for those most directly affected by the critical incident.

Thus, communities trying to recover from a traumatic critical incident need to allow the community's social structure to operate as normally as possible. Private, public, and parochial networks need to do what they do well: Private networks support and comfort us, public networks provide resources and expertise needed for reestablishing a community's infrastructure, and parochial networks bring people together as they perform the activities of everyday life.

## ACKNOWLEDGMENTS

This material is based upon work supported by the National Science Foundation under Grant No. 0735471 and Grant No. 0825662. Any opinions, findings and conclusions, or recommendations expressed in this material are those of the authors and do not necessarily reflect the views of the National Science Foundation.

## REFERENCES

Academy for Critical Incident Analysis (ACIA). 2009. The Critical Incident Concept, http://www.aciajj.org/critical_incident_concept.html (accessed July 1, 2009).

Airriess, C. A., W. Li, K. J. Leong, A. C. Chen, and V. M. Keith. 2008. "Church-based Social Capital, Networks and Geographical Scale: Katrina Evacuation, Relocation, and Recovery in a New Orleans Vietnamese American Community." *Geoforum* 39: 1333–1346.

Alexander, J. C. 2006. *The Civil Sphere.* New York: Oxford University Press.

Baker, P. 2005. "FEMA Director Replaced as Head of Relief Effort." *Washington Post,* September 10, http://www.washingtonpost.com/wp-dyn/content/article/2005/09/09/AR2005090900795_2.html.

Barry, J. M. 1997. *Rising Tide: The Great Mississippi Flood of 1927 and How It Changed America.* New York: Simon & Schuster.

Barton, A. H. 1969. *Communities in Disaster: A Sociological Analysis of Collective Stress.* Garden City, NY: Doubleday.

Bates, F. L., and W. G. Peacock. 1987. "Disasters and Social Change." In *The Sociology of Disasters,* ed. R. R. Dynes, B. DeMarchi, and C. Pelanda, 291–330. Milan, Italy: Franco Angeli Press.

BBC News. 2008. "Finland Memorial for Gun Victims." September 28. http://news.bbc.co.uk/1/hi/world/europe/7640221.stm (accessed October 22, 2009).

Bin, O., and B. Edwards. 2009. "Social Capital and Business Giving to Charity Following a Natural Disaster: An Empirical Assessment." *The Journal of Socio-Economics* 38: 601–607.

Bolin, R. 1982. "Long-term Recovery from Disaster." Boulder, CO: University of Colorado, Institute of Behavioral Science.

———. 1993. "Household and Community Recover after Earthquakes." Boulder, CO: University of Colorado, Institute of Behavioral Science.

Bolin, R., and L. Stanford. 1998. *The Northridge Earthquake: Vulnerability and Disaster.* London: Routledge.

Brehm, J., and W. Rahn. 1997. "Individual-level Evidence for the Causes and Consequences of Social Capital." *American Journal of Political Science* 41: 999–1023.

Bridge, G., R. Forrest, and E. Holland. 2004. *Neighbouring: A Review of the Evidence.* Bristol: University of Bristol, Centre for Neighbourhood Research.

Brinkley, D. 2006. *The Great Deluge: Hurricane Katrina, New Orleans, and the Mississippi Gulf Coast.* New York: HarperCollins.

Bugliosi, V., and C. Gentry. 1974. *Helter Skelter: The True Story of the Manson Murders.* New York: Norton.

Bursik, R. J., and H. G. Grasmick. 1995. "Neighborhood-based Networks and the Control of Crime and Delinquency." In *Crime and Public Policy: Putting Theory to Work,* ed. H. Barlow, 103–130. Boulder, CO: Westview Press.

Campbell, C. D. 2000. "Social Structure, Space and Sentiment: Searching for Common Ground in Sociological Conceptions of Community." In *Research in Community Sociology,* ed. D. A. Chekki, 21–57. Stamford, CT: JAI Press.

Carretero, P., and E. Angel. 2003. "Social Reaction in the Face of Catastrophe: An Interpretation of Communal Response to an Ecological Crisis." *Societies* 3, 91–103.

Carroll, Matthew S., Patricia J. Cohn, David N. Seesholtz, and Lorie L. Higgins. 2005. "Fire as a Galvanizing and Fragmenting Influence on Communities: The Case of the Rodeo-Chediski Fire." *Society and Natural Resources* 18: 301–320.

Carroll, M. S., L. L. Higgins, P. J. Cohn, and J. Burchfield. 2006. "Community Wildfire Events as a Source of Social Conflict." *Rural Sociology* 71, 261–280.

Carter, S., M. Kang, and R. Taggart. 1999. "An Interdisciplinary Approach to a Critical Incident Course." *Journalism and Mass Communication Educator* 54 (2): 4–14.

Catalano, R., and T. Hartig. 2001. "Communal Bereavement and the Incidence of Very Low Birthweight in Sweden." *The Journal of Health and Social Behavior* 42: 333–342.

Center for the Advancement of Health. 2003. "Report on Bereavement and Grief Research." Washington, DC: Center for the Advancement of Health.

Chavis, D. M., and A. Wandersman. 1990. "Sense of Community in the Urban Environment: A Catalyst for Participation and Community Development." *American Journal of Community Psychology* 18: 55–81.

Collins, R. 2004. "Rituals of Solidarity and Security in the Wake of Terrorist Attack." *Sociological Theory* 22: 53–87.

Cullen, D. 2009. *Columbine.* New York: Twelve.

da Cruz, J. 1993. *Disaster and Society: The 1985 Mexican Earthquakes.* Lund, Sweden: Lund University Press.

Dash, N., W. G. Peacock, and B. H. Morrow. 1997. "And the Poor Get Poorer: A Neglected Black Community." In *Hurricane Andrew: Ethnicity, Gender, and the Sociology of Disasters,* ed. W. G. Peacock, B. H. Morrow, and H. Gladwin, 206–225. New York: Routledge.

Drabek, T. E. 1986. *Human Responses to Disaster.* New York: Springer-Verlag.

Durkheim, E. 1915. *The Division of Labor in Society,* trans. George Simpson. New York: Free Press, 1966.

Dynes R. R. 1970. *Organized Behavior in Disaster.* Lexington, MA: Lexington Books.

Dynes, R. R., and E. L. Quarantelli. 1971. "The Absence of Community Conflict in the Early Phases of Natural Disaster." In *Conflict Resolution: Contributions of the Behavioral Sciences,* ed. C. G. Smith, 200–204. South Bend, IN: University of Notre Dame Press.

Erikson, K. 1976. *Everything in Its Path: Destruction of Community in the Buffalo Creek Flood.* New York: Simon and Schuster.

Ferman, B., and P. Kaylor. 2001. "Building the Spatial Community: A Case Study of Neighborhood Institutions." *Policy Studies* 18: 53–70.

Fradkin, P. L. 2005. *The Great Earthquake and Firestorms of 1906: How San Francisco Nearly Destroyed Itself.* Berkeley: University of California Press.

Franklin, D. 1995. "The FEMA Phoenix: Reform of a Federal Emergency Management Agency." *Washington Monthly* (July/August), http://www.washingtonmonthly.com/features/2005/0509.franklin.html (accessed October 17, 2009).

Giddens, A. 1990. *The Consequences of Modernity.* Stanford: Stanford University Press.

Gilbert, K. 2009. "Having a Heart for Service." *Virginia Tech Magazine* (Spring), http://www.vtmagazine.vt.edu/spring09/feature1.html (accessed October 29, 2009).

Girard, C., and W. G. Peacock. 1997. "Ethnicity and Segregation: Post-hurricane Relocation." In *Hurricane Andrew: Ethnicity, Gender, and the Sociology of Disasters,* ed. W. G. Peacock, B. H. Morrow, and H. Gladwin, 191–205. New York: Routledge.

Gotham, K. F., and M. Greenberg. 2008. "From 9/11 to 8/29: Post-disaster Recovery and Rebuilding in New York and New Orleans." *Social Forces* 87: 1039–1062.

Granello, P. F. 2001. "A Comparison of Wellness and Social Support Networks in Different Age Groups." *Adultspan Journal* 3: 12–22.

Granovetter, M. 1973. "The Strength of Weak Ties." *American Sociological Review* 78: 1360–1380.

———. 1983. "The Strength of Weak Ties: A Network Theory Revisited." *Sociological Theory* 1: 201–233.

Harden, B., and S. Vedantam. 2005. "Many Displaced by Katrina Turn to Relatives for Shelter." *Washington Post,* September 8.

Hawdon, J. 2009. "Communal Bereavement." In *The Encyclopedia of Death and the Human Experience,* vol. I, ed. C. Bryant and D. Peck, 213–216. Thousand Oaks, CA: Sage.

Hawdon, J., and J. Ryan. 2008. "Campus Violence: Exploring a Community's Response to Tragedy." Washington, DC: National Science Foundation.

———. 2009. "Social Capital, Social Control, and Crime." *Crime and Delinquency* 55: 526–549.

Hawdon, J., J. Ryan, and L. Agnich. 2010. "Crime as a Source of Solidarity: A Research Note Testing Durkheim's Assertion." *Deviant Behavior* 31 (8): 679–703.

Hawdon, J. E., J. Ryan, and C. Mobley. 2000. "An Evaluation of Community Anti-violence Programs: Implications of Some Preliminary Findings." *Journal of Community Research* 10: 163–185.

Hawdon, J., J. Ryan, P. Räsänen, and A. Oksanen. 2009. *Social Relations and Community Solidarity: An International Comparative Analysis.* Washington, DC: National Science Foundation.

Hawkins, C. A., and P. N. Rao. 2008. "CEDER: A Social Development Response to the Tsunami Recovery in Tamil Nadu, India." *Social Development Issues* 30 (1): 29–46.

Hunter, A. J. 1985. "Private, Parochial, and Public Social Orders: The Problem of Crime and Incivility in Urban Communities." In *The Challenge of Social Control: Citizenship and Institution Building in Modern Society,* ed. G. D. Suttles and M. N. Zald, 230–242. Norwood, NJ: Ablex Publishing.

Hurlbert, J., J. Beggs, and V. Haines. 2000. "Social Networks and Social Capital in Extreme Environments." In *Social Capital: Theory and Research,* ed. N. Lin, K. Cook and R. Burt, 209–231. New York: Aldine de Gruyter.

Kasapoglu, A., and M. Ecevit. 2003. "Impact of the 1999 East Marmara Earthquake in Turkey." *Population and Environment* 24: 339–358.

Kirby, E. 2012. "A Conceptual Model for Critical Incident Analysis." In *Handbook of Critical Incident Analysis,* ed. Richard W. Schwester, 21–34. Armonk, NY: M.E. Sharpe.

Klinenberg, Eric. 2002. *Heat Wave: A Social Autopsy of Disaster in Chicago.* Chicago: Chicago University Press.

Lochner, K., I. Kawachi, and B. P. Kennedy. 1999. "Social Capital: A Guide to Its Measurement." *Health and Place* 5: 259–270.

McMillan, D. W., and D. M. Chavis. 1986. "Sense of Community: A Definition and Theory." *Journal of Community Psychology* 14: 6–23.

Miki, H. 2002. "After the Great Earthquake: An Emergent Prayer for Humanity Supreme." Brisbane, Australia: International Sociological Association.

Morrow, B. H. 1997a. "Disaster in the First Person." In *Hurricane Andrew: Ethnicity, Gender, and the Sociology of Disasters,* ed. W. G. Peacock, B. H. Morrow, and H. Gladwin, 1–19. New York: Routledge.

———. 1997b. "Stretching the Bonds: The Families of Andrew." In *Hurricane Andrew: Ethnicity, Gender, and the Sociology of Disasters,* ed. W. G. Peacock, B. H. Morrow, and H. Gladwin, 141–170. New York: Routledge.

Morrow, B. H., and W. G. Peacock. 1997. "Disasters and Social Change: Hurricane Andrew and the Reshaping of Miami?" In *Hurricane Andrew: Ethnicity, Gender, and the Sociology of Disasters,* ed. W. G. Peacock, B. H. Morrow, and H. Gladwin, 226–242. New York: Routledge.

Nelson, C. 2009. *Rocket Men: The Epic Story of the First Men on the Moon.* New York: Viking.

O'Neill, A. 2009. "Manson's Lasting Legacy: 'Live freaky, die freaky.'" CNN, August 10, http://www.cnn.com/2009/CRIME/08/10/california.manson.murders/index.html (accessed October 6, 2009).

Ochberg, F., S. Cinti, E. Goodman, E. Kirby, H. Melia, S. Prior, and R. Taggart. 2007. "The Critical Incident Model." Presentation at the meeting for Critical Incident Analysis at John Jay College of Criminal Justice (CUNY), January 4.

Olson, R. S. 2000. "Toward a Politics of Disasters: Losses, Values, Agendas, and Blame." *International Journal of Mass Emergencies and Disasters* 18: 265–287.

Orr, M. 1999. *Black Social Capital: The Politics of School Reform in Baltimore, 1986–1998.* Lawrence: University Press of Kansas.

Peacock, W. G., and A. K. Ragsdale. 1997. "Social Systems, Ecological Networks and Disasters: Toward a Socio-political Ecology of Disasters." In *Hurricane Andrew: Ethnicity, Gender, and the Sociology of Disasters,* ed. W. G. Peacock, B. H. Morrow, and H. Gladwin, 20–35. New York: Routledge.

Peacock, W. G., B. H. Morrow, and H. Gladwin (eds). 1997. *Hurricane Andrew: Ethnicity, Gender, and the Sociology of Disasters.* New York: Routledge.

Pijawka, K. D., B. Cuthbertson, and R. S. Olson. 1987. "Coping with Extreme Hazard Events: Emerging Themes in Natural and Technological Disaster Research." *Omega* 18: 281–297.

Portes, A., and A. Stepick III. 1993. *City on the Edge: The Transformation of Miami.* Berkeley: University of California Press.

Putnam, R. D. 1993. *Making Democracy Work: Civic Traditions in Modern Italy.* Princeton, NJ: Princeton University Press.
———. 2000. *Bowling Alone: The Collapse and Revival of American Community.* New York: Simon and Schuster.
Quarantelli, E. L., and R. R. Dynes. 1976. "Community Conflict: Its Absence and Its Presence in Natural Disasters." *Mass Emergencies* 1: 139–152.
Räsänen, P., A. Oksanen, J. Hawdon, and J. Ryan. 2008. *Paikallistason yhteisöllisyys massaväkivallan jälkeen Suomessa ja Yhdysvalloissa.* [Community Cohesion after Mass Violence in Finland and the United States]. *Sosiologia* 45 (4): 347–353.
Reckdahl, Katy. 2007. "Social Services." In *City Adrift: New Orleans Before and After Katrina,* ed. J. Bergal et al., 59–70. Baton Rouge: Louisiana State University Press.
Rekers, George. 1995. *Susan Smith: Victim or Murderer.* Lakewood, CO: Glenbridge Publishing.
Risk Management Solutions. 2006. "Managing Tsunami Risk in the Aftermath of the 2004 Indian Ocean Earthquake and Tsunami," http://www.rms.com/Publications/IndianOceanTsunamiReport.pdf (accessed September 5, 2009).
Rivera, J. D., and D. S. Miller. 2007. "Continually Neglected: Situating Natural Disasters in the African American Experience." *Journal of Black Studies* 37: 502–522.
Ruzek, J. I., M. J. Brymer, A. K. Jacobs, C. M. Layne, E. M. Vernberg, and P. J. Watson. 2007. "Psychological First Aid." *Journal of Mental Health Counseling* 29: 17–50.
Ryan, J., and J. Hawdon. 2008. "From Individual to Community: The 'Framing' of 4–16 and the Display of Social Solidarity." *Traumatology* 14: 43–52.
Ryan, J., M. Hughes, and J. Hawdon. 1998. "Marital Status and General-Life Satisfaction: A Cross-National Comparison." *International Journal of Comparative Sociology* 39: 224–236.
Ryang S. 2003. "The Great Kanto Earthquake and the Massacre of Koreans in 1923: Notes on Japan's Modern National Sovereignty." *Anthropological Quarterly* 76: 731–748.
Sampson, R. J. 2001. "Crime and Public Safety: Insights from Community-level Perspectives on Social Capital." In *Social Capital and Poor Communities,* ed. S. Saegert, J. P. Thompson, and M. R. Warren, 89–114. New York: Russell Sage Foundation.
Sampson, R. J., S. Raudenbush, and F. Earls. 1997. "Neighborhoods and Violent Crime." *Science* 277: 918–924.
Savage, A., and L. A. Russell. 2005. "Tangled in a Web of Affiliation: Social Support Networks of Dually Diagnosed Women Who Are Trauma Survivors." *Journal of Behavioral Health Services and Research* 32: 199–215.
Scarisbrick-Hauser, A. M., and J. M. Lewis. 1990. "Theoretical Approaches to Societal Reactions to the Hillsborough Soccer Disaster." Washington DC: American Sociological Association.
Scorr, S. L., S. McSpirit, S. Hardesty, and R. Welch. 2005. "Post Disaster Interviews with Martin Country Citizens: 'Gray Clouds' of Blame and Distrust." *Journal of Appalachian Studies* 11: 7–29.
Shrum, W. 2007. "Hurricane Stories, from Within." *Social Studies of Science* 37: 97–102.
Siegel, J. M., L. B. Bourque, and K. Shoaf. 1999. "Victimization after a Natural Disaster: Social Disorganization or Community Cohesion?" *International Journal of Mass Emergencies and Disasters* 17: 265–294.
Smelser, N. J. 2004. *Collective Trauma and Collective Identity.* Berkeley: University of California Press.
Sweet, S. 1998. "The Effect of a Natural Disaster on Social Cohesion: A Longitudinal Study." *International Journal of Mass Emergencies and Disasters* 16: 321–331.
Tedford, Deborah. 2009. "Obama, Soldiers Gather to Mourn at Fort Hood." NPR.org, November 10, http://www.npr.org/templates/story/story.php?storyId=120277295&ft=1&f=1001&sc=YahooNews (accessed November 10, 2009).
Tierney, K. 2007. "From the Margins to the Mainstream? Disaster Research at the Crossroads." *Annual Review of Sociology* 33: 503–525.

Tierney, K., C. Bevc, and E. Kuligowski. 2006. "Metaphors Matter: Disaster Myths, Media Frames, and Their Consequences in Hurricane Katrina." *The Annals of the American Academy of Political and Social Science* 604: 57–81.

Tuan, Y. F. 1987. *Space and Place: The Perspective of Experience.* Minneapolis: University of Minnesota Press.

Turkel, G. 2002. "Sudden Solidarity and the Rush to Normalization: Toward an Alternative Approach." *Sociological Focus* 35: 73–79.

U.S. Department of Homeland Security (U.S. DHS). 2006. *Katrina Hurricane Disaster Weekly Report.* Washington DC: Office of the Federal Coordinator of Gulf Coast Rebuilding.

U.S. Geological Survey (USGS). 2009. *Most Destructive Known Earthquakes on Record in the World: Earthquakes with 50,000 or More Deaths,* http://earthquake.usgs.gov/regional/world/most_destructive.php (accessed August 10, 2009).

Wisner, B., P. Blaikie, T. Cannon, and I. Davis. 2003. *At Risk: Natural Hazards, People's Vulnerability and Disasters.* New York: Routledge.

CHAPTER 6

# Modeling Critical Incidents to Improve Response

ROBERT C. TILL, MATTHEW DURAK, AND ERIK D. GOODMAN

An important question that arises in the modeling of critical incidents is, "What is a model?" The *New Oxford American Dictionary* calls a model "a simplified description, especially a mathematical one, of a system or process, to assist calculations and predictions." The state of mathematical modeling capabilities with regard to representing the types of human behavior that occur during critical incidents is still rather primitive.

Critical incident analysis is a relatively new field of study. Fourteen distinct definitions of "critical incident" have been identified to date (Schwester, Horning, and Dank 2012). For our purposes, a critical incident is defined as "a relatively brief occurrence involving injury, loss, conflict, discovery or change of significant proportion, usually unscripted and unanticipated, with the potential to alter existing societal norms. Critical incidents are usually traumatic, threatening the bonds of trust that bind communities, but may be positive, initiating historic consequents" (Ochberg, as cited in Schwester, Horning, and Dank 2012.

## BACKGROUND: CRITICAL INCIDENT MODELS

Three models will be described in the course of this chapter:

1. A generalized framework for critical incidents.
2. An information retrieval model for critical incidents.
3. A more detailed model of a critical incident in which city dwellers use their cars (modeled as agents) to flee a chemical spill.

None of the models described here (or to the authors' knowledge described anywhere else) is able to predict the complete outcome of a critical incident at the participant level. At this time the main reason for modeling critical incidents is to explore generalizations about them. By determining how they are similar to one another we can move toward the development of models that have predictive capability, at least at the strategic level.

Critical incident models have many variables and are complex. Those that have predictive capability at the participant level will likely take the form of agent-based models (ABMs).

By recognizing the similarities of critical incidents we will also be able to index or categorize them and their components and apply the lessons learned in one critical incident to another incident that may not at first seem similar. In summary, we will explore:

1. *Generalization:* recognizing similarities (and differences) among incidents by fitting them into a common, "high-level" modeling framework, increasing our ability to take a broad view of the problem, compare outcomes of various types of responses, and understand the aftermath of the incidents.
2. *Information Retrieval:* establishing a classification scheme according to which data about critical incidents may be archived and subsequently searched according to keywords that are relevant to their analysis.
3. *Generating Agent-Based Models:* exploring bottom-up, ABM models to learn how to generate plausible rules to explain the behaviors of participants that produce higher-level "emergent" behaviors that agree with observed behaviors in the situations modeled. While these models are usually agent based, they can often include other elements as appropriate to representing the incident.

Let's look at each of these in turn.

## GENERALIZATION

By the term "generalization," we mean recognizing similarities (and differences) among incidents by fitting them into a common, "high-level" modeling framework. This increases our ability to take a broad view of the problems facing the individual, compare outcomes of various types of responses, and evaluate the possible aftermath of the incidents. Critical incidents are diverse and no two incidents share every feature, but there are sometimes similarities that may be helpful to notice; even among seemingly very dissimilar incidents. Establishing a high-level model forces the identification of roles, settings, and types of communication, attitudes, and many other attributes of any incident. Some incidents will have important attributes that are not shared by others, but it seems desirable to have a general model that is broad enough to include the most important attributes of most of what we would like to analyze.

It is useful to pursue a general model that helps in structuring the wealth of information available about any critical incident. Such a model allows discussion of an incident's similarities to and differences from other incidents. It seems possible that this process might yield insights that lead to changes in the way future incidents are handled by first responders, by the press, by subsequent interveners, by governmental entities, or by other individuals.

## Example: The ACIA Model: A Generalized Modeling Framework for Critical Incidents

At the inaugural meeting of the John Jay Academy for Critical Incident Analysis (ACIA) in January 2008, a model framework for critical incident analysis was presented (Goodman and Kirby 2008). This model was the latest of those developed by different individuals. The first models dated to 1978 (Ochberg 1978).

Implicit in this critical incident analysis framework is that a model of a critical incident requires the incorporation of the event itself, the demands it makes in the political arena, the influences of policy on the actions taken by the bureaucracy and adhocracy, and the resulting tactics that are fed back into the event. The political arena consists of interaction between the government and the governed, based primarily on trust. As an incident unfolds, an individual may feel either protected and supported or betrayed by the government, in various ways at various times. The bureaucracy and adhocracy consists of management and interveners, to whatever extent they are delegated responsibility. (The term "adhocracy" is used to describe whatever structure of interveners is active in dealing with "the event," whether by standing policy or by event-specific delegation by management.) Within the framework, the media are depicted as influencing all other elements described. These interactions are depicted in Figure 6.1.

The generalized CIA model highlights core uncertainties within critical incidents. The first question that arises is, "How exactly does the threat I'm considering fit into the framework?" The arrows and blocks within the CIA model represent questions an individual might ask, such as:

1. "How much do I know about the threat at a given time?"
2. "Is the government entity local or federal?" This question is tied into the question, "Is the response appropriate based on what I know about the threat?"
3. "How large a threat can be mitigated with the resources available?"
4. "What is the level of trust between the government and the governed? Will the government's response to this incident cause a significant change in that level of trust?"

Most of these questions are not answered in Figure 6.1, but Figure 6.2 adds labels for the relationships captured within each of the three shapes: those being labeled "threat" between violators and target, "trust" between government and the governed, and "delegation" between management and interveners. The generalized CIA model shown in Figure 6.2 identifies potential areas where data are readily available from other fields and can be incorporated into an overall model of critical incidents. It also identifies areas that need to be modeled in detail. Within an event, it is seen that there are violators and a target. The violators are the source of the threat to the target. While the word "violators" may need to be understood as sometimes describing inanimate sources of threat, such as earthquakes, it occupies a specific position in the model corresponding to the source of whatever threat the "victims" are subjected to. Interactions between the government and the governed are described as being based on trust—for example, responses of citizens

Figure 6.1 **The Model**

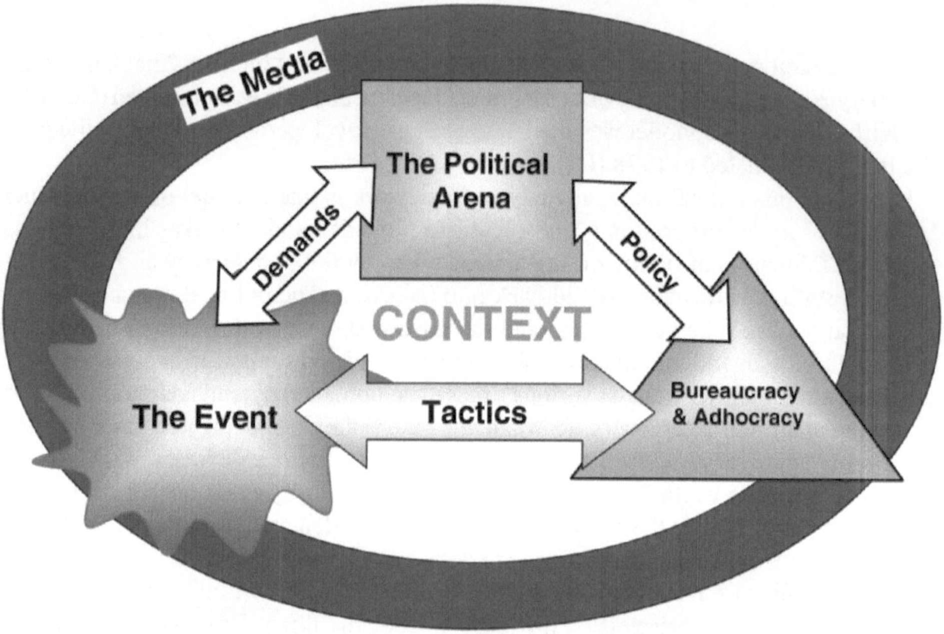


*Source:* Goodman and Kirby, 2008.

to government actions may depend strongly on the level of trust between the citizens involved and the particular government agency involved. The relationships within the bureaucracy and adhocracy are also clearly delineated. This model shows qualitative relations of interest—threat, trust, and delegation. The generalized model provides a context in which to discuss the critical incident.

A model such as this one, which structures the wealth of information available about any critical incident and delineates the similarities and differences between incidents, is extremely valuable. The process of analyzing a critical incident in terms of such a model generally yields insights that may lead to changes in the way future incidents are handled.

## INFORMATION RETRIEVAL

Information retrieval (IR) as defined by Manning and colleagues is "finding material (usually documents) of an unstructured nature (usually text) that satisfies an information need from within large collections" (usually stored on computers) (Manning, Raghavan, and Shutze 2008). At the current time it is very difficult to retrieve information about critical incidents that are similar to an ongoing incident or an incident that is the subject of an academic study. While it is possible to use a commercial search engine such as Google, the returned results are often unsatisfactory, as these engines are not designed to

Figure 6.2 **Model Interactions**

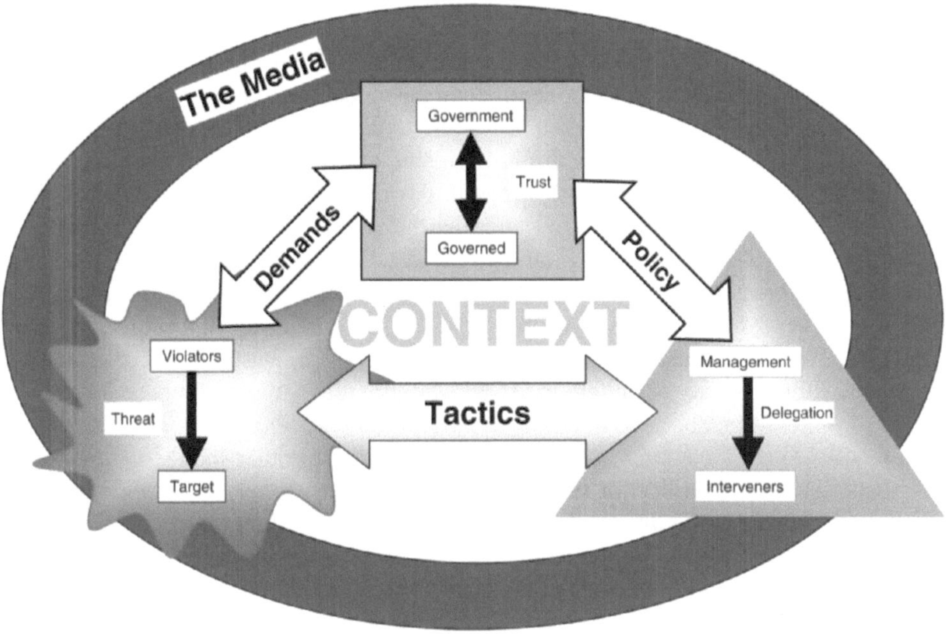


*Source:* Goodman and Kirby, 2008.

meet the specific requirements of critical incident analysis. This section will discuss the challenge of developing a classification scheme according to which data about critical incidents may be archived and subsequently searched according to keywords that are relevant to their analysis.

## Classification Model

Development of a classification model according to which data about critical incidents may be archived and subsequently searched according to *index terms* that are relevant to their analysis is one important way to meet this challenge. An index term or *keyword* is defined as "a pre-selected term which can be used to refer to content in a document. Usually, index terms are nouns or noun groups. In the Web, however, some search engines use all the words in a document as index terms" (Baeza-Yates and Ribeiro-Neto 1999).

The classification scheme to be used in critical incident analysis should be derived from the high-level critical incident model discussed previously—or some similar, unified model—but is a much more detailed, rigorous, and concrete entity, since it is embodied within a computer-based system for data capture, storage, and retrieval. The structure of a typical database is not, and cannot be, the same as that of the currently published critical incident analysis model discussed in Chapter 2. Databases are generally *relational* systems and capture data based on relationships defined among the data they contain. High-level models may seem to satisfy that description as well, but they are not forced to contain as precise definitions of those relationships as are explicit in a relational database.

Clearly, using the CIA model to frame a critical incident might include retrieving information about each of the areas identified in the model for comparison purposes. Currently, data is not necessarily indexed according to the syntax identified within the CIA model, so this would be extremely difficult. In fact, existing models of critical incidents may not be particularly helpful in this regard. For example, the standard "intensity" or "severity" indexes provided for hurricanes, earthquakes, and other natural phenomena would be of only very limited use. Standard indices alone are generally not a good measure of the severity of the incident, as a Richter scale measurement of 7 in a metropolis would have very different effects from one of the same intensity in a rural area, and a category 5 storm out at sea may not even make landfall, or may strike as a category 1 or 2, causing little damage. No single index will capture all we are seeking, so an array or vector of indices must be used in order to have relevance.

A method to combine indices or classifiers in a vector or other such *structured* index is critical to maximal utility for study of critical incidents. It may be important to provide a standardized (according to the CIA model, for example) set of descriptors that are recognized as synonyms or more generic terms for the specific indices or terms used to describe specific types of incidents in the general literature. For example, a term such as "severity" might be explicitly recognized to subsume such terms as "category X" in the case of a hurricane or "Richter Y" in the case of an earthquake, or even some combination of such physical measurements with impacts on human populations.

A major goal of having a critical incident archive is to allow easy retrieval of many incidents (potentially of varied types) that share some combination of features that are captured in a query. To enable this, content-bearing words (terms) must be marked when an incident is indexed. Once this indexing is accomplished, a method needs to be developed to retrieve the appropriate document when it is needed.

### Search Method: Data and Document Retrieval

An information retrieval (IR) system specific to critical incidents could be constructed using many different methods. Singhal (2001) provides an excellent overview, although entire texts have been written concerning technical aspects and implementation (Baeza-Yates and Ribeiro-Neto 1999; Grossman and Frieder 2004; Manning et al. 2008). There are several models proposed for the document retrieval process. Currently the three most commonly used are (1) the vector space model; (2) the probabilistic models (Grossman and Frieder 2004); and (3) the inference network model (Singhal 2001). An exhaustive discussion of these methods is beyond the scope of this chapter. The Boolean model of data retrieval will be discussed to provide context for how information has historically been stored and retrieved, and how such databases must be included in any document retrieval system. The vector space model of document retrieval will also be discussed to point out a possible way forward.

### *Boolean Model of Data Retrieval*

An early model for accessing data that is still used today is the simple Boolean model. A query is usually of the form CONTAINS X AND Y BUT NOT Z. There are drawbacks to this method. For example, only exact matches will be returned, so if the appropriate search terms are not used, no results will be found. Using simple Boolean models, information also is not rated by relevance, which is expected by most users of modern information retrieval systems such as those available on the Internet. Complex expressions are also usually difficult to understand (Manning et al. 2008).

The National Fire Incident Reporting System (NFIRS) (USFA 2008) is an example of a database system that may contain references to critical incidents. Boolean information retrieval is commonly used to extract information from the NFIRS database. The database contains information about individual fire incidents, however, and usually not specific document references. As such it provides output, depending on the query, concerning specific incidents such as the number of fires that have occurred in certain kinds of construction or ignited in a certain way.

It is clear that while data from NFIRS usually does not identify specific documents, it could assist the researcher by providing context for what constitutes a critical incident as well as information about a specific critical incident. In this case the database itself is a very large document that should be identified by a document retrieval system. There may be other databases of similar import that may also need to be identified in any document retrieval system concerned with critical incidents.

### *Vector Space Method of Document Retrieval*

When the information in the data collection is less structured, as is the case when one is storing a collection of documents, the problem of determining the relevance of a particular document to a query becomes an issue. Modern IR systems will estimate the usefulness of a document to a user query and assign the document a rank, usually a numeric score, and present the documents to the user according to this score.

The vector space model is one way to meet the challenge of developing an IR system specific to critical incidents and determining document similarity. It is one of the predominant techniques used in search engine design (Berry and Browne 2005). An example of the vector space model is found in Figure 6.3 (Polyvyanyy and Kuropka 2007).

The figure clearly shows a simplification of the process that may be used to determine similarities in documents pertaining to critical incidents. Such documents might have axes such as:

1. lethality
2. iconicity
3. unfamiliarity
4. cost
5. damages

Figure 6.3 **Example of a Vector Space Model**

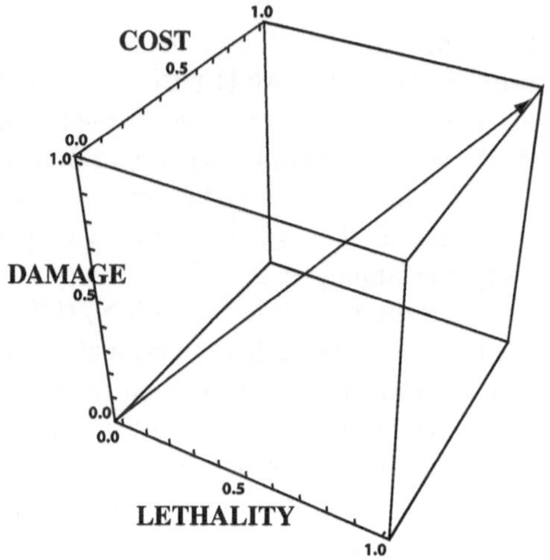


*Source:* Adapted from Polyvyanyy and Kuropka, 2007.

All of these phrases are or should be associated with the critical incident lexicon, so there could be hundreds or even thousands of vectors in the index that a query would be compared to. Other terms that would be appropriate indices can be taken directly from the generalized CIA framework.

An indexing scheme for critical incidents should be sophisticated enough to recognize that terms specific to a type of incident, such as "Richter" or "category X," should in turn generate a "count" for an index term such as "severity." This would make it possible to retrieve an entry about severity of an earthquake when the word "severity" does not appear but the word "Richter" does. Similar mechanisms should be introduced for all of the important terms used in the CIA model or whatever model is to guide the organization of the data.

## Information Retrieval Summary

It can be seen that vector-based and possibly other models, such as probabilistic or inference network models, may ultimately need to be combined in a large-scale IR system of critical incident documents. While modern search engines such as Google are useful for some critical incident research, data they search are typically not properly indexed to be of maximum use to the researcher. It is our hope to begin to remedy this problem by developing search engines and databases specific to critical incidents. A vector space–based engine would be an appropriate beginning. Ultimately, simultaneous access to multiple data sources, such as older Boolean-type databases, is desired. This may make the user interface more complex. It is hoped that this complexity can be avoided, because as interface complexity increases, usability may suffer.

## GENERATING AGENT-BASED MODELS

High-level models, representing flows of materials, energy, and so on, among a set of interacting entities work very well for representing many physical systems. However, neither individual people nor companies tend to act very much like flowing electricity or flowing water. In many cases, the behavior of individuals and of corporations, agencies, segments of the public, and so forth are much more successfully modeled by treating them as individual *agents,* each of which behaves according to a set of rules that determines its behavior in any particular situation. That is, faced with a particular situation, one or more rules "fires," determining the response of that agent in that situation. Each agent's set of such rules may have developed through time as the agent attempted to optimize its responses to various situations, or may represent an innate response over which the agent has minimal control. The situation is determined by the other participating agents, the surrounding environment, and perhaps the history of the agent. So, for example, a decision to lower a price on a product produced by a company might be determined by a rule that weighs the cost of producing the product, the recent sales history, and the current prices of competing products from other agents. The aggregate behavior of a large set of individual agents, each acting according to its own rule set, is often very difficult to predict simply by examining the rules. It is much better understood by calculating the behavior of the agents in a particular scenario and observing the overall behavior that *emerges* as a function of the behaviors of the many agents. A lack of agreement between the emergent behavior of a set of agents in a given scenario and what has been observed in the real world indicates that there is an error, either in the rules governing the agents or in the environment as presented to the agents, or that some significant factor has been omitted from the environment or the conditions to which the agents' rules are sensitive. However, when such a problem is uncovered, the act of discovery helps us to learn something about the underlying phenomenon; namely, that it cannot be adequately represented using the current model. Therefore, this agent-based model (ABM) drives us to discover missing or erroneous elements in the model, increasing our understanding and capture of the real-world occurrence. It is also useful as a means for testing what may be overly simplistic explanations for observed phenomena. When someone asserts that a certain behavior is the inevitable result of a particular set of interactions, it becomes relatively simple to test that assertion in this framework. If the observed behavior agrees with the original assertions, it does not mean they are true, but they have not been invalidated and the case is still "open." If, however, they do not agree, then it is clear that the explanation was not adequate to explain the behavior observed.

As North and Macal state (2007), ABMs are appropriate in the following instances:

- When the problem has a natural representation as consisting of interacting agents
- When there are decisions and behaviors that can be defined discretely, that is, with well-defined boundaries
- When it is important that agents adapt and change their behavior
- When it is important that agents learn and engage in dynamic strategic behavior

- When it is important that agents have dynamic relationships with other agents, and agent relationships form and dissolve
- When it is important that agents form organizations, and adaptation and learning are important at the organizational level
- When it is important that agents have a special component to their behaviors and interactions
- When the past may be a poor predictor of the future
- When scaling up is important, and scaling up consists of adding more agents and agent interactions
- When process structural change needs to be a result of the model, rather than the input to the model

All of the these are appropriate to critical incidents.

Simple example models can provide a context in which the benefits of generalizing, developing ABMs, and indexing can be described. It is clear that extensive time and resources are required to develop the generalized CIA model into a full agent-based model or into a set of such models. However, it is possible to begin to attack pieces of the model using ABM, showing its possible role in critical incident analysis. To illustrate this application, we shall describe and model a simulated critical incident requiring evacuation of cars from a simulated urban area in the event of a toxic chemical spill.

**An Urban Evacuation Model: Example of an ABM and Its Application**

A relatively easy example is an ABM model of the routing of automobile traffic away from a chemical spill. Considerable attention has been paid by many researchers to the study of possible urban evacuation strategies, much of it making use of traffic models or simulations of various sorts. Chen (2005) cites and summarizes much of that early work. Much of the traffic simulation work has not made use of the ABM methodology, has not included modeling the effects of the hazard on the evacuation process, and has not allowed explicit optimization of the response strategy, but rather only comparison of a few prespecified strategies. Chen (2005) describes four common urban evacuation strategies adopted by various U.S. cities: (1) setting signals to flash; (2) having police direct traffic at critical intersections; (3) using P.M. peak (generally outbound) light timings; and (4) setting timing plans to maximum cycle length on the evacuation routes, giving the majority of green time to the major roadways. The methodology used in our research differs from Chen's study in that it allows individual agents to have much more complex behaviors, including intermediate goals (stops) prior to evacuation. Our model includes the threat, which is allowed to act on the evacuation process, and the traffic signal control strategy is optimized over a broad range of options. However, Chen's study considers many phenomena not modeled here, such as inbound emergency vehicles, accidents, roadside parking, and public transportation.

Interesting behavior may be observed even when modeling only one or a few distinct agent types in this case. Even given an ABM that is deemed to model the behavior of the participating agents acceptably for the purpose at hand, it may still be difficult to find optimal strategies for emergency responders to follow in order to save lives when faced with such scenarios. This section will present an example of one type of approach that may be used to test and optimize strategies for interveners to use in dealing with the event.

Furthermore, in order to test each possible strategy, one or more computer simulation runs may be needed, each of which may take significant time and/or computing expense. It becomes impractical to test the many combinations of parameters that would determine all possible strategies in an exhaustive manner, so a more efficient search method must be used. In this case, it is advantageous to use an optimization tool aimed at global optimization, such as stochastic search software.

*An Agent-Based Traffic Model: Individuals in Vehicles as the Target*

The agent-based traffic model is written in NetLogo (Wilensky 1999). It was originally based on the "traffic grid" model (Wilensky 2003), but has been heavily modified and greatly extended. The premise of the model is that cars drive throughout a city, going to destinations determined at random before and during the simulation. In the context of the generalized CIA model, the drivers represent the target of the incident. After some time, an emergency (in this case, a chemical spill) occurs on one side of the city and drivers are instructed to move away from the spill, primarily using sirens to indicate an emergency and radio announcements to describe the problem. The model focuses on only a portion of the city, represented as a grid of eight "streets" by eight "avenues." Prior to the emergency, cars can enter and leave the visible portion of the city freely because under normal conditions the world is toroidal (cars leaving one side reappear on the opposite side), thereby retaining a constant traffic density in the area without creating new car "agents." This behavior changes during an emergency, when cars leaving an edge of the simulated area actually "stick" at that edge and must come back to the visible eight-by-eight world to complete their evacuation, unless that edge represents a direction of safe evacuation.

The eight-by-eight urban area has five thousand moving cars during normal conditions (Figure 6.4). Each car is an individual agent with variables and behaviors. These cars have random goals, which are represented by going to an intersection or leaving the visible area in a certain direction. Cars each have different target driving speeds in a Gaussian distribution around the speed limit.

When the emergency occurs, only half of the cars will immediately begin to obey the commands to leave the city. The noncompliant agents model the behavior of people who would ignore orders in such a situation because they decide to pick up a family member or friend in the immediate area first, and people who simply would not immediately become aware of the emergency directive. The model has the remaining cars learning the nature of the emergency on a probabilistic basis as time goes on.

Another source of cars during an evacuation is from parking garages. These garages are placed with exits in predetermined but random positions (but not within 100 feet of intersections), with fifty to two hundred cars in each garage. In the agent-based car evacu-

Figure 6.4 **Traffic Model "Snapshot"**

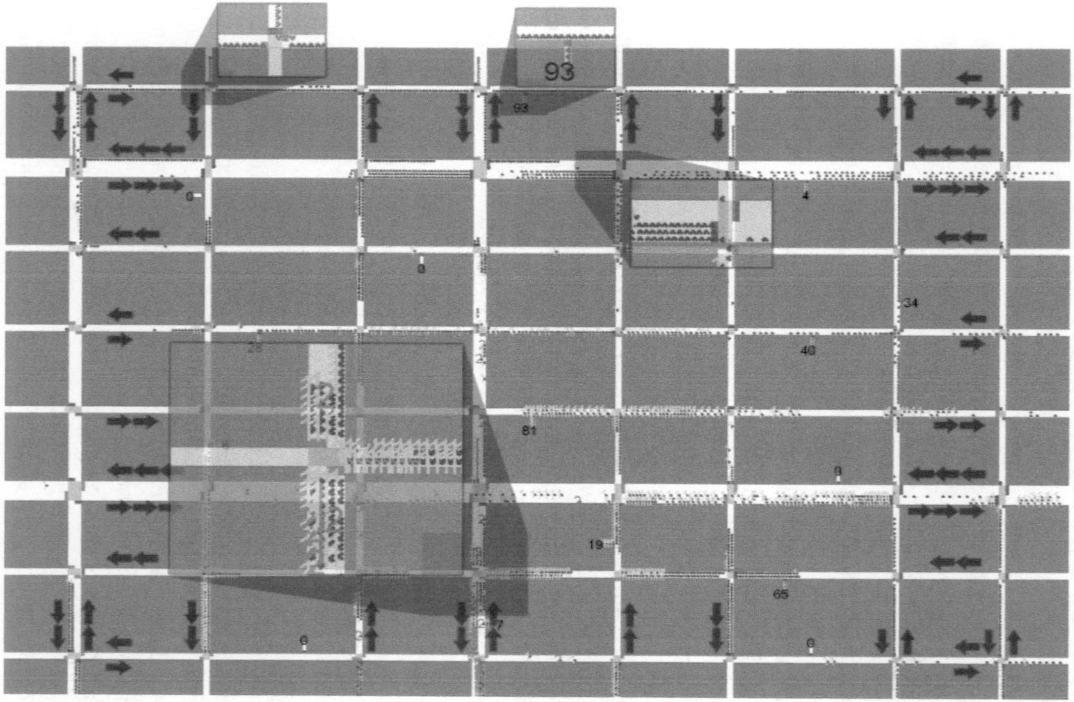


*Note:* Traffic model "snapshot" during a run in NetLogo with arrows indicating direction of traffic and number of lanes. Some portions of the map are enlarged to show detail. The black numbers on the map represent the number of cars in each parking garage. Cars that have reached AEGL-1 have a yellow "1," those in AEGL-2 have an orange "2," and the cars in AEGL-3 have a red "3."

ation model, all cars in the garages pull out and make a right turn onto the road (unless they are on a one-way road and must turn left). Before the emergency, cars do not enter or leave the garages (or equivalently, entry and exit are balanced).

The form of intervention by the "bureaucracy and adhocracy" of the ACIA model that is being tested with this ABM is control of traffic signals during an emergency. Officials can attempt to assist drivers in the agent-based car evacuation model by changing the signal timings of traffic lights when the emergency occurs, to a timing strategy preoptimized for emergencies of various sorts and locations in runs of an ABM. Finding an optimal signal control strategy for a toxic plume emergency just south of the simulated urban area is the task presented in this chapter to a commercial stochastic optimization package. Use of such a model and optimization may also be helpful to judge the usefulness of a proposed intervention strategy such as a change in traffic signal timings.

*Plume Model: The Threat*

The emergency in this model is a chlorine gas leak from a railroad tank car. Such a chlorine release resulting from an actual train wreck is documented in Dunning and Oswalt (2007), which led to a no-notice evacuation. In the urban situation modeled in this chapter, the leak occurs 1,100 feet south of the visible portion of the city. For the simulations performed here, it is assumed that officials respond to the accident by alerting drivers to the danger and activating the emergency evacuation traffic signal timing one minute and fifteen seconds after the incident begins (i.e., the time to the initial response) to try to speed evacuation of the area.

The plume itself is modeled using the ALOHA (U.S. EPA 2007) "threat at a point" tool. The plume is modeled as a chlorine gas leak from a horizontal cylindrical tank. The tank is 9 feet in diameter and 32 feet long, containing 140,000 pounds of chlorine (Streit 2002). The leak is a one-inch hole at the bottom of the tank. The parameters set for atmospheric conditions assume a partly cloudy day at 70 degrees Fahrenheit over an urban area. The wind is set to blow 3 knots from the south. The threat is measured at every 550 feet going north and every 880 feet to the east and west (the average size of a city block in the model). The outdoor (ground-level atmospheric) concentrations are used throughout, and are calculated at each intersection when the plume first enters the map and at 5, 10, 15, 20, 30, 40, 50, and 60 minutes thereafter. Once this data is collected in a spreadsheet form, the concentration for every 30 seconds is estimated with linear approximations. Then this data is linearly approximated between recorded points in the model in order to create a gradient of the chemical that updates every 30 seconds (60 steps in the model). A visual representation of the plume is shown in Figure 6.5. Here, the darker shades indicate chlorine gas and white represents no chlorine.

Individual vehicle exposures are calculated at each time step using a rate of seven air exchanges per hour in a car (Ott 2007). This assumes car windows are closed. The exposure is calculated by keeping track of the concentration in the car at any point in time and adding the difference of the concentration in the car and outside multiplied by the number of air exchanges per step in the model.

Figure 6.5 **Visual Representation of the Plume**

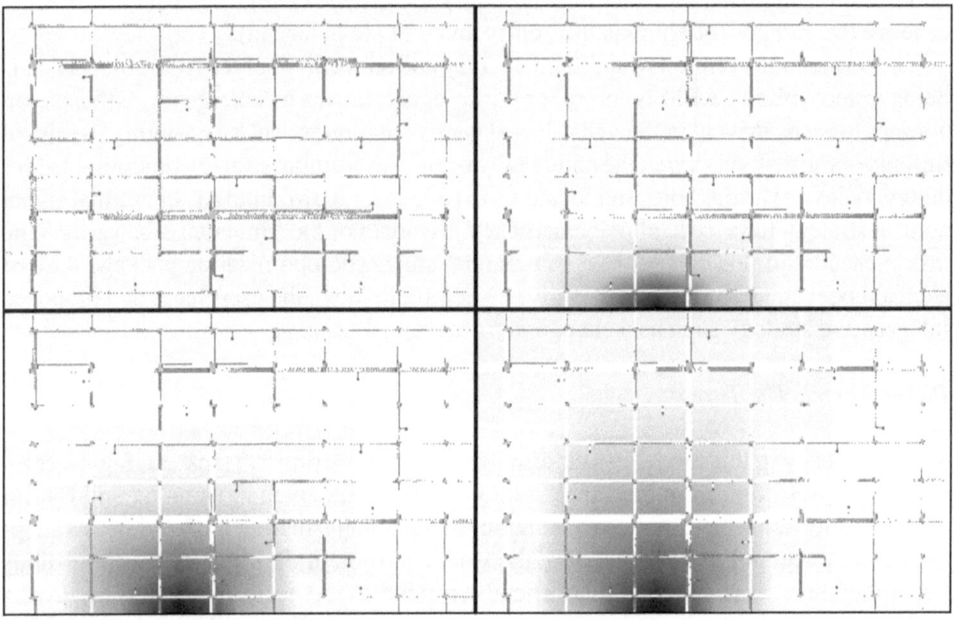

*Note:* The concentration of chlorine in the air is shown on the model at time intervals of 5,10,15, and 20 minutes (from top left to right, then bottom left to right). Darker shades indicate higher chlorine concentration.

In order to look at how the concentration of gas in the car affects the driver, acute exposure guideline levels (AEGL) for chlorine were used (Subcommittee on AEGL 2004). According to the guidelines, AEGL-1, the lowest level of exposure detected by a subject, occurs at approximately 0.5 parts per million (ppm) and the symptoms are simply detection of the gas; no disablement occurs at this stage. AEGL-2 symptoms are disabling asthma-like attacks, and AEGL-3 is lethal. In order to determine which exposure level an individual is subjected to, the calculation $C^{1.5} \times t = k$ was used, where $C$ is the concentration in parts per million, $t$ is the time in seconds, and $k$ is a constant. This approximation was chosen because some research has stated $C^2 \times t = k$. Others have used $C \times t = k$. In order to determine the constant $k$, studies of 50 percent lethality in humans ($LC_{50}$) were used and then time-scaled to data for different time and concentration levels.

The problem with determining the value of $C^{1.5} \times t$ for a driver is that the concentration level is changing constantly. The exposure is calculated by summing $C^{1.5} \times 0.5$ for every time step, where 0.5 seconds is the length of the step. This way, a constant concentration will yield the same AEGL level after a given time and a variable concentration will be averaged. Each driver/car combination also has a level of susceptibility, normally distributed around 50 percent with a standard deviation of 20 percent. This susceptibility is multiplied by the $C^{1.5} \times t$ value for each AEGL level so that some cars are more susceptible than others to the gas but the $LC_{50}$ value still holds.

Based on the experimental symptoms of chlorine gas at levels AEGL-2 and AEGL-3, drivers who experience AEGL-2 are considered incapacitated and will stop driving. Individuals who pass out while driving will cause congestion in the traffic and could cause other drivers to get stuck long enough to also be affected by the gas. Because drivers sensing chlorine gas in the air are likely to respond by being more willing than normal to break traffic rules, cars in the model that encounter traffic blockage in their direction of travel will enter opposing lanes to pass a sequence of stopped cars, even though that would not occur under normal circumstances.

*Control Strategy (Chromosome) Design and Optimization Goal*

This model of evacuation of an urban area will be used to try to develop a scheme for speeding evacuation in a particular direction by revising the cycles of the traffic lights in the area. The system is assumed not to have traffic sensors in the control loop, so the new cycle is simply installed in the lights when the evacuation order goes into effect. The strategy for development of the evacuation control scheme will be to run many simulations of the evacuation scenario, with different light cycles, and determine which one is most effective in minimizing loss of life and incapacitation. Rather than testing against random patterns, commercial optimization software will generate the traffic light cycles to be tested, learning about successful patterns as the optimization progresses.

An example of a normal cycle (except for phase length adjustments) for a single light is shown in Figure 6.6. The traffic simulation also uses an input file with instructions for how traffic lights should be timed when an emergency occurs. Because the optimization tool used, HEEDS (Hierarchical Evolutionary Engineering Design System), has an evolutionary algorithm foundation, the specification of any traffic light control strategy is called a chromosome, which denotes a specification of all necessary numerical parameters for a particular strategy. Independently for each street (horizontal road), the control strategy specifies the length of the green light, including a fixed (standard) time for the yellow light, the length of green for all vertical roads (avenues) crossing that road, and a positive timing offset for the start of horizontal green from the start of green of the light immediately to the left of it, so that lights can be synchronized in the horizontal (evacuation) direction. Thus each horizontal row of streets has the same phase duration for all vertical green lights and a second (different) time for horizontal green lights, as shown in Figure 6.7.

The unit offset on the chromosome is multiplied going right to left in the model. For example, a unit offset of 25 would set the offset of each intersection going right to left as 0, 25, 50, 75, . . . , $25 * n$. In this way, lights to the left will begin at a point in time further in the cycle so that, ideally, by the time cars driving to the right reach the next light, it is still green. This left-to-right synchronization is chosen because that is the direction that will be used for evacuation (to the right). This signal timing will synchronize to the desired pattern regardless of the light timings during nonemergency operation. Roads are grouped by row because cars evacuate from left to right so that, ideally, these traffic

Figure 6.6 **Example of Traffic Light Timing**

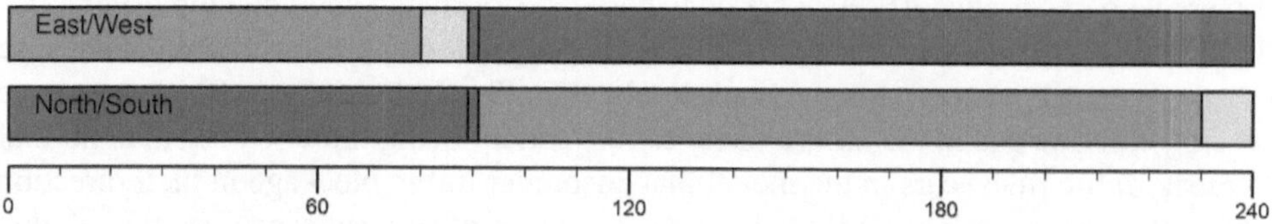


*Note:* The top bar represents the color of the light for traffic headed east or west, and the lower bar is for north/south traffic.

Figure 6.7 **Layout of the Chromosome Subjected to Optimization**

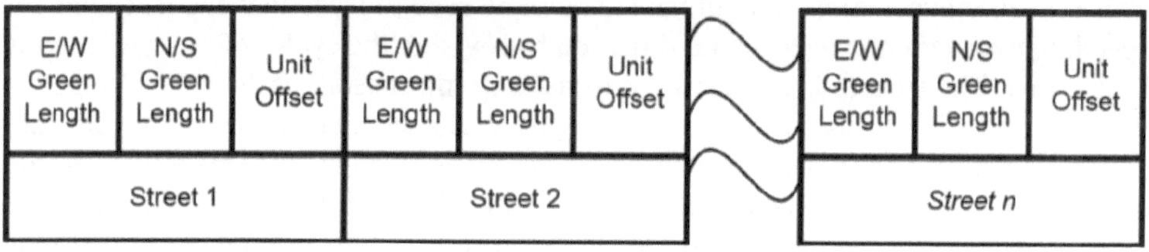


lights will be timed to let cars hit as few red lights as possible while preserving reasonable flow of the traffic in the other directions.

The goal is to keep traffic smoothly flowing out of the city. The value that HEEDS attempts to minimize is the number of drivers killed plus 10 percent of the number incapacitated, after the model is run for 7,200 ticks or one hour. This time is chosen because it is the longest length of time for which ALOHA provides data about the gas plume. Comparing all runs for the same length of simulated time provides a better measure of which strategies performed better than others.

*Optimization of Traffic Light Control: Intervention*

Because there are many "control variables" (light timing parameters) that can be changed to attempt to speed the evacuation of this urban area, and because changes in any one of them can indirectly affect the way the others actually work, this agent-based model presents a challenging optimization problem. The commercial optimization package, HEEDS (of which one of the authors is a developer), is used to perform the optimization, using its proprietary SHERPA search algorithm, which does not require any tuning parameters. Evaluation of each potential solution requires running of a significant ABM simulation, so it is important to minimize the number of total evaluations needed. This motivated the selection of SHERPA, which distributes the evaluation of individual evacuation strategies over a set of machines in a computer cluster or grid to significantly reduce the time to solution. In this exercise a total of 143 processors were used. Note that this application is "stretching" the NetLogo platform—there are other ABM platforms that could run this traffic simulation in a small fraction of the time needed by NetLogo, requiring many fewer processors. NetLogo was used here for ease and simplicity of development and the excellent user interface already available for depicting results.

*Preliminary Results*

This study is still in progress, so only preliminary results are presented here. The traffic model has been optimized using HEEDS with 8,004 total strategies evaluated. Figure 6.8 shows the results of the optimization. The values plotted are the fitness, calculated as the number of cars that reached AEGL-2 (disabled) plus ten times the number of cars that reached AEGL-3 (casualty). HEEDS will run designs multiple times if the design performs well, by running the same design but using a different random seed in the ABM model, producing different distribution and behavior of the agents. This effectively tests the better designs to be sure that they will work well even under different random initial configurations and decisions that are expected in the real world. The numbers plotted are the averages of these fitness values.

The downward trend of the plot indicates that HEEDS did a good job in finding better solutions as the optimization ran. However, this also indicates that given more computing time, HEEDS may have been able to find even better solutions. The unusual-looking spike around 7,500 evaluations on the graph is not an error—it is caused because the HEEDS software noticed that the optimization progress was slowing, so it introduced additional diversity to try to make the search more robust. It was then able to obtain slightly better fitness after only a few evaluations beyond that point.

Given only the data from the 8,004 initial evaluations, the best average fitness was taken as the one with the lowest value among all of those for which at least ten random seeds had been tested. This light timing strategy was then tested with ten additional random seeds to collect individual data. Average fitness of 3,865.5 was based on an average of 424.5 cars in AEGL-2 and 344.1 cars in AEGL-3. The standard deviations of these values were 328.4, 60.1, and 30.2, respectively.

In order to have something to compare the optimized strategy against, a standard light timing was also designed and tested with the ten random seeds used for the optimized strategy. This design took a simple approach of setting the traffic light offsets to random values to assume no synchronization (since when not evacuating, signals on two-way streets with nonuniform block lengths cannot be synchronized for traffic in both directions). The "standard" strategy used the duration 120 ticks (1 minute) for all east–west greens, 90 ticks (45 seconds) for north–south greens with two lanes of traffic, and 120 ticks (1 minute) for north–south greens with four lanes of traffic. The results are plotted on the graph as a straight line representing the average fitness of this design, which was calculated as 6,687.6, representing 337.6 cars in AEGL-2 and 635 cars in AEGL-3. The standard deviations were 391.1, 70.6, and 40.5 respectively. We can see that HEEDS was able to design a strategy that saved an average of 175 lives, with a smaller decline in the number of people disabled. Also, the design optimized by HEEDS performed slightly more consistently than the standard strategy. The difference in the mean fitness of the best optimized design and the standard case was highly significant ($t = 17.4731$,

Figure 6.8 **Average Fitness vs. Time**

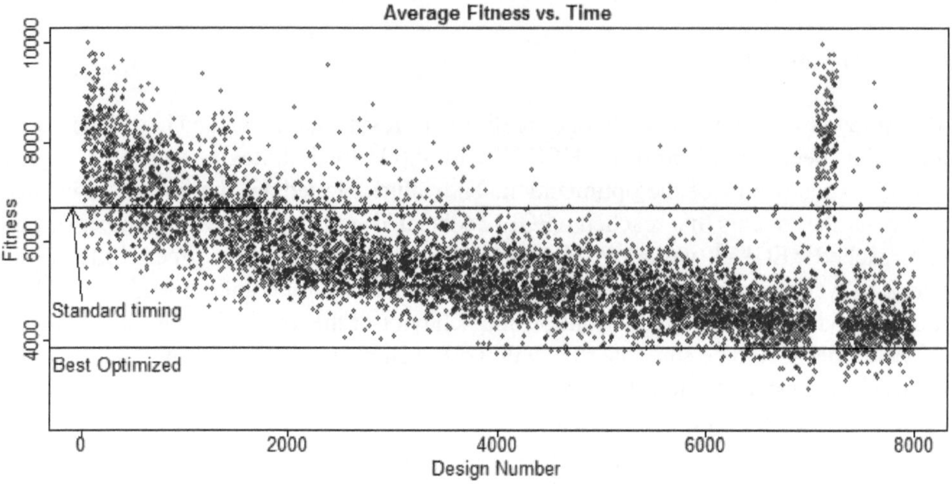


*Note:* This plots the optimization versus time. The values plotted are the calculated fitnesses, averaged over all evaluations for any designs that have been evaluated multiple times with different random seeds. The average fitness of the standard solution is presented as a line. The line for the best optimized design has been retested so some designs appear below it on the graph.

$p = 1.653e{-12}$). Note that because Figure 6.8 plots all designs evaluated, there are some points below the line presented as the "best" optimized design. That is because the seemingly better designs were evaluated fewer than ten times, so may be designs that were simply "lucky" in the small number of evacuations on which they were evaluated. Of course, extending the run further would eventually allow any designs among them that were actually better to prove their superiority after they were tested on more evacuations (i.e., with different random seeds).

These results are only an initial optimization, but they successfully demonstrate that HEEDS can optimize an ABM to outperform a simple standard timing strategy for evacuating an area. Further computing time will be required to demonstrate the full effectiveness.

*Next Steps and Summary*

This evacuation scenario is being developed as part of an ongoing study. The ABM run to evaluate one evacuation strategy can take anywhere from thirty minutes to two hours, so even though many simulations are done in parallel on the cluster, an optimization study can still take several weeks to complete. (As mentioned earlier, using a more efficient ABM platform than NetLogo could speed these evaluations dramatically.)

This is not presented as an example of performing critical incident analysis, which has not been done, but to illustrate how an agent-based model can be useful in formulating or testing a potential form of intervention—in this case, modification of the normal cycles controlling traffic lights to cope with an emergency evacuation. That is, it explores ways in which the bureaucracy and adhocracy in the triangle of the CIA model can use models to develop strategies for better response in critical incidents. Additional reflection may allow the reader to see how many other aspects of the drivers' behaviors could be added to the model, producing a richer array of behaviors and perhaps representing more closely the human response to a particular scenario. Given that, the bureaucracy and adhocracy might choose other ways to notify people of an emergency or to control the flow of traffic, for example. Given a variety of outcomes predicted by the model, augmented by some sort of model for the adverse consequences suffered by those not successful in evacuating in time, and capturing the costs (multidimensional) of performing the evacuation per se, one might be able to begin posing questions about the longer-term implications of an incident requiring evacuation on the people's attitude toward the governmental agencies supervising it, as influenced by the way the media reported it. At that point, one might be said to be performing critical incident analysis.

The current ABM predicts that there could be about 635 casualties with the traffic lights running under normal conditions and roughly 345 casualties with the traffic lights optimized. This suggests that at least 290 lives could be saved by properly timing the traffic lights in a city during an emergency, even without the availability of traffic sensors. Clearly these provide a plausible range that allows one to determine if optimization is something that is worth an investment in time and money to examine in greater depth. This will, of course, depend on the speed with which the threat can be recognized and responded to by city officials. If this response is not adequate, another method for separating the target populace from the threat may need to be developed. The flexibility of ABMs would very likely allow other methods to be explored, often with only small modifications to the model. Other efficiencies may be discovered upon further investigation of this particular model. This demonstration model clearly shows that ABM can be used to identify tradeoffs and efficiencies within a system, and also provides insight into the outcomes for individual agents.

Future versions of the traffic light control strategy optimization will need to look at more factors affecting traffic flow and analyze the distribution of states of knowledge and goals of the drivers. There are important behaviors that have been left out of the model in order to obtain these initial results. For instance, many drivers might not obey the law and sit at a red light for two minutes if they sense immediate danger. The current model allows drivers to use opposing traffic lanes in an evacuation if the same-direction lanes are blocked, but does not include other forms of emergency behaviors. In real-world emergencies, drivers may also travel at higher speeds, possibly resulting in more traffic accidents. In order to create a more realistic model, it is important to explore how people behave under emergency situations and how they may deviate from their normal patterns.

Further study should also be done on a larger variety of situations that may occur. For instance, the current experiment finds optimal strategies for a single specific scenario. This scenario defines 5,000 cars initially, and an additional 1,500 cars in garages. It also specifies the chlorine spill location and the wind speed and direction in order to estimate casualties. The optimized results in this case may not apply if the traffic is much heavier or lighter. It may also be more difficult to optimize for situations in which the location of the spill and the resulting plume are different. A more thorough optimization should test multiple scenarios, such as high and low levels of traffic and different chlorine plumes, to prepare for a wide variety of situations. The purpose of the present model is to illustrate the effectiveness of optimizing response strategies using an ABM, and the given results appear to do that.

## CONCLUSIONS

Three main reasons for modeling critical incidents are (1) to provide an ability to generalize critical incidents to recognize their similarities and differences; (2) to facilitate indexing or categorizing critical incidents to enable applying the lessons learned in one critical incident to another incident that may not at first seem similar; and (3) to allow use of simplified ABMs that can mimic incident dynamics and provide insights to outcomes of "what if" scenarios.

Critical incidents are incredibly complex, especially at the participant level we are trying simulate. None of the models described in this chapter is able to predict the outcome of a critical incident for the individuals involved. However, given the flexibility of ABMs and the increasing speed and capability of computers, there is little doubt that progress can be made in this area in determining distributions of outcomes in some critical incident scenarios.

Currently, critical incident models may be useful in classifying incidents, in helping to illuminate the dynamics that are important to study, and even in analyzing outcomes or potential outcomes for specific "what if" scenarios. Because the predictive capabilities of a model of any complex critical incident are currently limited, the outcomes provide a good basis for discussion about the particular situation being modeled but probably will not account for the final outlook for individuals being modeled after the incident occurs.

## REFERENCES

Baeza-Yates, R., and B. Ribeiro-Neto. 1999. *Modern Information Retrieval.* New York: Addison Wesley.

Berry, M. W., and M. Browne. 2005. *Understanding Search Engines: Mathematical Modeling and Text Retrieval*, 2d ed. Philadelphia: Society for Industrial and Applied Mathematics.

Chen, M. 2005. "Traffic Signal Timing for Urban Evacuation." Master thesis, Dept. of Civil Engineering, University of Maryland.

Dunning, A. E., and J. L. Oswalt. 2007. "Train Wreck and Chlorine Spill in Graniteville, South Carolina." *Transportation Research Record: Journal of the Transportation Research Board* 2009: 130–135.

Goodman, E., and B. Kirby. 2008. "Conceptual Model for Critical Incident Analysis." In *Proceedings of the Academy for Critical Incident Analysis 2008.* John Jay College of Criminal Justice (CUNY), January.

Grossman, D. A., and O. Frieder. 2004. *Information Retrieval: Algorithms and Heuristics.* The Information Retrieval Series. Norwell, MA: Springer.

Manning, C. D., P. Raghavan, and H. Shutze. 2008. *Introduction to Information Retrieval.* Cambridge: Cambridge University Press.

North, M. J., and C. M. Macal. 2007. *Managing Business Complexity: Discovering Strategic Solutions with Agent-Based Modeling and Simulation.* New York: Oxford University Press.
Ochberg, F. 1978. "The Victim of Terrorism." *Practitioner* 220 (1316): 293–302.
———. 2008. "Critical Incident Concept." The Academy for Critical Incident Analysis, http://aciajj.org/the-critical-incident-concept/ (accessed July 23, 2011).
Ott, W. R. 2007. "Secondhand Smoke in Motor Vehicles Study." TobaccoSmoke.org, http://tobaccosmoke.exposurescience.org/shs-in-motor-vehicles (accessed October 27, 2009).
Polyvyanyy, A., and D. Kuropka. 2007. "A Quantitative Evaluation of the Enhanced Topic-based Vector Space Model." Hasso Plattner Institute. University of Potsdam, http://kuropka.net/files/HPI_Evaluation_of_eTVSM.pdf (accessed July 23, 2011).
Singhal, A. 2001. "Modern Information Retrieval: A Brief Overview." *IEEE Data Engineering Bulletin* 24 (4): 35–43.
Streit, S. 2002. "17,360 Gallon Chlorine Tank Car: HO Scale." *Railroad Model Craftsman,* June, http://www.atlasrr.com/Reviews/17krmc.htm (accessed July 23, 2011).
U.S. Environmental Protection Agency (EPA). 2007. *ALOHA® User's Manual.* Washington, DC, http://www.epa.gov/osweroe1/docs/cameo/ALOHAManual.pdf (accessed July 23, 2011).
U.S. Fire Administration (USFA). 2008. *National Fire Incident Reporting System 5.0 Complete Reference Guide.* Washington, DC: National Fire Data Center, U.S. Department of Homeland Security.
Wilensky, U. 1999. NetLogo, http://ccl.northwestern.edu/netlogo (accessed July 23, 2011).
———. 2003. "Traffic Grid." NetLogo Models Library, http://ccl.northwestern.edu/netlogo/models/TrafficGrid (accessed July 23, 2011).

CHAPTER 7

# Predicting Critical Incidents

NORMAN GRONER

Scientific empirically based work about critical incidents often concerns prediction, whether explicitly stated or not. Applied or action research is the primary agenda in critical incident studies, where the goal is to foster positive change that will prevent or mitigate the negative effects of critical incidents. Prediction is implicit in most scholarly discussions about critical incidents because generalizations are made based on past critical incidents and projected onto similar incidents that might occur in the future.

This chapter reflects the author's viewpoint advocating for the empirically based scientific investigation of prediction as it relates to critical incidents. I take the broadest possible view of the breadth of methods that can be used to analyze data. Still, naïve predictions that are not objectively based on good empirical evidence are important because these perceptions guide people's preparations for and reactions to critical incidents. The concluding part of this chapter concerns perceptual biases that affect subjective predictions, and how these are potentially relevant to an understanding of critical incidents.

I review the considerable difficulties in predicting both the onset and evolution of critical incidents, based on the premise that such prediction is possible within limits and desirable because it provides opportunities to prevent and mitigate the negative outcomes that often coincide with critical incidents.

## THE ROLE OF PREDICTION IN CRITICAL INCIDENT ANALYSIS

Prediction is implicit any time there are plans to improve outcomes in preventing, planning for, responding to, or recovering from the negative effects of critical incidents. Whenever we try to affect outcomes, we are making an implicit prediction that our intervention will cause improvements. In this chapter, I discuss the issue of prediction and divide it into two general topics: empirically—or scientifically based—and intuitive predictions. An illustrative example of the importance of considering both scientific and naïve predictions is the recent controversy about the efficacy of critical incident stress debriefing as a means to mitigate the effects of post-traumatic stress disorder. The positive effects of conducting a debriefing where emergency responders and victims retell their experiences immediately following very stressful events was nearly universally accepted and mostly left unexamined scientifically. Recent research (Roberts et al. 2009) has cast serious doubt on how well the approach alleviates negative symptoms and has provided some evidence that debriefing may exacerbate them in some people. Moreover, recent empirical studies indicate that alternative approaches, notably trauma-focused cognitive behavioral therapy, are potentially valuable in mitigating the psychological impact of severely stressful events (Ekeberg et al. 2008). The point here is not to review the validity of the therapeutic approaches and the research that examines it, but rather to use it as an example of the importance of skepticism about commonly held beliefs, and the considerable value of rigorous empirical investigation.

## THE EMPIRICAL BASIS FOR PREDICTION: GENERALIZING FROM THE PAST TO THE FUTURE

On the face of it, it may seem senseless to ask why it is useful to predict critical incidents. However, people and the cultures that they represent can be fatalistic, questioning both the possibility and the value of predicting critical incidents. Critical incidents in particular may seem predestined and beyond anyone's ability to control, as evidenced by labeling them "acts of God." From this standpoint it may even be counterproductive to predict potentially catastrophic events, serving only to increase anxieties rather than improve outcomes.

"Generalizability" is a primary goal of scientific endeavor. Researchers seek to discover and explain empirically supported statements that are true for all similar phenomena. Unless our interest in critical incidents is strictly historical, we will want to project on events that have not yet occurred. This goal implies prediction as a primary goal of many critical incident analyses.

The action research or applied research model underlies much of the research community's interest in critical incidents (Greenwood and Levin 2000). Modern scientific cultures are based on an instrumental view that purposeful action can prevent and mitigate negative outcomes. Based on our scientific approach, we have made substantial progress in some areas. Perhaps our greatest predictive success is advanced weather models that enable us to predict the general progress and intensity of hurricanes. Despite our increasingly accurate predictions, these storms still cause extensive damage and loss of life because people increasingly put themselves in the paths of such destruction, choosing to live in shoreline and low-lying areas that are most affected by these storms. Still, without the predictive models, the damage, suffering, and loss of life would certainly be much greater.

Relative to our success in predicting the paths of tropical storms, our ability to predict the precipitating and cascading events associated with critical incidents seems very limited. We have had limited success in accurately predicting the precise locations and times for earthquakes and tornadoes. Beyond natural disasters, the prediction of technological and intentional catastrophic events is, to date, well beyond our capabilities. Nonetheless, the value in prediction, given the monumental losses that can occur, is inherently worth pursuing, even considering the difficult and sometimes insurmountable obstacles to our ability to predict the onset of catastrophic events and the ways in which they play out.

## THE CASE STUDY APPROACH

Case studies predominate in research about critical incidents because critical incidents involve inherent boundaries that make it easy to identify them as cases. The boundaries are relatively clear; there is a start and an end, and the effects are usually geographically circumscribed. Stallings (2007) identifies the field research case study approach as the predominant method used in disaster research. Of course, these boundaries can be fuzzy, but relative to most phenomena, they are relatively easy to define.

A "case" is, by definition, a representative of some broader category of which it is a member. The purpose in studying a case is to increase our understanding of the collection of cases to which this instance belongs. Thus, when we perform a study of some particular critical incident, our interest also extends to some broader category of critical incidents that are somehow similar. Critical incidents are typically multidimensional complex events, the entirety of which is beyond the scope of interests of any particular researcher. Depending on our interests, cases may be assigned to more specific categories, such as floods, external threats to national security, school shootings, incidents involving violations of social trust, and incidents that induce instances of stress-related psychological problems.

Case studies may simply have intrinsic value, but more often they have instrumental value (Stake 2000); that is, the case is studied because it provides a better understanding of the class to which the case belongs. Case studies and action research have traditionally been closely tied, as expressed by George Santayana: "Those who do not learn from history are doomed to repeat it" (1905, 284). The underlying motivation is to predict how negative effects of critical incidents can be mitigated to the extent the case study is similar to future incidents along the parameters of interest to the researchers.

Critical incidents are often exceedingly complex events, so the researcher is typically limited to a topic of interest. Research studies may involve a wide variety of topics circumscribed by the case. In case study research, there is a tradeoff between scope and specificity. Killian (2002) explains that the testing of hypotheses requires specificity; that is, variables must be identified and operationalized. This certainly applies where the predictive validity of the research needs to be supported. The following sections covering predictive models focus on the type of research that is relatively specific. However, Killian also encourages researchers to collect and examine exploratory or relatively open-ended research agendas that can yield novel findings and hypotheses that would otherwise have been undetected.

The case study approach is not a methodology per se, but only a convenient way to parse the phenomenon of interest (Stake 2000). The specific methods used to describe a case study may be qualitative, quantitative, or some hybrid of the two. Specific analytic methods should be chosen according to how well they allow the researcher to accurately represent the available data.

## METHODS FOR CONSTRUCTING PREDICTIVE MODELS

The Merriam-Webster online dictionary defines "model" as "a system of postulates, data, and inferences presented as a mathematical description of an entity or state of affairs." However, as used here, a model as defined as "a representation containing the essential structure of some object or event in the real world" (Stockburger 1996). As such, models can be qualitative, that is, they can be derived from qualitative data without the use of mathematics as long as they are structural representations of phenomena in the real world (in our discussion, the real world as it relates to critical incidents).

### Qualitative Models

Case studies of critical incidents may use qualitative, quantitative, or combinations of qualitative and quantitative data. Ryan and Bernard (2000) describe qualitative data as text (e.g., newspapers, government documents, e-mail) and narratives (i.e., people telling about their experiences). Qualitative data can be analyzed using quantitative as well as qualitative methods. For example, qualitative data can be coded into categories, and the frequencies of occurrence can be associated using nonparametric statistical methods; that is, methods that are not based on assumptions requiring that variables use ordinal or interval level of measures.

Perhaps the best-known qualitative method for analyzing qualitative data is so-called grounded theory. Researchers who use this approach carefully examine data and allow categories and themes to emerge (Chamaz 2000). The process is iterative; as themes emerge, the models are increasingly refined and data that was examined earlier is reexamined for additional insights. The researcher is expected to always keep the findings "grounded" in the data. There is considerable controversy among grounded theory researchers, who take the positivistic view of scientific endeavor that objectivity is possible, and constructivist theorists, who believe that researchers' preconceived notions invariably influence the analysis and that objective analysis is impossible (Chamaz 2000).

### Survey Research

Survey research has been used repeatedly to investigate critical incidents (Bourgue, Shoaf, and Nguyen 2003). As the term is used here, "survey research" relies on close-ended items and is distinguished from interview research that uses open-ended items that yield qualitative data. In the realm of critical incidents research, surveys are usually employed as part of a case study approach. However, it is also possible to use the same, or similar, survey instruments or items to generalize across "cases."

## APPROACHES TO PREDICTIVE QUANTITATIVE MODELING

Quantitative models use mathematical representations of the real-world phenomena. Quantitative predictive models can be roughly divided into models that are empirically or rationally derived.

## Empirical Models

An empirical model is solely constructed based on data. Empirical models are descriptive, not inherently explanatory. As an example, empirical models have been developed to predict the spread of wildfires (Perry 1998). No theory or a priori notions about the model are used to build an empirical model. While explanations about the relationships between predictor (i.e., independent) and criteria (i.e., dependent) variables may be sought after the model is derived from empirical models, such theoretical relationships among variables play no role in constructing the models. Testing a theoretical hypothesis is not required to generate a purely empirical model. Some researchers have argued that purely descriptive research is not of much value in social science research about disasters (Stallings 2003).

Inferential statistical analysis is closely associated with both the empirical and the rational approaches to model building, although the role of analysis is somewhat different. In developing an empirical model, inferential statistics are used to determine whether the observed relationships used to build the model are very unlikely to have occurred by chance. Such models are often based on relationships among many variables. When testing relationships among many variables, some may appear to be statistically significant even though the relationships result from random variations. As such, tests of significance when building empirical models should be very conservative.

Exploratory analysis is an important type of empirical model building that serves as a bridge to rational modeling. An exploratory analysis looks for structure in data sets for the purpose of building a rational model that can be used to generate hypotheses for further testing. Inferential statistics are used to determine whether the relationships detected in the data analysis are unlikely to have occurred by chance. Exploratory research is quasi-empirical because the data drive the model building, but the process of understanding the meaning of the analysis to develop hypotheses is rational.

Linear regression is the simplest and most straightforward means to build an empirical predictive model. A linear correlation appropriate to the level of measurement (e.g., Pearson product moment coefficient for interval data, Spearman rank order correlation for ordinal data) is computed, and a prediction is made about an unknown value of an event that has not yet occurred based on a known value for the same event. For example, a researcher finds the correlation between how quickly individuals prepare to evacuate and their scores on a media credibility scale. Assuming that there is a statistically significant linear correlation between these two variables, the researcher can use the model to predict how quickly a person is likely to prepare to evacuate based on his or her score on the media credibility scale.

When investigating most topics related to critical incidents, predictive models based on linear relationships between two variables are too simple to yield adequate predictive validity. Taking the example of predicting how quickly people prepare for an evacuation, the perceived credibility of media is only one among a great many potentially influential predictor variables, including but not limited to the availability of transportation resources, the support of family and friends, having a clear destination, past experiences with similar events, and so forth (Sorenson and Sorenson 2007). Fortunately, there are many extensions to the use of linear regression to create predictive models. Several predictor variables can be associated with a single criterion variable in a multiple regression model. The model need not be linear. A nonlinear regression can be built where Z exponential functions, logarithmic functions, trigonometric functions, or power functions are likely to yield a better predictive model.

## Rational Models

Empirically derived models are relatively rare in the study of critical incidents; rational models predominate. Rational models are tested using postulates or hypotheses about the associations or causal relationships among the variables of primary interest in the models. The quality of the scientific endeavor largely depends on how successfully researchers can operationalize the variables used in the rational models. To the extent that researchers can reliably and validly measure variables, they will be successful in testing hypotheses generated by rational models.

In a rational modeling approach, the researcher seeks to test hypotheses that are developed prior to analyzing the data. Again, the inferential statistics are used to test whether the hypothesized relationships between variables are very unlikely to have occurred by chance. Hypothesis testing is always based on rational models.

As noted earlier in this chapter, critical incidents are often too complex to understand holistically. Mini-theories are circumscribed sets of postulates that attempt to address a narrow range of phenomena, and mini-theories that are relevant to critical incidents are abundant, especially from the fields of public health and disaster sociology. Examples of mini-theories may be used to predict the onset of wars and humanitarian disasters based on competition for resources and the availability of superior weaponry. Similarly, mini-models predict the onset and development of certain specific types of critical incidents. Examples are quantitative epidemic models that are used to predict where and how rapidly infectious diseases are likely to spread. Another example is graphical information systems models that are used to calculate the spread of toxic materials from accidental or intentional releases of biological, chemical, or radiological toxins.

Critical path analyses have been used to develop predictive models in disaster sociology. The method requires the researcher to develop a rational model about the causal relationship between variables. As an example, a critical path analysis was used to test a model for predicting evacuation delays among building occupants. The analysis used telephone survey data collected by the National Institute of Standards and Technology in its investigation of the World Trade Center disaster of September 11, 2001 (Averill et al. 2005): "The estimated parameters of the models for WTC 1 and WTC 2 revealed that the model had a very high degree of success in explaining evacuation initiation delay, pre-evacuation actions, seeking information, and perceived risk in both towers" (131).

## CATEGORIZING PREDICTIVE MODELS OF CRITICAL INCIDENTS

It is of value to briefly examine some of the many predictive models that are relevant to predicting the onset and develop of critical incidents. This list is hardly exhaustive. The idea is to provide a sample of relevant models. Some scheme for categorizing predictive models is valuable, and in this section the scheme is organized around the traditional phases used in the emergency management field: preparedness, response, and mitigation. These categorizations of predictive models are shown in Figure 7.1, where the traditional phases of emergency management are related to the general model provided in Chapter 2.

- *Preparedness* always occurs prior to an incident, and is a nexus between how the incident is managed and the policy-driven planning by public and private entities meant to cope with the incident.
- *Response* occurs after the onset of an incident. Once an incident has begun, the mode switches to response, when management of the incident attempts to cope directly with the threat.
- *Mitigation* always occurs before the incident. In emergency management, mitigation is often defined as referring to long-range or durable approaches for reducing the impacts once a threat or hazard has materialized. In practice, mitigation generally refers to physical features that are installed to lessen the impacts of threats. As such, mitigation is a function of the nexus between policy-driven changes in infrastructure and building practices used by public and private entities to lessen the impacts after the onset of an incident.

### Preparedness Models

Preparedness is the nexus between policy and management of an incident. Policy dictates the structures that enable and constrain the management of incidents. Examples at the federal level in the United States include the National Response Framework and the National Incident Management Systems. Prediction is implicit in policy theory and models; well-conceived policy is believed to improve outcomes.

### Response Intervention Models

Responses to disasters have been the focus of a great deal of social science research, and that research has generated many mini-theories and limited-scope models. Two areas that have generated significant contributions are geographic information systems and emergency management systems.

#### *Geographic Information Systems*

Geographic information systems are increasingly being used as a support tool for responses during many critical incidents. These computer-based systems provide information that is

Figure 7.1 **Traditional Emergency Model Processes Related to the Critical Incident Analysis Model**

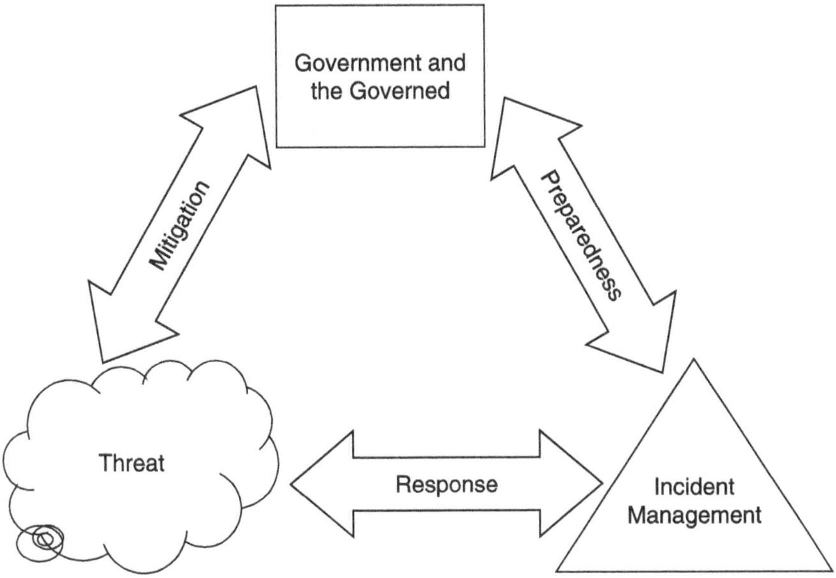


spatially organized using maps, including information about transportation routes, physical structures, topography, and the locations of resources. Features that make predictions about the spread of hazards are of particular value in responding to critical incidents. These include the anticipated plumes related to airborne and waterborne contaminants and flooding.

### *Emergency Management Systems*

Emergency management systems are rational models that underlie government (and increasingly private sector) efforts to respond effectively to critical incidents that involve threats to lives and tangible property.

People often seek to control the outcomes associated with critical incidents. Command-and-control models, whether implicit or explicit, guide efforts to control outcomes by gathering and trying to understand information about events (control processes) and by directing the actions of participants based on that understanding (command processes). Command-and-control models have an exceedingly long history, dating back to ancient Chinese and Roman writings concerning the conduct of warfare. Recent innovations largely concern decentralizing command-and-control to enable more adaptive responses by avoiding the time and administrative overhead required when information is centrally transferred and processed and instructions from the resulting decisions are later disseminated.

The Incident Command System (ICS) is a seminal emergency management model that is being universally incorporated into United States government response plans, from local public safety agencies to the federal government. The ICS model provides a good example of a contemporary command-and-control model designed with the intent of promulgating more positive outcomes to emergencies, critical incidents included.

The ICS model divides response activities into functional categories or sections, each of which has a clearly defined role: operations, logistics, documenting financial variables, and planning or forecasting future events during the incident. The source of authority resides in an incident commander, who delegates authority to sections that function autonomously as long as they adhere to the ICS framework. The ICS approach, and all the allied command-and-control models, are based on the assumption that response actions coordinated using the approach will lead to *predictably* better outcomes.

## Mitigation Models

In the emergency management framework, the term "mitigation" has a circumscribed meaning. The Stafford Disaster Relief and Emergency Assistance Act is U.S. federal legislation that defines the relationships among the emergency responses at different levels of government. The act defines *hazard mitigation* as "any action taken to reduce or eliminate the long-term risk to human life and property from natural hazards." The reference to "long term" is ambiguous, but mostly refers to physical modification and features that remain in place over an extended period of time before the incident for the purpose of reducing its negative effects. Mitigation is distinguished from "preparedness," which generally refers to planning the short-term actions taken during the incidents.

More than any of the other categories, modeling risk plays a central role in mitigation models. Risk models are used to decide how resources can be most cost-effectively deployed. Although models for risk vary, they are generally variations on the following equation: Risk equals Probability that a threat will occur multiplied by the Consequences if the threat does occur.

However, risk models break down for low probability/high consequence where events converge in unforeseeable ways. Risk models are often based on normal distributions of hazardous occurrences. However, normal distributions may substantially underrepresent the real probabilities of very low- and high-consequence events. Higher than expected probabilities for rare events may occur because risk models omit important factors and interactions that are too complicated to incorporate, an argument discussed later in this chapter in the section on uncertainty. The problem is of particular concern in financial markets where it is variously known as the Black Swan theory (Taleb 2007) and the fat tail problem (Rosenberg and Schuermann 2006).

## Business Models

Prediction (often labeled as "forecasting" in this context) is of great interest to businesses. To the extent that they can predict future events, they are more likely to earn a higher return on their endeavors. Perhaps risk models are the most relevant of business models to the study of critical incidents. Although various terms are used to define "risk," quantitative risk models are generally based on calculating the product of the probability that an event will occur multiplied by the consequences if the event does occur. Such models are the basis for predicting actuarial losses in the insurance business, and provide generally accurate results when applied to events that are frequent enough to provide a good record of data on the probability and consequences of events involving loss. However, risk models can yield unreliable results when applied to low probability/high consequence events. The inherent hazards of using risk models have been recently publicized by the meltdown of financial markets. In this instance, events converged in ways that the risk models were unable to predict, resulting in exceedingly large and unforeseen losses (Taleb 2007).

Predicting customer and client behaviors is a frequent quest for business analysis. Models include attempts to forecast narrowly defined events such as where to direct marketing resources and which customers are more likely to engage in fraudulent and delinquent behaviors. Other models are used to forecast the economic trends and the associated performances of investment portfolios. These approaches often rely on regression techniques where a variety of predictor variables are statistically associated with criterion variables. Risk underwriting is another area of predictive modeling where businesses seek to use a predictor variable to identify persons or other entities where the risk is not commensurate with the expected insurance premiums.

### Engineering Models

Engineering models are intended to avoid or guide the design of mitigating measures in technological systems. Engineering models are another area in which predictive models are used to prevent and mitigate negative outcomes. For example, buildings and critical structural components are designed to withstand earthquakes and to prevent fires from spreading into conflagrations. Levees are designed to limit flood damage, and building facades may be designed to limit the damage caused by explosions. An extensive array of quantitative mini-theories allows engineers to predict the probabilities that their designs can resist insults of various intensities. Engineering approaches are also used in predictive simulations of a variety of dynamic processes, including pedestrian and traffic movement, a topic explored in Chapter 6.

## UNCERTAINTY

Uncertainty concerns the limits to predictive and forecasting models. Given the nature of critical incidents, models of uncertainty are also relevant to the problems of predicting the precipitating and cascading events that comprise critical incidents. Theorists distinguish between aleatory and epistemic uncertainty (Daneshkhah 2004).

Aleatory uncertainty is caused by the inherent randomness or stochasticity of events. Because aleatory uncertainty concerns inherent natural fluctuations, it is irreducible; that is, any increase in our understanding of the behavior of a system will not increase our ability to predict those events. Chaos theory is a related concept of interest in critical incident analysis.

Stochastic models incorporate representations of randomness when predicting events. The approach involves running numerous computerized iterations of the model while randomly choosing values from a distribution of the criteria or independent variables. Each iteration is a deterministic application of the model, but the result is a distribution of values for the criterion or dependent variable that is thought to represent the naturally occurring fluctuations in the criterion variable. Markov models can be used when dealing with a series of systems states, each of which involves probabilities that the system will change to subsequent states. While our current understanding of how critical incidents evolve from one state to another is inadequate to the task, Markov models provide a potentially valuable approach to predicting the onset and cascading nature of critical incidents.

Chaos theory postulates that events are not random per se, but that initial conditions and complex interactions make events impossible to accurately model, even without postulating any randomness. Stated differently, it is our inability to accurately model the way that events play out that makes them unpredictable. Epistemic uncertainly is caused by limitations in the models that we use to predict events and our lack of understanding about the behaviors of a system. As we improve our models, we can reduce epistemic uncertainty, thereby improving our ability to predict how systems behave. As epistemic uncertainty is reduced, our models more accurately predict the onset of critical incidents or mitigate negative outcomes from critical incidents. Advances in modeling weather phenomena are good examples. As the models improve, that is, as epistemic uncertainty is reduced, predictions have become increasingly accurate. What once appeared to be aleatory uncertainty, or chaotic or random weather fluctuations, was later understood to be epistemic uncertainty, which, in turn, was reduced as models (and the computing power needed to drive the models) improved.

Critical incidents sometimes evolve unpredictably from relatively minor precipitating events, a view consistent with chaos theory. Charles Perrow (1999) argues that it is inherently impossible to prevent catastrophic systems accidents where technologies are so complex that external and internal events can interact in ways that cannot be predicted. The argument can be extended to other domains where tightly coupled, complexly interactive systems can yield unexpected and severe events. Recent problems in financial markets are illustrative.

The problems of predicting the occurrence of a precipitating event differ depending on whether the event is intentional or unintentional. Intentional events involve adversaries who try to anticipate the actions of their opponents. Prediction becomes a function of success in "playing the game." To the extent that adversaries are rational in their approaches to maximizing their outcome, the epistemic uncertainty can be modeled using a game theoretic approach. However, game theory is limited to the extent that adversaries are irrational; that is, they are not concerned with maximizing their outcomes. For example, a suicide bomber is not going to be concerned with earthly gains.

## PERCEPTIONS OF PREDICTABILITY AND PROBABILITIES

Researchers have identified several perceptual biases and heuristics that affect the perceived predictability of events and the perceived probabilities that events will occur. Perceptual biases are important in understanding the outcomes of critical incidents, because such biases impact the likelihood that people will prepare for and respond appropriately to such events.

Naïve predictions about critical incident predictions are common to many cultures, from the apocalyptic predictions in the Bible to the belief that environmental insults could cause the catastrophic events depicted in books and movies, from *Godzilla, King of the Monsters!* (1956) to the depiction in *The Day After Tomorrow* (2004) of global weather catastrophe.

## PERCEPTIONS OF THE LIKELIHOOD OF CRITICAL INCIDENTS

Communicating risk is one of the core agendas in action research on critical incidents (Rodriguez et al. 2007). To the extent that experts are able to effectively communicate the probabilities associated with natural hazards such as severe weather and earthquakes, the belief is the people will prepare accordingly. Considerable research has been devoted to examining the credibility and modalities of risk communicators and messages. However, this section focuses on the heuristics and biases that influence the recipients of such messages.

Biases that lead people to underestimate the probabilities that catastrophic events are likely to occur, leading them to not adequately prepare for such events, are of particular interest in an action research agenda. At the most fantastical level, there always seem to be some people who believe in predictions of eschatological end-of-the-world events and prepare accordingly.

Confirmatory and illusory correlation biases may account for biases where people make erroneous judgments about the link between causes and effects. The confirmation bias involves the tendency of people to seek and find more credible information that confirms their hypotheses or preconceived notions. Illusory correlations result when people access only information about the co-occurrence of events, while ignoring instances where one event occurred in the absence of the other event.

The "availability heuristic" means that the easier it is to remember a threat, the greater the subjective estimate of its probability (Tversky and Kahneman 1973). The ease of recall may, in turn, be influenced by "such factors as the concreteness, drama, familiarity, recency, relevance, similarity or vividness of instances" (Dube-Rioux and Russo 1988, 223). This is strongly influenced by both personal experiences and by the media. For example, people who watch local news are likely to have a different view of probabilities than people who do not, including, for example, traffic accidents, tragic fires, robberies, and murders and assaults. Contemporary news broadcasts that provide vivid depictions of catastrophes may enhance an availability bias by making the events particularly memorable (Kahneman, Slovic, and Tversky 1982).

Emotions appear to mediate perceived risk. Persons who react to the same event with fear tend to have greater subjective probabilities of occurrence for similar events than people who react with anger (Lerner et al. 2003).

Discounting remote risks means that people treat them as having a zero probability, even when the consequences of such an event would be severe (Kornreuther and Pauly 2004). For example, it is not likely to matter to people if the probability of a life-threatening asteroid striking the earth in a single year is 1 in 10,000 or one in 10,000,000 even though the difference is 1,000 times greater in the latter case. In either situation, most people simply treat the event as though the probability of occurrence is zero; that is, if the probability is small enough, they simply behave as if the event will never happen, even if the consequences would be catastrophic.

The "gambler's fallacy" is the belief that after a string a bad luck, luck is bound to change. For example, if you toss a coin and it comes up "heads" five consecutive times, it would seem to make sense that your luck would change and the next toss would likely come up "tails." But for each coin toss, the probability that a coin toss will be heads-up never changes. To the extent that the gambler's fallacy affects perceptions of probability, one who gambles and suffers a string of losses is more likely to continue, feeling that his or her luck will surely change. But the probabilities remain the same for each and every new gamble. The same might apply to some persons who are victims of hurricanes; now that they have suffered the loss, they believe they are less likely to experience another hurricane. But everything else being equal, the chance of experience remains the same.

The hindsight bias leads people to the erroneous conclusion that events were more predictable than they really were. In retrospect, all of the other possible ways in which events could have evolved are not apparent, causing a sense of inevitability where none existed at the time. In particular, this cognitive bias plays a role in perceptions of catastrophic events with people expressing the view that victims and public officials "should have seen it coming" (Fischhoff and Beyth 1975).

## CONCLUSION: WHY BOTHER IF PREDICTION IS SO UNCERTAIN?

Given the high levels of uncertainty and the innately chaotic manner in which critical incidents play out, it is reasonable to ask, "Why bother?" Critical incidents are often exceedingly complex, and the high levels of uncertainty impose severe limits on the accuracy of predictions. Nonetheless, theory still provides for significant advances in our abilities to mitigate negative outcomes and perhaps to avoid the onset of critical incidents. The difficulties are more epistemic than practical. While we cannot predict events with much precision, our understanding allows us to constrain events. Guided by theory, we can prevent many negative outcomes; we may not know what will happen, but we can sometimes know that it will not happen in a certain way.

A useful example is the use of construction practices to mitigate incidence of casualties from earthquakes. To date, we have had little success in developing models for predicting the onset of earthquakes, their precise locations, and the associated ground movements (Geller

1997). Nonetheless, we have had great success in mitigating the effects of earthquakes by designing better buildings. The same applies to advances in the social and behavioral sciences, although our successes are less tangible. Disaster sociology, in particular, has generated an extensive body of theory in the form of postulates that await sets of predictive validity (Drabek 1986; Tierney, Lindall, and Perry 2001). Based on this information, we have made significant advances in constraining outcomes by applying predictive theory. We know now that it is unnecessary to over-allocate resources toward preventing civil disturbances following natural disasters, that we should control the content of emergency communications to provide consistent and accurate messages, and that we should not simply ask victims to recount their traumatizing experiences following a critical incident.

## REFERENCES


Averill, D. J., D. S. Mileti, R. D. Peacock, K. D. Erica, N. Groner, G. Proulx, et al. 2005. "Occupant Behavior, Egress, and Emergency Communications (NIST NCSTAR 1–7)." In *Final Report of the Federal Building and Fire Safety Investigation of the World Trade Center Disaster* (8 vols.). Gaithersburg, MD: National Institute of Standards and Technology.

Bourgue, L. B., K. I. Shoaf, and L. H. Nguyen. 2003. "Survey Research." In *Methods of Disaster Research,* ed. R. A. Stallings. Philadelphia, PA: Xlibris.

Chamaz, K. 2000. "Grounded Theory: Objectivist and Constructivist Methods." In *Handbook of Qualitative Research* (2d ed.), ed. N. K. Denzin and Y. S. Lincoln, 509–535. Thousand Oaks, CA: Sage.

Daneshkhah, A. R. 2004. "Uncertainty in Probabilistic Risk Assessment: A Review." August 9, http://www.shef.ac.uk/content/1/c6/03/09/33/risk.pdf.

Drabek, T. 1986. *Human System Responses to Disaster: An Inventory of Sociological Findings.* New York: Springer-Verlag.

Dube-Rioux, L., and J. E. Russo. 1988. "An Availability Bias in Professional Judgment." *Journal of Behavioral Decision Making* 1: 223–237.

Ekeberg, Ø., L. Weisaeth, H. Kornør, D. Winje, I. Kirkehei, K. Johansen, et al. 2008. "Early Trauma-focused Cognitive-Behavioural Therapy to Prevent Chronic Post-traumatic Stress Disorder and Related Symptoms: A Systematic Review and Meta-analysis." *BMC Psychiatry* 8: 81.

Fischhoff, B., and R. Beyth. 1975. " 'I knew it would happen': Remembered Probabilities of Once-Future Things." *Organizational Behavior and Human Performance* 13: 1–16.

Geller, R. J. 1997. "Earthquake Prediction: A Critical Review." *Geophysical Journal International* 131 (3): 425–450.

Greenwood, D. J., and M. Levin. 2000. "Reconstructing the Relationships between Universities and Society through Action Research." In *Handbook of Qualitative Research*, 2d ed., ed. N. K. Denzin and Y. S. Lincoln, 85–106. Thousand Oaks, CA: Sage.

Kahneman, D., P. Slovic, and A. Tversky. 1982. *Judgments under Uncertainty.* New York: Cambridge University Press.

Killian, L. M. 2002. "An Introduction to Methodological Problems in Field Studies in Disasters." In *Methods of Disaster Research,* ed. R. A. Stallings, 49–93. Philadelphia, PA: Xlibris.

Kornreuther, H., and M. Pauly. 2004. "Neglecting Disaster: Why Don't People Insure against Large Losses." *Journal of Risk and Uncertainty* 28: 5–21.

Lerner, J. S., R. M. Gonzalez, D. A. Small, and B. Fischhoff. 2003. "Effects of Fear and Anger on Perceived Risks of Terrorism: A National Field Experiment." *Psychological Science* 14 (2): 144–150.

Perrow, C. 1999. *Normal Accidents: Living with High-Risk Technologies.* Princeton, NJ: Princeton University Press.

Perry, G. L. 1998. "Current Approaches to Modelling the Spread of Wildland Fire: A Review." *Progress in Physical Geography* 22 (2): 222–245.

Roberts, N. P., N. J. Kitchiner, J. Kenardy, and J. I. Bisson. 2009. "Multiple Session Early Psychological Interventions for the Prevention of Post-Traumatic Stress Disorder." *Cochrane Database of Systematic Reviews,* DOI: 10.1002/14651858.CD006869.pub2.

Rodriguez, H., W. Diaz, J. M. Santos, and A. Benigno. 2007. "Communicating Risk and Uncertainty: Science, Technology, and Disasters at the Crossroads." In *Handbook of Disaster Research,* ed. H. Rodriguez, E. L. Quarantelli, and R. R. Dynes, 476–488. New York: Springer.

Rosenberg, J. V., and T. Schuermann. 2006. "A General Approach to Integrated Risk Management with Skewed, Fat-tailed Risks." *Journal of Financial Economics* 79 (3): 569–614.

Ryan, G. W., and H. R. Bernard. 2000. "Data Management and Analysis Methods. In *Handbook of Qualitative Research,* 2d ed., ed. N. Denzin and Y. Lincoln, 769–802. Thousand Oaks, CA: Sage.

Santayana, G. 1905. *The Life of Reason; or, the Phases of Human Progress.* New York: Scribner's Sons.

Sorenson, J. H., and B. V. Sorenson. 2007. "Community Processes." In *Handbook of Disaster Research,* ed. H. Rodriquez, E. L. Quarantelli, and R. R. Dynes, 183–199. New York: Springer.

Stake, R. 2000. "Case Studies." In *Handbook of Qualitative Research,* 2d ed., ed. N. K. Denzin and Y. S. Lincoln. Thousand Oaks, CA: Sage.

Stallings, R. A. 2003. "Methods of Disaster Research: Unique or Not?" In *Methods of Disaster Research,* ed. R. A. Stallings, 21–44. Philadelphia, PA: Xlibris.

———. 2007. "Methodological Issues." In *Handbook of Disaster Research,* ed. H. Rodriguez, E. L. Quarantelli, and R. R. Dynes, 55–82. New York: Springer.

Stockburger, D. W. 1996. "Models." In *Introductory Statistics: Concepts, Models, and Applications,* http://www.psychstat.missouristate.edu/IntroBook/sbk00.htm (accessed July 23, 2011).

Taleb, R. 2007. *The Black Swan: The Impact of the Highly Improbable.* New York: Random House.

Tierney, K. J., M. K. Lindall, and R. W. Perry. 2001. *Facing the Unexpected: Disaster Preparedness and Response in the United States.* Washington, DC: John Henry Press.

Tversky, A., and D. Kahneman. 1973. "Availability: A Heuristic for Judging Frequency and Probability." *Cognitive Psychology* 5: 207–232.

CHAPTER 8

# Praxis, Applications, and Problems

## Data Mining for Critical Incident Analysis

VINCENZO A. SAINATO

Never before has so much data from so many disparate sources been accessible to researchers or practitioners of critical incident analysis (CIA). The construction of massive data arrays has the capacity to yield a bounty of riches for the analysis of incidents. That said, emergency managers at all levels need tools to facilitate decision making in a host of issues, including evacuation directives, management of near-the-scene logistics, triage at the scene and at hospitals, use of volunteers and off-duty personnel, reduction of telephone traffic congestion, and integration of response with second- and third-level responders from other jurisdictions (Larson, Metzger, and Cahn 2006).

The Academy for Critical Incident Analysis Database (ACIAD) at John Jay College was designed to help fill a necessary gap: to provide a centralized depository for data, specifically designed to facilitate CIA. The ACIAD makes it possible for practitioners to contribute directly to active research in areas that can help them in their roles as the responders to critical incidents (CIs).

This abundance of possibility gives way to an emerging issue: The excess of data can be so overwhelming that useful intelligence and patterns may be overlooked. Knowledge discovery in databases, or data mining (DM), is the process of discovering meaningful new correlations, patterns, and trends by sifting through data stored in repositories such as the ACIAD. Data mining is an interdisciplinary area, drawing upon research in statistics, databases, pattern recognition, machine learning, data visualization, optimization, and high-performance computing. However, until recently only those with quantitative and programming expertise have been able to apply DM. Now the necessary tools are accessible to others, as familiar applications with familiar interfaces (such as Excel and SPSS) have acquired the ability to apply DM analysis via third-party add-ons.

The purpose of this chapter is to assess the praxis, applications, and problems related to DM and CIA. The first section focuses on the praxis (the practical application of theory). It describes DM and related technologies and provides a walk-through of the different kinds of DM analytical tools available to practitioners and researchers. This is followed by a review of the problems inherent in DM, with a description of how these concerns relate to the design or collection of CIA data. The final section provides brief illustrative examples of how DM can be useful with respect to the Virginia Tech shooting incident of 2007.

## PRAXIS: WHAT IS DATA MINING?

### Databasing and Data Mining: A Symbiotic Affair

Data management technologies permit the storage, sorting, modeling, and analysis of a given data set. Data management can be broken into two broad subclasses: databasing and data mining. The focus of this chapter is the discovery (or mining) of information patterns that exist within large tracts of data; however, an appreciation of both the problems and the potential of data mining require some understanding of databasing.

### Databasing: The Acquisition and Storage of Data

Data acquisition technologies include all devices that can record, monitor, obtain, or digitize data about our legal, illegal, or illicit behaviors, choices, or preferences (Farmer and Mann 2003). In the ACIAD context, data acquisition includes a complex web of data, including personal statements, official reports, and media coverage.

Advanced acquisition systems, commonly referred to as distributed or intelligent, involve the acquisition and integration of digitized data from many individualized sources or disparate systems (Remagnino 2002; Velastin and Remagnino 2006) (see Figure 8.1). Distributed systems have grown in popularity among agencies in large urban areas. For example, since 2001 the New York Police Department has been incrementally implementing a citywide surveillance strategy called Operation Sentinel, which is modeled after London's Ring of Steel (Coaffee 2004; Lisberg 2007). Among the many acquisition technologies that the agency plans to adopt is an interconnected web of three thousand closed-circuit televisions in lower Manhattan alone (Baker 2008; Kramer 2008). Once completed, the system will be integrated with surveillance systems from government agencies within a 50-mile radius of the city. This distributed system will maintain continuous tracking of automobiles into and out of New York City by automatically scanning licenses and searching for suspicious travel patterns throughout the metropolitan area.

Databasing (or data warehousing) refers to technologies that can aggregate, store, and sort massive amounts of digital data. A database is typically a computer that serves as a repository for data. However, any device that can permanently store information is, strictly speaking, also a database. And, given the storage capacity of "everyday" acquisition technologies such as iPods and cell phones, the lines between these categories have become decreasingly "bright" and increasingly "fuzzy."[1] That said, for the purpose of this research, databases are technologies that are expressly designed and organized to aggregate and store digitized data for later retrieval.

Figure 8.1 **Architecture of a Typical Distributed DST System**

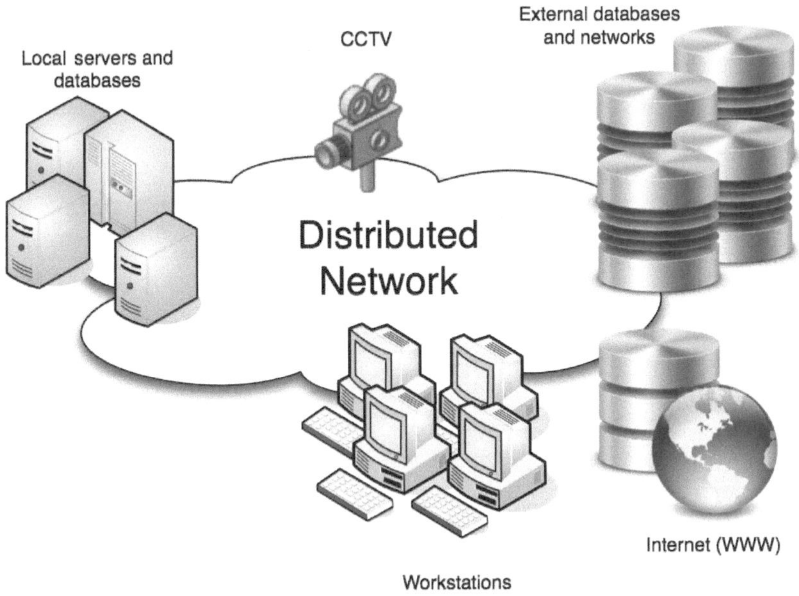


The modern development of databases is a fascinating tale that has been outlined elsewhere (Garfinkel 2001; O'Harrow 2005; Packard 1964; Rosenberg 1969; Rule 1974; Solove 2004; Westin and Baker 1972). Databases have proven themselves to be efficient at organizing and sorting vast tracts of data.

It is not only the government that has found value in taking advantage of the capabilities of databases; both government and nongovernment entities generate and gather a tremendous amount of data in the routine and critical functioning of these institutions. Increasingly, businesses and government entities (large and small) are not merely tracking exchanges or service deliveries but attempting to acquire as much detailed information as possible (Andrejevic 2007; Clippinger 2007; Privacy Rights Clearinghouse 2007; Rule 2007; Solove 2004, 2007; Wood 2006).

The biannual WinterCorp (2007) survey of government and commercial databases shows an astounding rate of change in the size of databases (see Table 8.1). Between 2001 and 2005 the size of the largest-known commercial or government databases increased by 900 percent and, for the first time ever, exceeded the 100 terabyte (TB) barrier.

Another way to assess the magnitude of these databases is to consider the following: The Yahoo! database, at 100 TB, has the storage capacity to fit about 2 billion books. That is roughly 120 times the number of books held in the world's largest collection, the Library of Congress.

That all of this data is being collected begs the obvious question: "How could any individual, group, or entity possibly sort through all of the digitized data and makes sense of it?" One potential answer might be found in the market for data analysts—perhaps

Table 8.1

**Winter Corporation's Annual Survey of Database Size and Performance**

| 2001 | | 2003 | | 2005 | |
|---|---|---|---|---|---|
| Company/Organization | Size(TB) | Company/Organization | Size(TB) | Company/Organization | Size(TB) |
| Telstra | 10 | France Telecom | 29 | Yahoo! | 100 |
| British Telecom | 8 | AT&T | 26 | AT&T[1] | 94 |
| United Parcel Service | 8 | SBC | 25 | KT IT-Group | 49 |
| Experian | 3 | Anonymous | 16 | AT&T | 27 |
| US Customs Service | 3 | Amazon.com | 13 | LGR-Cingular Wireless | 25 |
| Korea Telecom (KT ICIS) | 2 | Kmart | 13 | Amazon.com | 25 |
| Dacom System Tech. | 2 | Claria Corporation | 12 | Anonymous | 20 |
| CheckFree | 1 | Health Insurance Review Agency | 12 | UPSS | 19 |
| Centrelink | 1 | FedEx Services | 10 | Amazon.com | 19 |
| LG TelCom | 1 | Vodafone D2 GmbH | 9 | Nielsen Media Research | 18 |

*Source:* WinterCorp, 2007.
*Note:* Size in Terabytes (TB). 1 TB= 1000 gigabytes
[1]This is a collection of databases acquired through mergers and acquisitions.

there has been a significant increase in demand by businesses to hire them. However, Grossman (2001) finds that there is an increasing "data gap" in the United States, as the demand for new analysts has been constant since 2000. Also, the complexity of the data systems and volume contained therein has dramatically increased.

Given the scale and complexity of collecting data, it stands to reason that many valuable insights are "hidden" in the data patterns. Imagine conducting a factor analysis on a matrix that (at the extreme) might contain as many as 150 trillion cells. Even a modest array could take weeks for the typical human analyst to uncover anything useful; or, more simply, the enormity and complexity of the data would leave much of it unanalyzed.

### Data Mining

Science has developed a solution to the problem of analyzing massive collections of digitized data: data mining. This process has many alternative names[2] and definitions. For our purposes data mining is defined as

- nontrivial extraction of implicit, previously unknown, and potentially useful information from data; and
- the exploration and analysis, by automatic or semiautomatic means, of large quantities of data in order to discover meaningful patterns.[3]

Simply stated, data mining is using a computer application to comb through, sort, assess, analyze, categorize, and explore a vast array of data. Data mining is not merely looking up a number in a phonebook; rather, it is the discovery that the (specific) last names Kelley, O'Doules, and McSweeney, for example, are more common on the west side (along I-90) of Cleveland, and Spino, LoPresti, and Zingaro are more common on east side (but west of I-271). At the most basic level, data mining is taking raw data (such as the phonebook for the Greater Cleveland metropolitan area) and turning it into useful insights or, more simply, information.

As a scientific operation, data mining, has roots in the late 1980s and early 1990s and is a synthesis of developments in overlapping and related, but somewhat indistinct areas, including:

- *Statistics:* theory based and focused primarily on testing hypotheses
- *Machine learning:* focused on improving performance of a learning agent (computer)
- *Knowledge discovery:* integrates theory and heuristics focus to further the process of knowledge discovery, including data cleaning, learning, and integration and visualization of results[4]

In and of itself data mining represents a great leap forward with respect to taking raw data and turning it into useful intelligence; nonetheless, it has two broad limitations. First, like many operations involving statistical analysis, data mining may be able to indicate corollary patterns and relationships, but it is not probative. Second, data mining will help uncover the hidden patterns in the data, but it cannot tell the analyst the usefulness or significance of its findings. In other words, data mining detects the latent patterns but cannot determine what they mean. While the "heavy lifting" of the computations is accomplished with analytical software, creating or selecting of the model or algorithm employed and interpreting the outputs are necessarily human functions.

## Central Data Mining Methods and Tasks

Data-mining methods are divided, at the highest level, into two categories. The first category is prediction, which is a collection of methods that uses variables to predict unknown or future values of other variables. The most common predictive task is classification. Other classification tasks include traditional regression analysis and deviancy (outlier) detection.

Classification is the process in which a model is inferentially developed from the given data for the purpose of uncovering the previously unknown or hidden records. Common uses of classification include direct marketing, fraud detection, and consumer behavior analysis.

The second broad category of data-mining methods is description, in other words, finding the human-interpretable patterns that describe the data. Three data-mining tasks are descriptive. First, clustering is used to partition data into related subsets. Common clustering endeavors include document clustering or market segmentation. Second, association-rule discovery is the process of creating dependency rules that can predict an occurrence (Lin and Brown 2003). A common use of this task is the positioning of products in a grocery store. A classic illustration is the finding by retailers that when men are sent out to purchase diapers, they also purchase a six-pack of their favorite beer, probably as a sort of reward. Conclusion: Place the beer and the diapers near each other to maximize sales.[5]

Another task, sequential pattern discovery, assesses variables that have their own timeline of events to find rules that predict the sequential patterns (Brown and Gunderson 2001). This is distinct from the association rule, which is focused on correlations rather than timing. An example of this would be the discovery by retailers that people who typically purchase a baseball glove and conditioning oil in February are likely in mid-March to purchase cleats and batting gloves; however, these marketing data may also show that the typical baseball consumer is more price-conscious than retailer-loyal. With this insight, the sports store might give those customers an incentive to come back in March for their next purchase.

### *Cross-Method Tasks*

Some DM tasks are not inherently grouped into the methodological categories described earlier. Frequently, there is a need to distill vast arrays of data into more discrete arrays. Data-reduction techniques provide the researcher with the ability to consolidate data into manageable groupings (Shmueli, Patel, and Bruce 2007). An extension of data-reduction is data exploration. Shmueli and his colleagues (2007) liken this to a detective who examines a crime scene. The detective surveys the scene in order to answer a specific set of questions pertaining to the identity of the likely offender; however, some data at the crime scene can be consolidated and grouped for simplification. Likewise, with data exploration, the researcher can reduce the scale of variables or combine variables or other dimensions of the array to allow for more focused attention on a particular aspect of the data.

Data visualization is an approach to exploring data in which the graphing of relationships is critical to analysis. Visualization approaches are uniquely capable of illustrating the relationships within and between individual variables, as well as groups of variables. Traditional statistical data visualizations include histograms, pie charts, and box plots. Network visualization allows researchers to assess data as a constellation of relationships that can be especially helpful, as is frequently the case with crime analysis, when the data are incomplete or inconsistent (Xu and Chen 2005b). Network visualizations (Figure 8.2) integrate quantitative and qualitative techniques in order to map the data to a graph and subsequently measure the discovered patterns into meaningful narratives (see de Nooy, Mrvar, and Batagelj 2005). This kind of visualization task is inherently labor intensive. Researchers frequently work within networked data environments that allow the creation of competing network visualizations of the same array; thus novel approaches to automating visualization comparisons have been proposed (Yang, Li, and Zhang 2008).

The converse of the visualization of networked data would be the measuring of patterns associated with visual or other "rich media" content. Undoubtedly, the proliferation of Web-based media and a wide array of mobile media devices provide researchers with a wealth of new opportunities to gain valuable data for mining purposes. For example, blogs can provide real-time insights with respect to the detection of terrorism or other threats such as cyberterrorism (Tsai and Chan 2007; Yang and Ng 2007). That said, a nearly endless number of media types, technologies, and sources present unique challenges, such as the analysis of content and indexing of distributed data, semantic modeling, and privacy (Datta et al. 2007; 2008; Ghafoor 2004).

## Analytics Software

The capacity to conduct DM, until recently, was strictly under the purview of sophisticated computer engineers and scientists. It required an individual to directly code or script the analysis and algorithms by hand—as in the pre-Macintosh[6] interface. What this meant was that the individual had to have a deep understanding of both the math and the intricacies of data acquisition and management.

Over time, more powerful computers and software became accessible to everyday users and greater demands by customers led some manufacturers to innovate.[7] Today, both practitioners and researchers have options. For example, all the major database management software packages (i.e., Microsoft SQL) contain DM tools. The target audience is those who are responsible for maintaining the databases themselves. These also require some programming knowledge; however, these tools have the DM algorithms and some graphical presentation functions built in.

Figure 8.2 **Network Visualization of the 9/11 Hijackers and Their Accomplices**

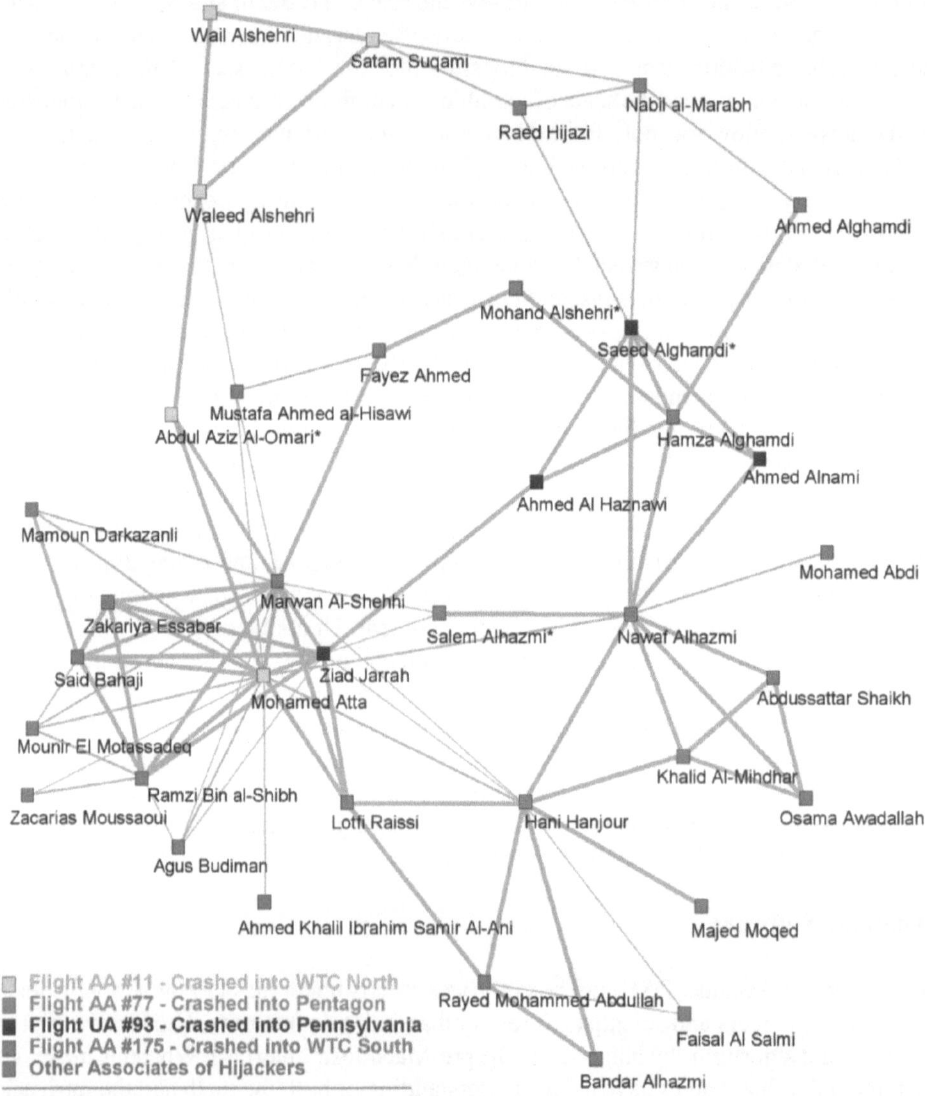


*Source:* Krebs 2002.

## Good News—Many Options!

New DM tools tend to fall into one of two categories: stand-alone software suites or add-ons to existing software packages. These new tools enable greater participation among a wider audience (both statisticians and analysts) and tend to be based on a familiar graphical interface (see Figure 8.3).

Stand-alone DM tools can be formulated around a specific algorithm. Each DM operation (i.e., creating a decision tree) is derived from a specific algorithm, many of which (such as C.5 or CART) are standardized. Some tools, such as SAS DataMiner, include an extremely broad selection of options, are very powerful, and can be purchased as an extension of familiar software. However, the core program, SAS, is very expensive, and the features of DataMiner must be purchased separately at a premium. The upside is that this analytics software is familiar to many, is customizable, and is supported by customer service and documentation. The downside is an extremely high cost that puts this tool out of reach for many. SPSS Clementine extends the capacities of SPSS in much the same way and with similar strengths (familiarity, service) and weaknesses (costs).

Others have created DM tools that work within popular applications, such as Microsoft Excel. While not as robust and powerful as the more expensive and customizable competitors, DM in Excel can suffice for many purposes. Excel, due to certain computing limitations, may not be able to analyze more than 60,000 rows in a DM operation; however, by using software such as XLMiner, the user is provided with a familiar environment at a very low cost. This is a fairly new software product, and its documentation is limited. That said, one of the very few DM texts that has been produced for practitioners (rather than advanced computer engineers) is based on XLMiner (Shmueli et al. 2007).

Last, but certainly not least, some of the most powerful and useful commercial software applications for DM are free. Weka is a collection of machine-learning algorithms for data-mining tasks that has been developed by researchers at the University of Waikato (New Zealand). It contains tools for the most common DM-related tasks, including data preprocessing, classification, regression, clustering, association rules, and visualization. Weka has a highly customizable interface that is cross-platform compatible (PC, MacOS, Linux) and was designed using open-source software. This means that sophisticated users who want to customize the analytics can do so. The downside is that it was designed and written for serious DM experts. The interface is learnable over time and with direction, and it does not require programming knowledge; however, the available documentation assumes at least a moderate level of knowledge and the texts that incorporate Weka are for DM sophisticates (Witten and Frank 2005).

Pajek is, in many ways, the opposite of Weka. It focuses on data visualization and social network analysis. It has a clumsy interface, but it is accessible and, with some trial and error, easy to learn. The engineers who crafted it are responsive to user inquiries, and the textbook written by the creators of the software is sufficiently low level that DM and social network neophytes are not deterred (de Nooy et al. 2005).

### You Push the Button and the Computer Does the Rest? Not Yet . . .

Despite the many options available to those who are interested in analyzing CIA data sets, there is still an outstanding problem: how to acquire the data from the ACIAD and get it into the analytics software of choice. Ideally, a researcher would be able to query the ACIAD directly, define parameters, and observe the progress of data streaming to the computer.

Sadly, the ability of a user to push a button and let the computer do the rest is a step ahead of where things now stand.

As far as the software is concerned, most software packages are not designed to handle this kind of inquiry (from the software to a separate database), and those that are do so because someone has invested in developing an application programming interface (API). An API is a suite of functions that allows two or more discrete systems, applications, networks, and so forth to transfer data back and forth. Without an API that allows data to travel directly between the specific applications (i.e., SPSS), the researcher has to export from one program and import into the other.

The second issue pertains to data preprocessing. Properly conducting DM requires "cleaning" the data; moreover, it is likely that preprocessing (or cleansing) of downloaded data from ACIAD would be done via an application that is separate from the analytics. The reality of this varies based on the type of media or data being analyzed and the particulars of the analysis. Issues specific to the data are listed in more detail in Figure 8.3.

## APPLICATIONS OF DATA MINING TO CIA

There is no shortage of creative, useful, and novel applications of data mining. From retailers trying to sell more beer and Pampers (Whitehorn 2006) to the detection of hidden criminal transactions (Bache et al. 2007b), the tools are available. What follows is a description of specific DM research that involves core topics related to CIA. The review has been organized based on the current sub-subdivisions in the ACIAD.[8]

### Data Mining Applied to Critical Incident Analysis: A Review of the Literature

#### *Terrorism*

Since 9/11, the demand for security, risk reduction, and crime prevention has resulted in the development of a range of strategies related to the pervasive growth and adoption of data acquisition efforts and technologies to surveil populations and events. This has resulted in a massive increase in the amount of data collected by entities for the purpose of risk analysis or incident prevention (Shaikh et al. 2007). Visualization systems, such as CCTVs (Cameron 2004; Cole 2004; Gras 2004; Monahan 2006; Norris and Armstrong 1999; Norris, McCahill, and Wood 2004; Ruegg, November, and Klauser 2004) or X-ray cameras (Adey 2003, 2006; Electronic Frontier Foundation 2004; Electronic Privacy Information Center 2008) are increasingly being networked (DARPA 2008; Shachtman 2003). Haggerty and Ericson (2000) refer to these massive collections as data "assemblages." Because of the real and perceived harm associated with terrorism, the demand to integrate and make sense of assemblages has consequently led a greater level of resources and higher expectations to be directed toward information technology in general and data mining in particular (Ahsan and Shah 2008).

Terrorism, however defined, involves the commission of extreme outlier acts.[9] Since 9/11 a tremendous amount of scholarship has developed with the intention of developing

Figure 8.3 **The Weka Visual Interface Allows You to Visually Construct the Step-by-Step DM Activities**

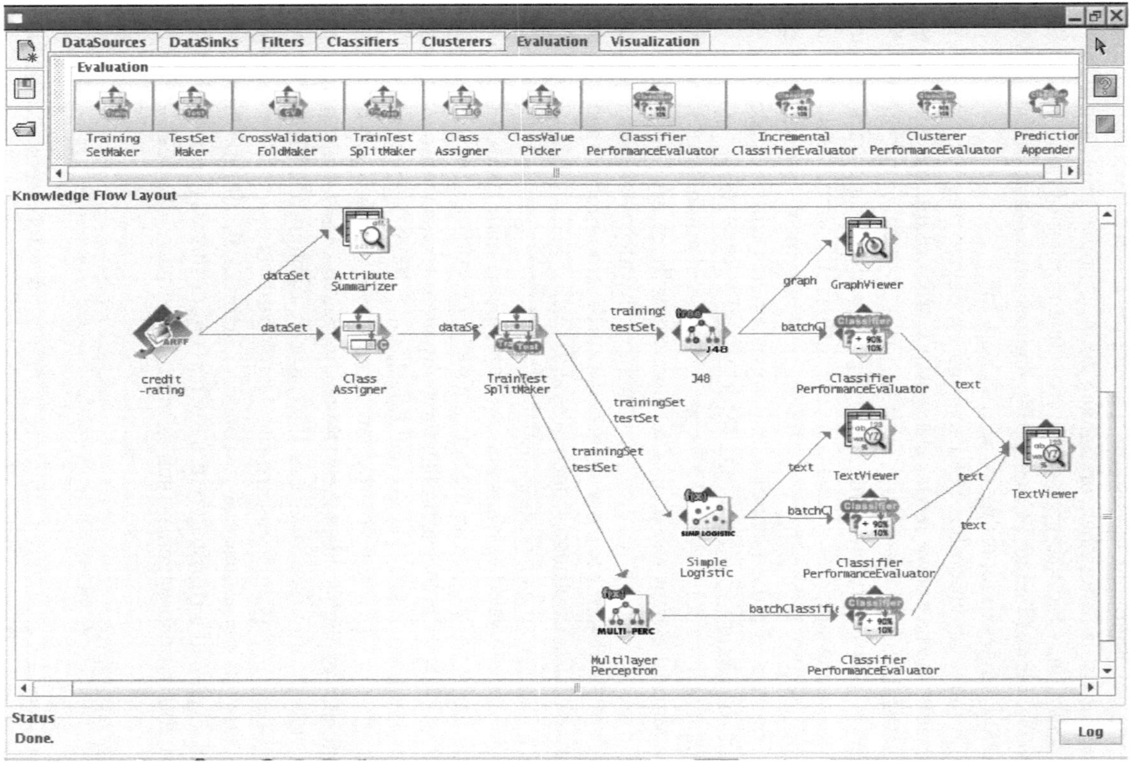

models for detecting ongoing terrorist acts or anticipating terrorist activity. Frequently, the plotting of terrorist acts is not obvious until it is too late, because the needed data is spread over many disparate systems. This leads to an information problem that typical DM algorithms cannot accommodate (Jonas and Harper 2006); however, there are promising developments in this area that employ novel applications of "fuzzy" logic and sequence analysis (Cao, Zhao, and Zhang 2008). Another approach involves anticipatory event detection, which comprises the real-time mining of news feeds (He, Chang, and Lim 2007), blogs (Tsai and Chan 2007), and text-messaging streams (Shen et al. 2006) for subtle changes in reporting or activity patterns.

While much attention has been focused on developing ways to monitor inherently monitorable media (news feeds, etc.), others are looking for ways to assess real-time data related to suspicious patterns of behavior in physical environments (Memon, Hicks, and Harkiolakis 2008; Yan et al. 2007). In line with this approach are those developing ways of creating adaptive systems that are sufficiently simple to operate and can provide powerful insights (Chandra et al. 2007); thus, a typical law enforcement agency can take advantage of these technological advancements without needing a virtual army of scientists on twenty-four-hour duty (McCue and Parker 2003).

Memon and his colleagues (2007) provide a useful reminder of why it is a challenge to analyze terrorist groups and their related networks: Terrorists (along with organized crime groups)[10] are reticent to share their membership lists with government organizations, let alone researchers. Responding to the scarcity of data to develop effective analytics requires creative integration of data from numerous sources and creative analytics choices (Memon et al. 2007) as well as the recasting of existing modeling analytics (Ozgul, Erdem, and Aksoy 2008). One approach, demonstrated by Yang and Ng (2007), is to mine blogs for topic-specific content and their links to other sites for the inherent social network relationships. Another involves an attempt to build psychological profiles through pattern-text analysis of e-mail content (Balamurugan and Rajaram 2008; Galitsky and Kovalerchuk 2006). Hackworth and Treleaven (2007) go beyond most DM researchers by developing novel approaches to the information scarcity problem; they explicitly adapt their DM analytics to take advantage of relevant political and sociological theories (see McGinnis 1991; Roach, Ekblom, and Flynn 2005).

Some researchers are focused on analyzing patterns of activity consequent to terrorism events, such as the efficacy of responses (Larson et al. 2006). Others are investigating particular types of terrorist activities, such as online threats (Ertoz et al. 2003); bio-attacks (Lateef 2007), or stratagems, such as multistage attacks (Li et al. 2007).

### *Natural Disasters*

Much of the DM literature pertaining to natural disasters trends toward modeling early warning systems. The modeling of river flows (Zhang, Zhou, and Li 2008), flood prediction (Chen and Yu 2007; Damle and Yalcin 2007), and forecasting water flood recovery (Weiss et al. 2001) inspires some; Reinhardt (2008) applies a combination of GIS, quantitative simulation, and spatial data mining to assess landslide vulnerabilities in light of flood, forest fire, and earthquake risks.

Some are concerned with simulating response systems in emergencies (Albores and Shaw 2005), with an emphasis on creating tools for decisions makers by modeling preparation and response factors (Larson et al. 2006). Others have directed their efforts toward novel applications of visualization and media analysis. One example includes a combination of geospatial information science (GIS) mapping of real-time video news feeds to assess correlations (Pan and Faloutsos 2002). Another example, which has greater intrinsic value, involves posthurricane satellite image mining to detect and classify concentrations of damage, as well as to facilitate real-time response operations by identifying blocked routes and staging areas (Barnes, Fritz, and Yoo 2007).

Meteorological events, because of the vast combination of geological and other factors, are notoriously complicated to model (Major and Mangano 1995). However, data-mining analysis has led to advancements in spatial analysis of outlier metrological events. New techniques allow researchers to build multidimensional data points and create highly sophisticated prediction models for occurrences and trajectories of natural events such as forest fires, tornadoes, and hurricanes (Trafalis and White 2003; Zhao, Lu, and Kou 2003).

The mining of extraterrestrial data is essential to CIA. One principal issue concerns sunspots and other solar events. This year the National Oceanic and Atmospheric Administration (NOAA; 2008) announced that we had entered an eleven-year cycle of heightened activity. According to NOAA (2007, 2008), this will result in increased risks to power grids and communication systems (wired and wireless), as well as an increase in storm events. Various approaches to solar DM have been proposed (Bezir et al. 2007; Li and Ye 2002; Lopez-Ariste et al. 2001). Researchers are still exploring ways to link space-effect modeling with terrestrial effects. Given the expected structural changes to solar activity, the models may have to be adjusted. Related predictive DM work has been developed on the simulation of asteroid activity (Burl et al. 2006; Thuillot et al. 2006). The nature of astronomical analysis requires a massive integration of data from many sources. One approach researchers have developed to resolve this need is the linking of ground-based telescoped images, space-based satellites, and DM to create virtual observatories (Allan et al. 2006; Jaenisch et al. 2004; Liu, Wang et al. 2006; Voisin et al. 2004).

### *Public Health*

Governments and NGOs have argued for global changes to modernize health-care information infrastructures (HII) (Knoop 2006). One particularly vexing problem is the complex and fragmented nature of national HIIs, which results in tremendous inefficiencies and introduces sources of error. In fact, some question whether it is possible for the benefits of HIIs to outweigh the cost to invest in them (Hauben, Patadia et al. 2005). Others are concerned about a variety of problems inherent in mining health data from distributed networks. Conversely, some people have proposed approaches to knowledge management aimed at homogenizing the distributed data within the HIIs (Abidi 1999, 2001; Pagnoni, Parisi, and Lombardo 2001).

Research that attempts to create useful systems for public health planning and decision support in the face of critical events is ongoing both in the United States and abroad

(Abidi 2001; Lavrac, Bohanec et al. 2005; 2007; Mirhaji et al. 2004; Pur et al. 2005). So-called pharmacovigilance decision support systems have been developed as a way to identify patterns of adverse drug reactions in a population. In the past, this required real-time data and in-depth domain-expert knowledge (Almenoff et al. 2007; Banks et al. 2005; De Castro et al. 2007; Hauben, Madigan et al. 2005). However, advancements in mapping knowledge domains and procedural checks to detection systems have helped increase the deployment of these systems (Boyack 2004; Young Moon et al. 2001).

Another area that has received a lot of attention from DM and network analysts has been the spread of disease. Some have developed analysis tools for influenza surveillance at global (Moore et al. 2008), national (Germann et al. 2006), and regional (Kaufman, Wong et al. 2007) levels. Others have looked at ways of predicting the spread of dangerous and drug-resistant nonsocomial infections (Brossette et al. 1998; 2000; Lamma et al. 2006).[11] Rapid outbreaks, consistent with bioterror attacks, present a challenge for researchers when analyzing densely populated and distributed data arrays (Buckeridge et al. 2005).

Epidemiological network analysis is critical for predicting patterns pertaining to the spread of infectious diseases. Therefore, identifying links and relationships among at-risk populations is of prime concern to practitioners. Different diseases operate or "move" differently among populations. Consequently, researchers have developed social network analysis tools for the analysis of HIV/AIDS (Barbosa, et al. 2000; Klovdahl 1985; Wallace 1993; Yoshikura 2005), tuberculosis (Fitzpatrick et al. 2001; Sterling et al. 2000), smallpox (Barrett, Eubank, and Smith 2005), and syphilis (Rothenberg et al. 1998).

Disease outbreaks are affected by communication, health, government, and other infrastructures. Moreover, models that predict and assess the various response levers have been proposed (Carley et al. 2006; Kaufman et al. 2007). Certain viruses or bacteria have been identified as potential bioterrorism weapons (i.e., smallpox), but other diseases have unknown potential for harm. Applying knowledge discovery and text-pattern analysis to a vast array of biomedical literature, some researchers inductively assessed and rank-ordered the associated probabilities. Thus, policymakers could adapt existing warning or other management programs (Hu et al. 2006).

Other public-health problems that researchers have assessed with DM include monitoring food safety by detecting patterns in consumer complaints (Dubrawski et al. 2006), food terrorism (Yoshida 2008), outbreaks related to the sharing of food (Trostle et al. 2008), detecting disease in food stocks (e.g., cattle) (Dick and Bracho 2007), environmental effects on human health (Sokolova and Fernandez-Caballero 2009), and patterns of illness outbreak in light of water quality measures (Gopinath and Tamjis 2008).

### *Technology*

"Technology" refers to variety of mutually interdependent and interlinked sectors of the critical infrastructure (CrI). These sectors include our energy, transportation, and other resources[12]—all of which facilitate the deepening integration of societies by providing the means by which systems (regulatory, economic, etc.) can be linked and made more efficient. The orchestration of the various CrI industrial sectors makes it possible for both government and free market operators to simultaneously provide essential goods and services (i.e., water), interconnectedness and operability (i.e., telecommunications), and public safety and security (U.S. DHS 2003).

Threats to the CrI, whether posed willfully (e.g., terrorism) or not (e.g., human error) are prima facie, but it is the information infrastructure that makes the interconnections possible. Thus, an attack against the information infrastructure can have contagion effects, which can be felt spatially, temporally, and across all sectors (Perry 2006; Zimmerman and Horan 2004).

Consequently, there is a considerable amount of analysis concerning various aspects of information assurance (IA) or the detection of threats against the information infrastructure (Srivastava 2002). There have been attempts, on a macro scale, to detect evolving patterns of Internet traffic dynamics. Some have developed analytics for assessing service demands and related traffic patterns (Xu, Zhang, and Bhattacharyya 2008); others have focused on particular types of attacks, such as denial of service (Sanford, Parish, and Sanford 2007a, b; Wang, Deng et al. 2003) or worms and viruses (Ishibashi et al. 2005; Yamanishi, Takeuchi, and Maruyama 2005). Still others are concerned about the fragmentation of macro traffic data and question the efficacy of current techniques and propose technical solutions (Andersen and Feamster 2006; Patcha and Park 2007).

Conversely, microlevel approaches in which a firm or agency is the unit of analysis appear more frequently in the relevant literature. Network traffic logs are often analyzed for developing micro models, which is useful for developing intrusion detection systems (IDSs) customized to detect the behaviors of the entity under study (Thuraisingham 2004; Tsai and Chan 2007; Yamanishi et al. 2005). The standard IDS approach of reviewing raw log data is insufficient[13] for preventing novel attacks that exploit unknown or undisclosed computer network or application vulnerabilities, known as "0-day attacks" (Song et al. 2007). To combat them, some researchers have developed systems that detect anomolies using different data-mining techniques, including inductive-rule generation for infrequent events (Lee and Stolfo 1998), deriving fuzzy rules from each network behavior and using those initial rules to further define fuzzy rules for detecting attacks (Dickerson and Dickerson 2000), "self-aware" artificial intelligence (Balducelli et al. 2007); tracking land classifying nodal or information packet behavior(s) or within the network (Fugate and Gattiker 2003; Jaenisch et al. 2003; Ramadas, Ostermann, and Tjaden 2003), and using clustering algorithms to discover outliers within probability-distribution densities (Ertoz et al. 2003; 2004; Pawling, Chawla, and Madey 2007).

Data mining can be instrumental in protecting modern power systems, however, predicting a blackout is notoriously complex. It requires the ability to model and analyze data that is multidimensional, temporally varied, fragmented, and distributed across expansive networks with many nodes (Xue 2006; 2007). One approach, developed by Zhao and his colleagues (2003), is designed to help network operators detect potentially unstable system components as well as network stability. In many areas, the power grids are made up of a mixture of old and new components. The modern components have the capacity to provide operators with refined system data, whereas the older systems do not. Thus, operators are frequently operating with incomplete information, which causes a problem for modeling the network (Taft 2006; U.S. DOE 2003). Researchers have proposed workarounds, including using neural network analytics to approximate sufficient data refining (Morais et al. 2007), forecasting power loads in advance (Wang, Ma, and Li 2005), using probabilistic models of power grid stability controls (Wehenkel et al. 1999), and detecting the presence of cascading factors (Venkatasubramanian, Kezunovic, and Vittal 2008).

Hydroinformatics is the use of DM to assess data as it relates to the efficacy of water distribution (Savic and Walters 1999). Increasing global and regional populations, as well as industrial uses, results in growing demands on water resources—a serious disruption of which can destabilize the social and political economy of a state. Hydroinformatic tools have been developed to facilitate a host of water-related problems, including water management in arid climates (Anderson et al. 2008; Hewett, Leuchner, and Carvalho 2001; Tadesse, Brown, and Hayes 2005; Uddameri 2007), water asset management (Babovic et al. 2002; Bessler, Savic, and Walters 2003; Chang and Chang 2002; Wang and Yin 2007; Wang et al. 2004; Yin et al. 2006), and pollution, sewage, and treatment (Chen and Chang 2007; Korving and Ottenhoff 2008; Tachibana and Ohnari 1999; Vaseashta et al. 2007; Waissman et al. 2000; Yuan, Wang, and Morris 2000).

Air, land, sea, and telephone networks are vital to the national economy both in terms of everyday economic functions and in times of crisis. The efficacy of transporting people, goods, services, and communications networks is even more critical given the increased reliance on just-in-time logistics and the resulting interdependancies constructed to meet global demands. For example, in 2000, a still-unexplained cut to the world's longest subsea cable system severely disrupted communication services and data transmissions both into and out of Australia, Singapore, Indonesia, the United States, and Europe (Tangredi 2002). Losses due to business interruption totaled millions of dollars daily before repairs could be completed (Brewer 2000).

Land transportation systems analysis includes automobile and railway networks. This comprises a dizzying array of different types of subsystems and applications. On the cutting edge are intelligent transportation systems, which encompass an approach to roadway management using DM. These systems try to balance factors that typically are at cross purposes, such as vehicles, loads, and routes, in order to improve safety, transportation times, and fuel consumption (Luo 2008; Pande and Abdel-Aty 2006; Weigang et al. 2004; Wong and Woon 2008; Yang and Jia 2004; Zhou and Lu 2005). Others are focused on modeling drivers' route-choice behaviors under various circumstances by analyzing global positioning system (GPS) data (Gong, Gao, and Wang 2005; Ishak 2003). One serious problem for which DM researchers have attempted to develop models is that roadway data tends to be either highly detailed or sparse (Chen, Hu et al. 2006; Pang and He 2007; Tilak et al. 2007). Others have attempted to develop systems that fuse the heterogeneous data types into a framework to allow decision making by creating context-sensitive emergency management analysis tools (Chen, Chang et al. 2006; Lu et al. 2005; Pollard et al. 2004). As Chen and his colleagues (2006) argue, the patterns of roadway usage and associated risks are vastly different in light of different CrI events, such as hurricanes and pandemics.

Previous studies have tended to emphasize roadway analysis, but, given the increased demand for public rail and tramway transportation, it is no surprise that researchers have started turning their attention to those and related subject areas (Hale 2000; Yang et al. 2007). For example, Frederiksen and Schmid (2008) illustrate how previous models for predicting and responding to rail integrity are antiquated, as they are typically based on outmoded rail transport (slower trains with lower load capacities) and data collection technologies (analog versus digital). Another approach has been to create management tools for the maintenance and operation of trains to predict wheel failures (Yang and Letourneau 2005) as well as load safety (Yu, Gu, and Li 2004).

Air and sea transportation networks have come under greater scrutiny since 9/11. Air networks have been assessed by DM researchers for a wide range of modeling endeavors, including airport (node) and industry security and stability (Grubesic and Zook 2007; Koch 2004), the efficacy of routing air corridors ("Air transportation" 2003; Smith et al. 2000), and decision-making tools regarding physical safety equipment, such as the engines (Litt et al. 2000).

Every year more than 200 million containers are shipped around the world (Cudahy 2006). These containers are picked up, transported, and transferred among railroads, ships, and trucks, however, the linchpin is maritime shipping (Langewiesche 2004; Levinson 2006). Until recently this aspect of transportation analysis has received limited scrutiny by DM researchers, with the exception of logistics. Recent scholarship (Flynn 2006), public policy (U.S. DHS 2005a, 2005b), and the experiences of security concerns in Africa and Asia with maritime piracy (Langewiesche 2004; Ong 2002; Sinai 2004), have increased the demand for greater levels of DM modeling of maritime concerns. However, to date there has been limited public distribution of models in this area (Galdorisi and Goshorn 2006). In part, this is a function of both technical and data collection issues. Maritime shipping is arguably the most widely distributed global network, and much relevant data is not collected or is highly decentralized and fragmented (Galdorisi and Goshorn 2006). While some experiments have been developed (see Sickinger 2005; Tiburcio 2005), Horne (2008) proposes a suite of maritime security concerns that would benefit from DM analytics, including analysis of enemy courses of action and tactics; factors affecting the classification of time-critical targets, effects of policy changes on operations and vulnerabilities, and communication systems and networks among security agencies.

Law enforcement is the primary delivery mechanism for both safety and security in society. In many jurisdictions, the sifting and analyzing of data is still largely a manual process (Xu and Chen 2005a). Police agencies have, with varying success and levels of sophistication, developed a variety of methods to analyze and mine their rich collections of data (Thongtae and Srisuk 2008). In the United States, these systems are typically local and, mirroring the decentralized national organizational structure of police, are characterized by a limited amount of interagency data sharing (Holmes, Comstock-Davidson, and Hayen 2007). Beyond that, police in the United States face a number of challenges, not the least of which is a cultural distrust by Americans who equate a perception of increased police surveillance with a loss of civil liberties (Birrer 2005; Garfinkel and Smith 2006; Ghafoor 2004; Gunasekara 2007). Actions by the FBI and other agencies (Dunbar 2007;

Eggen 2005; Kramer 2008; Lisberg 2007; Office of the Inspector General 2002, 2005, 2006, 2007a, 2007b, 2007c, 2008; Singel 2007; Solomon and Gellman 2007) have only reinforced these concerns and, in some cases, resulted in a backlash against the use of serious data analysis systems (Kurlander 2005; O'Harrow 2005; Swartz 2004).

In spite of these concerns, police agencies have deployed a variety of DM systems that are designed to support their operations, including video surveillance for monitoring suspicious roadway patterns (Farmer and Mann 2003; Velastin and Remagnino 2006; Yuan, Sun et al. 2003); analysis of serial offenses (Bache et al. 2007a; Dahbur and Muscarello 2003; Ribaux and Margot 2001); patterns of crime occurrences such as money laundering (Zhang, Salerno, and Yu 2003), sex offenses (Adderley and Musgrove 2001a), and burglaries (Gunderson 2002), as well as more general crime pattern analysis (Adderley 2004; Lee and Phillips 2008; Lin and Brown 2003); and an analysis of offender networks (McAndrew 2000; McIllwain 1999; Oatley, McGarry, and Ewart 2006). Other researchers have focused on decision support for crime scene turnover (Adderley and Bond 2008; Adderley, Townsley, and Bond 2007), investigations (Adderley and Musgrove 2001b; Chen 2008; P.S. Chen, Chang et al. 2006; Ozkan 2004; Smith and Babiceanu 2004), and agency management (Adderley 2004; Chandra et al. 2007; Chen, Chung et al. 2004; Geradts and Bijhold 2002; Ku, Iriberri, and Leroy 2008; Li et al. 2006; McCue and Parker 2003; Oatley and Ewart 2003; Salasznyk et al. 2006).

### *Economic and Financial*

Economics—the art of how a society produces and distributes its resources—largely involves topics related to the critical infrastructure, whereas financial events involve capital investments and risk. One example of a DM analysis that pertains to finance would be a bank asking the following question: "I am the CEO of a large bank and I want to know what the probability is of a 'run' on my assets if. . . ." Financial institutions (FI) and businesses have used DM analytics for some time. For example, an FI has to assess risk and asset valuations and a retailer might want to determine how customers are likely to behave (Hormozi and Giles 2004). But these examples are not critical events per se; these are actually fairly mundane and benign everyday uses of data mining.

Analyses pertaining to finance often concern the attempt to detect fraudulent or corrupt behavior. Frequently, this involves the detection of behaviors that are cleverly camouflaged to look normal (Coderre 1999), such as insider trading (Donoho 2004; Fan, Yu, and Wang 2004; Safer 2002). However, DM as an analytical tool set for the detection and prediction of financial events comprises a wide array of behaviors and methods, including management fraud (Eining, Jones, and Loebbecke 1997; Fanning and Cogger 1998; Fanning, Cogger, and Srivastava 1995; Green and Choi 1997), insider fraud (Porter 2003; Virdhagriswaran and Dakin 2006),[14] or fraud associated with the misreporting of financial statements (Beasley 1996; Beasley et al. 2000; Bell and Carcello 2000; Kirkos, Spathis, and Manolopoulos 2007; Lin, Hwang, and Becker 2003; Persons 1995).

Frauds committed directly against or by government officials or entities have also been put under the microscope. For example, using DM, U.S. government officials discovered close to $1 billion in fraudulent payments following Hurricanes Katrina and Rita (Kutz and Ryan 2006). Another example would be the use of DM to determine the likelihood of voting fraud (Alvarez 2005) and campaign finance fraud (Cho and Gaines 2007). Some have developed models for sovereign risk analysis to detect patterns of corruption (Huysmans, Baesens, and Vanthienen 2008; Huysmans et al. 2006). Others have examined unusual circumstances such as agricultural (subsidy) fraud (Little et al. 2002), systemic fraud committed by schoolteachers (Jacob and Levitt 2003), and fraud by a sumo-wrestling league supported (in part) by the Japanese government (Duggan and Levitt 2002).

*Fires*

Regardless of the cause of any fire emergency, necessary and available resources must be optimally used to anticipate and respond to arising concerns. Ideally, relevant data and knowledge can be centralized, thereby allowing decision makers and managers to effectively coordinate their response to the event (Salasznyk et al. 2006). Analysis of fire events such as the Piper Alpha oil rig fire, the Mont Blanc tunnel fire, and the collapse of the World Trade Center towers revealed many managerial mistakes (Berry et al. 2005),[15] and DM researchers have proposed some useful tools for fire officials. That said, there is a paucity of analysis, and much work still needs to be done.

Much of the literature, with some exceptions (Kolesar and Walker 1974; Mosca, Giribone, and Bruzzone 1995), seems to focus on predicting the occurrence or direction of travel of wildfires (Douglas et al. 2006; Han et al. 2003; Mazzoni et al. 2007). This should not be surprising. In the United States there have been nearly 600,000 reported wildfires since 2003, which have destroyed more than 54 million acres (an area roughly twice the size of California) (National Interagency Fire Center 2008).

Arson can be a crime of violence, fraud, or vandalism, or it can be simply incidental to another crime (i.e., the cover-up of a murder) (U.S. Fire Administration 2004). Arson analysis by DM researchers includes the creation of offender profiles (Canter and Fritzon 1998) and offending profiles (Gunderson 2002). The United States has developed a collection of necessary data for conducting DM analysis. The U.S. Fire Administration (2004) argues that, with the possible exceptions of Florida and Tennessee, fire authorities around the country have been provided with insufficient guidance on what DM analysis can offer or how to go about employing it.

*Violence*

Riots, mass shootings, massacres, genocides, and a whole host of violent acts that constitute critical events are not everyday issues. In fact, school shootings, such as those at Columbine and Virginia Tech, can be characterized as the outliers of outlier events. The rareness of events and the challenge of collecting data probably explain why serious DM analysis in this general area has not occurred. It is far easier to upload a database of frequently occurring events, such as burglary, and analyze the patterns therein. The very nature and infrequency of these events requires extreme creativity, patience, and the capacity to integrate analytical frameworks from different sciences. These "on the ground" realities and the extreme lack of DM literature in this domain are powerful arguments in favor of the creation of the ACIAD database.

## PROBLEMS ASSOCIATED WITH DATA MINING

Data mining is not a perfect science. In fact, it is an artful assemblage of heavy computing power, robust statistical musculature, and the "sociological imagination." The problems associated with data mining are well documented, and we describe them here, followed by a review of specific concerns in the aggregation and analysis of critical incident data.

### Negative Spillover Effects Inherent in Mining Large Data Sets: Garbage In, Garbage Out

Negative spillover effects inherent in mining large data sets are unavoidable. The programmer's aphorism "garbage in, garbage out" (GIGO) refers to the *effect of the effects* outlined below—invalid inputs lead to invalid outputs. However, the real meaning of this phrase extends beyond the literal and into the "collective mythological wisdom" common among decision makers who rely on this technology and assume that whatever the machine says must be right. In addition, the technology has advanced beyond the literal meaning of GIGO, which implies that the problem with invalid results is merely a function of invalid inputs, but this is not so. The problem may be a combination of many kinds of "badness" that have a multiplier effect on each other. Sometimes it is possible to clean the data so that you have (valid) "roses," rather than garbage, coming in and *still* wind up with (invalid) garbage at the other end.[16]

The GIGO effects are neither new nor innovative to practitioners. Basically, the problem is that the ability to make increasingly larger "buckets" for the data (a mechanical dilemma) outstrips the ability to contain many of the effects, which are mostly nonmechanical. A common problem with many of the effects has to do with algorithms. An algorithm is the set of instructions that dictates the operations of computerized machines. The algorithms make it possible to do the things we do with data: calculate, clean, simplify, sort, store, analyze, and so on. The problem with many of the effects outlined here is that some of the fixes are insufficient to bridge the gap between naturally occurring errors in the data and the design of the algorithm; moreover, the problem multiplies as the databases grow.[17]

The "garbage" typically refers to biased results from the data-mining exercises, which constitute the "spillover effects." It is useful to refer to these effects in this context because, as one effect manifests, it typically spills over into another effect, thereby compounding the negative effects (i.e., a dynamic increase in errors or biased parameters).

A comprehensive assessment of effects, (potential) solutions, and workarounds that can create or reduce compounding error, bias, or unduly prejudiced results of the data mining have been covered elsewhere in greater detail.[18] Yang and Wu's (2006) survey of data-mining practitioners provides a parsimonious review of the key problems from the computer scientist's perspective. Seifert (2004) and the Technology and Privacy Advisory Committee (2004) reports focused on the public policy and oversight angles. Not surprisingly, the policymaker's and practitioner's perspectives differ in their focus. The differing foci provide a framework for segregating the categories noted here.

The first category, data-structured problems, is most closely associated with the practitioner who works directly with the data and the models. These problems are a function of the inherent complexities associated with assessing any large set of data. The second category, procedural problems, is most relevant to the policymaker. These problems relate to the procedural complexities of acquiring and managing the data.

## Data-Structured Problems

### *Dimensionality*

This refers to the number of separate variables and distinctive attributes that describe each case (or observation) in the data set. As dimensionality increases, so does the likelihood of biased parameters. A common solution is to preprocess the data into indexes to reduce the dimensionality.

### *Sparsity*

The majority of algorithms are best suited for situations where the data is sparse—that is, most of the variables are binary and the majority attribute is coded 0. The more robust the data, the greater the challenge to validate new or apply existing algorithms.

### *Scalability*

Data acquisition technologies vastly increase the quantity of data that can be simultaneously recorded from a single event. The number of variables per case (or event) can increase from several thousands to millions and even, in some cases, to billions. This contributes to the dimensionality problem and can result in a quantity of data that can confound algorithms; which is further complicated by the absence of sparsity.

### *Complex and Heterogeneous Data (Obtaining Complex Knowledge from Complex Data)*

Most search algorithms assume sameness of data type. Moreover, they are frequently designed with the assumption that the variables are identically distributed and are mutually independent. These assumptions are frequently violated and result in biased parameters. The identical distribution problem necessarily restricts the searching of complex data, such as sound, image, or graphics files. Also, much data is valuable *because* of its inherent relationship to other changing relationships, such as social or metabolic networks.[19]

*Data Quality*

This refers to the validity and reliability of the data as well as its completeness. Not all data sources operationalize their data in the same way, which is especially problematic since arrays for data mining are typically created from a multitude of sources. Completeness can refer to cases in which the information for a particular variable is not present, as well as a situation in which there are multiples of the same case. For example, an array with separate cases labeled "Bobby Chariot" and "Bobby Chariot III" might actually contain data from the same individual. Mined data typically originates from multiple sources whose original intent of the data collection varies. Consequently, the data might lack the ability to answer disparate questions accurately. In addition, and depending on the collecting entity, the respondent might not provide truthful answers.[20] Moreover, data is typically context-sensitive. Thus, an answer may be truthful in one circumstance but not in another. For instance, an individual might transact business with an online retailer using her personal credit card on behalf of her employer. Under such circumstances, it is truthful for her to state that her address is the business address (in that context).

*Compounding Error Rates*

Every collection of data has an accepted level of error. Which level is acceptable varies, based on what is being measured and how the error rate is being calculated. Thus, the error rate for the same group of data using one method can result in a rate as low as 6 percent and with another method it can approach 50 percent. Nonetheless, as you combine data from different sources you necessarily "take on" their differing errors, which can severely compound the aggregate error rates and bias the parameters.

*Conjoined Database*

Data mining can occur over a distributed network or simultaneously from multiple sources. It is not uncommon to have data that we know is, theoretically, linked, but for any number of reasons (technical issues or the scale of the data) cannot be put into a single database for mining (the interoperability problem). Nonetheless, when this does occur, it contributes to a compounding error rate problem.

*Resolution*

The scale at which the data is being measured can either hide or overrepresent deviations. For example, when viewed from a million miles away, the earth appears as a lovely orb; whereas from 105,000 million miles, it is indistinguishable from a speck of space dust.

*Special Data Problems: Time and Space*

Time and space (geographical coordinates) present special problems.[21] Both frequently exhibit autocorrelation, in which the proximity of one observation in space or time to another is indistinguishable or includes lag or leading-indicator bias. Simply put, this increases the "noise" in the data and lowers accuracy. Also, time-series data is extremely difficult for most data-mining applications to analyze, and it is frequently indexed into a category or another stationary measure. This can lead to resolution challenges.

*Timeliness*

Some data necessarily changes frequently and has a discrete shelf life. This is distinct from time-series data in which the thing being measured is being measured with respect to time, per se. Timeliness is basically the age of the data. Out-of-date data leads to patterns and models that are almost certainly biased.

*Special Data Problems: Skewed Distributions and Cost Sensitivity*

Most data-mining algorithms are designed with an assumption that the data is normally distributed or, at least, that the critical thing being measured is a "typical" case. This assumption is frequently broken. This is especially true when assessing scenarios of so-called fat-tail risk,[22] in which the outlier events are artificially weighted upward so that the machine properly detects the events and includes them in its model. This results in what amounts to a shoehorning of the inflated distribution into a normal distribution for analysis; consequently, it creates unpredictable results. Cost sensitivity refers to factors that should lead certain data to be weighted in some way. For example, errors in the data lower its value, thereby creating a cost. For example, data showing that certain customers cost retailers less to retain might indicate that these customers are "better"; however, the question typically left unanswered is how to transform the data in light of this. Frequently, it is too hard to accurately calculate what the costs (weights) are, so most models assume that all data costs are the same.[23]

## PROCEDURAL COMPLICATIONS

*Data Cleansing*

The problems inherent in the data are well known to practitioners. Thus, many have developed ad hoc prophylactic solutions to try to solve the various problems identified earlier. What is important to note is that there is no known method for completely "cleansing" the problems in a given data set, so there will always be some error. The real problem is when the actual error rate is unknown. Moreover, like the tainting of physical evidence in a police investigation because of some kind of contamination, there exists the very real possibility that the cleansing process will introduce more errors into the data.

*Interoperability*

This refers to the ability of different entities to share information with respect to the various data management standards and practices. For example, one database can code "no" as "no" or "0," or there may be incompatible standardization of data formats, such as "MMDDYYYY" versus "DDMMYY" versus "DD/MM/YYYY." In addition, issues can arise from the lack of interchangeability of data between entities due to differing technical standards of the software used for the mining and storage operations[24] of the databases.

*Dynamic Objectives*

Data that is collected or purchased by an entity for one reason is frequently repurposed. Moreover, the individuals whose data were originally captured might or might not have given their data initially if they knew about the secondary purpose. This, of course, is the main thrust of many civil libertarians and others who argue for stricter controls on government use of data, particularly on data mining (Stanley 2004; Stanley and Steinhardt 2003).

*Data Ownership and Distribution*

Even if a single entity has legitimate access to the data, it might be spread among different systems, which requires us to directly confront issues of preserving data integrity and security. With that come several interesting implications, the most visible of which has to do with data leakage. Data leakage is when a person, group, or entity inappropriately gains access to a data set or uses a legitimately gained data set in contravention to legal or procedural rules (e.g., government regulations).

*Algorithm Validation*

Whenever an algorithm or model is being developed, it is tested on multiple data sets. The results are never 100 percent the same. However, they have to be shown to be the same (at least) statistically. Even if this is shown to be true, there is always the possibility that, given the test data, the model might be internally valid but externally invalid. This means that the model is valid only on the existing test and validation data but might not be so when applied to a new data array. The most likely reason is biased test samples. If bias occurs, then the true validity of the model might be suspect at best, and its applications could result in increased error rates in the form of false positives or false negatives.

*Fragmentation*

Fragmentation is a variation on the incomplete data set problem mentioned earlier. It is a combination of ownership and distribution problems along with interoperability. This occurs when a large entity has several databases that, if combined, could make for a robust and sufficiently complete source, but technical issues or internal disputes over ownership prevent this from occurring.

### Data Mining Spillover Effect Summary

The previous sections have outlined a variety of problems, or effects, that can skew the validity of a data-mining operation. The point in including them all is not to try to prove that data mining is an invalid analysis tool but, rather, to address the myth of its infallibility. As has been shown, even if the data are good, there are a wide variety of problems that can negatively affect the final product.

The individual problems outlined, *ceteris paribus,* are cause for concern. But the reality is that they rarely occur in isolation.[25] The more realistic scenario is that their presence yields a negative spillover, which, if uncontrolled, can lead to a negative spiral or a multiplier effect. Simply put, the errors caused by one effect can cause errors in another, and then that can cause a reaction of error effects that result in a severely unreliable analysis. And, to make matters worse, the practitioners and those who rely on their analysis *might* not be aware of the degree to which the results are invalid.

The fact that these events can exist might not be such a big deal when you are marketing dog food or clothes. After all, if the error rate in a marketer's database is 5 to 10 percent, he might be quite happy as that implies that 90 to 95 percent of the "hits" are good leads—an excellent ratio. However, if the job is to detect patterns and profiles that might indicate a security threat, then a 5 to 10 percent error rate[26] can have tremendous impact. Error rates at this scale would likely result in a world-class waste of valuable resources and a very real concern about false negatives, false positives, and the likely disruptions of individual civil liberties.

### Potential Data-Mining Problems and the ACIAD

The inherent problems of data mining can be compounded when conducting CIA. However, the purpose of this section is not to undermine the rest of this chapter. There is a great deal of potential value to be gained by mining the ACIAD. Dr. Smith in the 1960s TV show *Lost in Space* did not whine about how we "are doomed." However, his friend the Robot tried to warn of the danger(s). Proceed with caution and know the potential pitfalls so that, whenever possible, you can design the problems out of your analysis.

To date, there is no consensus on how to design information collection or analysis systems with respect to critical incident response. However, the National Incident Management System (NIMS 2008) does give practitioners direction on how to collaborate and respond to disasters—natural or not. Much of the literature reviewed in this chapter includes proposed workarounds to address the various effects. There is no doubt that information systems can add a tremendous amount of value to practitioners, even in real time. Information systems can facilitate the effective deployment of resources, detect problems before they occur, and help authorities confront the lingering effects after an incident. But the fact is that data pertaining to events can be generated and held across jurisdictions and functional agencies. Moreover, because of either circumstance or the law, those agencies might not share information on an ongoing basis.

Arguments about the information problems that concern CIA and potential solutions to them have been proposed elsewhere (Chen, Wang, and Zeng 2004; Chen and Sharman 2005; Kim and Sharman 2005, 2006; Raghu, Jayaraman, and Rao 2004; Shen and Shaw 2004, 2005; Turoff 2002; Turoff et al. 2004). What follows is an attempt to describe the specific concerns of data collection, design, and analysis of which practitioners or scholars who employ CIA and emergency response data should be aware.

Each critical incident is unique. In some cases, like 9/11, an incident is made up of multiple distinct events. Particular events also determine which procedures and resources are involved. For example, a chemical fire includes EMT and other medical providers, firefighters, rescue, police, decontamination, and traffic and perimeter control resources, whereas a fire in a skyscraper has a host of different needs and resource allocations (Chen et al. 2007). Not only do different events call for different agencies to respond, but the sequence of who is called and when they are called varies as well.

Simply put, responses to events are time dependent (Horan and Marich 2006) and occur in phases (U.S. DHS 2004). The sequencing and time frames associated with who responds when affect the quality and relevance of data that is generated and collected by each responding agency for each individual event.

A related factor concerns the interdependency of agencies. This results in a decision and management framework that is complex and varied. Moreover, the actual decision making frequently changes during the lifecycle of an event (Beroggi and van Gent 2001; Beroggi and Wallace 1995). Given the fluctuating epicenter of action, the quality of the data captured can be compromised or, at least, biased. In addition, this increases the possibility of interaction effects. Thus, preplanning and streamlining interagency responses can reduce the likelihood of errors in the collected data.

Another factor is that different agencies value data differently and might not maintain digitized records (Carafano 2003). Moreover, given the lack of concern at some agencies, it is likely that whatever raw data they do have are incomplete and inaccurate (Carafano 2003; Turoff 2002).

Lastly, there is the question of data acquisition and appropriateness of collection measures/instruments. Basically, the problem has to do with the capacity and demands of different modes of data acquisition to acquire the data that is actually needed and in the necessary format. Wireless networks and wireless sensors are examples of instruments that generate these kinds of instrumentality concerns (see Derr and Manic 2007; Laiho et al. 2005; Liu, Choudhary et al. 2006; Manganaris et al. 2000; Peralta and Casas de Peralta 2002; Zhou and Lu 2005).

## AN ILLUSTRATIVE DISCUSSION: THE VIRGINIA TECH CASE

The previous sections described the many ways in which DM can contribute (and has contributed) to CIA, as well as the inherent problems associated with DM and CIA. The purpose of this section is to illustrate novel ways in which DM could be applied to the Virginia Tech case (see Massengill et al. 2007).

The Virginia Tech case (VTC) is an example of an extreme outlier violent critical event. Moreover, it is worth emphasizing that it is a single event, and the power of DM is its ability to sift through many observations or cases to identify useful patterns. Thus, data collected from events similar to the VTC would allow for more types of analysis and pattern assessment. However, the illustrative examples suggested here are meant to show how DM could be applied to a single case.

Perhaps the type of analysis that would be immediately understandable relates to various types of network visualizations. Given some of the concerns on how events were communicated and shared by responders and administrators, several concepts for analysis are suggested, including the following:

- An analysis of communications patterns among first responders. This involves tracing the communications that occurred within the first responder network. The value of this analysis is to understand which institutional or other bottlenecks affected the efficacy of the responses.
- An analysis of communications patterns among school administrators. This also involves tracing the communications that occurred, however, the interest here is in the timing and distribution of communications concerning the event(s) and institutional guidelines that can be proposed moving forward.
- An analysis of communications patterns among students and faculty. Who communicates with whom, how, and how quickly? This includes an analysis of cell phone and university landline calling patterns before, during, and after the crisis. Understanding how populations self-organize communications during crises can give responders and administrators ideas about how best to take advantage of "natural" collectives.

Another approach would be to mine the content of multimedia data mining and eyewitness testimony (text mining). Web pages, newsfeeds, blogs, and other forms of digital media are rich in mineable content. This would be useful for understanding how perceptions are formed and responded to in open exchanges online. Related to that kind of analysis is the mining of eyewitness account data. For example, it would be interesting to understand the (dis)similar patterns of language and meaning that are used by people of different demographics (age, sex, etc.) and roles (victim vs. responder).

## CONCLUSION

The creation of the ACIAD provides for an especially promising venue by which necessary and useful data concerning critical events can be aggregated and analyzed. This kind of database and the nature of the subject matter it houses can provide researchers with the ability to develop many new and innovative concepts and models for emergency and critical event-related operations research and decision-making guidance.

As researchers develop their research agendas and utilize the ACIAD, they will incorporate novel pattern detection techniques and develop workarounds to deal with the inherent shortcomings.

The purpose of this chapter has been to showcase what DM is and to illustrate the ways in which it has already been applied in various CIA domains. Many interesting concepts and research are available that will be useful for getting ideas. The ACIAD will allow for simultaneous analysis to a degree that has not been achieved to date. That is its promise. However, as we move toward realizing those potential gains, researchers must collaborate. To maximize the value of our research, domain and DM experts should "cross-pollinate" and leverage each other's expertise. The prospects of biased and poorly rendered (and useless) results, given the inherent problems of DM and CIA data, need to be accommodated. As we move forward, some trial and error will be required. And while these problems will never fully go away (they never do), as we become more experienced with using the ACIAD, we will be better equipped to manage and design around them.

## NOTES

1. For example, Apple's Web site boasts that the iPod with the largest capacity (160GB) can store, in some combination, up to 40,000 songs, 200 hours of video, or 25,000 photos (Apple 2011).

2. What we refer to as data mining has many different purposes. This has resulted in a plethora of nuanced monikers, including knowledge discovery (mining) in databases (KDD), knowledge extraction, data/pattern analysis, data archeology, data dredging, information harvesting, business intelligence (Grossman 2001; Guo and Grossman 1999).

3. See Frawley et al. 1992.

4. For more detailed discussions of the areas or operations listed, see Berry and Linoff 2000; Delmater and Hancock 2001; Hand, Mannila, and Smyth 2001; Tan, Steinbach, and Kumar 2006.

5. However, this specific illustration is nothing more than the data miner's version of an urban myth (for a detailed recounting, see Whitehorn 2006).

6. Or Windows, if you prefer.

7. For more detailed discussion on these topics, see Almenoff et al. 2007; Breyer 2000; Chen, Chung et al. 2004; Chen 2008; Edelstein 2007; Li et al. 2006; Oatley and Ewart 2003; Ribaux and Margot 2001; U.S. DHS 2006.

8. That said, many of the studies involve topics that explicitly or implicitly relate to other topics covered, thus, the inclusion of a topic in one area (i.e., terrorism) should not be interpreted definitively.

9. Outlier detection is a problem for data mining; this is covered with more clarity later in the chapter.

10. See McIllwain 1999.

11. A nosocomial infection is a disease that develops after admission to a hospital, and it is the consequence of a treatment, not necessarily surgical, by the medical staff (Lamma et al. 2006).

12. These sectors include Agriculture, Food, Water, Public Health, Emergency Services, Government, Defense Industrial Base, Information and Telecommunications, Energy, Transportation, Banking and Finance, Chemical Industry and Hazardous Materials, Postal and Shipping (U.S. DHS 2002, 2003).

13. Standard IDSs require frequent updates to their rule algorithms and signatures (Patcha and Park 2007).

14. Where the offender has privileged access inside the organization and either commits acts against the firm or uses the cover of the firm to commit acts against outsiders.

15. See also Drysdale, Macmillan, and Shillito 1992; Sylvester-Evans and Drysdale 1998; Torero, Quintiere, and Steinhaus 2002; Usmani, Chung, and Torero 2003.

16. Dr. Gary Weiss, personal communication, April 10, 2007.

17. Owners of some large databases have capped or scaled back the growth of the database in order to control this problem (Becla and Wang 2005).

18. See Cabena 1998; Kennedy et al. 1997; Tan et al. 2006; Weiss and Indurkhya 1998.

19. This is not to suggest that this problem is being ignored by scientists (see Hornick, Marcadé, and Venkayala 2007; Jensen and Neville 2002). However, these network analysis applications require sophisticated customizations to obtain validity and reliability.

20. For example, when retailers inquire about this author's address: 1313 Mockingbird Lane, the home address of the 1960s television family the Munsters.

21. While problematic, the "space" issue provides fewer challenges than time-series data. Li and Biswas (1995) illustrate a space-analysis method used by oil companies to prospect new fields.

22. The use of this term originates in finance and other risk-analysis literature (see Rosenberg and Schuermann 2006).

23. Turney (2000) provides a walkthrough of the variety of costs that can affect the data/model; see also McCarthy, Zabar, and Weiss 2005; Pan et al. 2007; Zadrozny and Elkan 2001.

24. In other words, different technologies can be used to construct and operate a database (i.e., Oracle, Microsoft). Transferring data between databases that are situated differently may be limited or impossible.

25. Dr. Gary Weiss, personal communication, April 10, 2007.

26. For a fuller conversation regarding the "1 percent" or "Cheney doctrine," see Suskind 2006; see also the 9/11 Commission's discussion on error rates in the federal government airport security program, CAPPS, and feasibility (National Commission on Terrorist Attacks upon the United States 2004); of course, there is the case of the former pop singer Cat Stevens's attempt to enter the United States after 9/11 (Donnelly 2004).

## REFERENCES

Abidi, S. 1999. "Applying Data Mining in Healthcare: An Info-Structure for Delivering 'Data-Driven' Strategic Services." *Studies in Health Technology and Informatics* 68: 453–456.

———. 2001. "Knowledge Management in Healthcare: Towards 'Knowledge-Driven' Decision-Support Services." *International Journal of Medical Informatics* 63: 1–2, 5–18.

Adderley, R. 2004. "The Use of Data Mining Techniques in Operational Crime Fighting." *Lecture Notes in Computer Science* 3073: 418–425.

Adderley, R., and J. Bond. 2008. "Predicting Crime Scene Attendance." *Policing* 31 (2): 292–305.

Adderley, R., and P. Musgrove. 2001a. "Data Mining Case Study: Modeling the Behavior of Offenders Who Commit Serious Sexual Assaults." In *Proceedings of the Seventh ACM SIGKDD International Conference on Knowledge Discovery and Data Mining*. New York: ACM.

———. 2001b. "Police Crime Recording and Investigation Systems: A User's View." *Policing* 24 (1): 100–114.

Adderley, R., M. Townsley, and J. Bond. 2007. "Use of Data Mining Techniques to Model Crime Scene Investigator Performance." *Knowledge-Based Systems* 20 (2): 170–176.

Adey, P. 2003. "Secured and Sorted Mobilities: Examples from the Airport." *Surveillance and Society* 1: 4.

———. 2006. " 'Divided We Move': The Dromologics of Airport Security and Surveillance." In *Surveillance and Security: Technological Politics and Power in Everyday Life*, ed. T. Monahan. New York: Routledge.

Ahsan, S., and A. Shah. 2008. "Data Mining, Semantic Web and Advanced Information Technologies for Fighting Terrorism." Paper presented at the International Symposium on Biometrics and Security Technologies.

"Air transportation." 2003. *Aerospace America*, 41(12): 24–25.

Albores, P., and D. Shaw. 2005. "Responding to Terrorist Attacks and Natural Disasters: A Case Study Using Simulation." In WSC '05 Proceedings of the Thirty-seventh conference on Winter simulation.

Allan, A.; A. Adamson; B. Cavanagh; F. Economou; S. Fraser; T. Jenness, et al. 2006. "What Do Telescopes, Databases and Compute Clusters Have in Common?" In *Proceedings of the SPIE,* vol. 6274, Advanced Software and Control for Astronomy, ed. Hilton Lewis and Alan Bridger.

Almenoff, J. S.; E. N. Pattishall; T. G. Gibbs; W. DuMouchel; S. J. W. Evans; and N. Yuen. 2007. "Novel Statistical Tools for Monitoring the Safety of Marketed Drugs." *Clinical Pharmacology and Therapeutics* 82 (2): 157–166.

Alvarez, M. 2005. "Voter Registration: Past, Present, and Future." National Commission on Federal Election Reform: CALTECH/MIT: Voting Technology Project.

Andersen, D. G., and N. Feamster. 2006. *Challenges and Opportunities in Internet Data Mining.* Pittsburgh: Carnegie Mellon University.

Anderson, L. O.; Y. Malhi; L. E. O. Aragao; and S. Saatchi. 2008. "Spatial patterns of the canopy stress during 2005 drought in Amazonia." Paper presented at the Geoscience and Remote Sensing Symposium, Barcelona, June 23–28, 2007.

Andrejevic, M. 2007. *iSpy: Surveillance and Power in the Interactive Era.* Lawrence: University Press of Kansas.

Apple.com. 2011. iPod classic. www.apple.com/ipodclassic/.

Babovic, V., J. P. Drecourt, M. Keijzer, and P. Friss Hansen. 2002. "A Data Mining Approach to Modelling of Water Supply Assets." *Urban Water* 4 (4): 401–414.

Bache, R.; F. Crestani; D. Canter; and D. Youngs. 2007a. "Application of Language Models to Suspect Prioritisation and Suspect Likelihood in Serial Crimes." International Workshop on Computer Forensics.

———. 2007b. "Mining Police Digital Archives to Link Criminal Styles with Offender Characteristics." *Lecture Notes in Computer Science.* In Proceedings of ICADL 2007: 493–494.

Baker, A. 2008. "City Would Photograph Every Vehicle Entering Manhattan and Sniff Out Radioactivity." *New York Times,* August 11.

Balamurugan, S.A.A., and R. Rajaram. 2008. "Learning to Classify Threaten e-mail." In Proceedings of Asia International Conference on Modelling and Simulation 2008: 522–527.

Balducelli, C.; S. Bologna; L. Lavalle; and G. Vicoli. 2007. "Safeguarding Information Intensive Critical Infrastructures Against Novel Types of Emerging Failures." *Reliability Engineering and System Safety* 92 (9): 1218–1229.

Banks, D.; E. J. Woo; D. R. Burwen; P. Perucci; M. M. Braun; and R. Ball. 2005. "Comparing Data Mining Methods on the VAERS Database." *Pharmacoepidemiology and Drug Safety* 14 (9): 601–609.

Barbosa, M. T.; M. R. Byington; and C. J. Struchiner. 2000. "Dynamic Models and Social Networks: A Review and Reflections on Their Contribution to Understanding the HIV Epidemic." *Cadernos de Saúde Pública* 16 (Suppl 1): 37–51.

Barnes, C. F.; H. Fritz; and J. Yoo. 2007. "Hurricane Disaster Assessments with Image-Driven Data Mining in High-Resolution Satellite Imagery." *IEEE Transactions on Geoscience and Remote Sensing* 45 (6): 1631–1640.

Barrett, C. L.; S. G. Eubank; and J. P. Smith. 2005. "If Smallpox Strikes Portland." *Scientific American* 292 (3): 54–61.

Beasley, M. S. 1996. "An Empirical Analysis of the Relation Between the Board of Director Composition and Financial Statement Fraud." *Accounting Review* 71 (4): 443–465.

Beasley, M. S.; J. V. Carcello; D. R. Hermanson; and P. D. Lapides. 2000. "Fraudulent Financial Reporting: Consideration of Industry Traits and Corporate Governance Mechanisms." *Accounting Horizons* 14 (4): 441–454.

Becla, J., and D. L. Wang. 2005. "Lessons Learned from Managing a Petabyte." Paper presented at the Conference on Innovative Data Systems Research, Asilomar, CA.

Bell, T. B., and J. V. Carcello. 2000. "A Decision Aid for Assessing the Likelihood of Fraudulent Financial Reporting." *Auditing* 19 (1): 168–184.

Beroggi, G. E. G., and D. van Gent. 2001. "Assessing Group Decision Support Systems for Emergency Response Using Gaming Simulation." Paper presented at the International Emergency Management Society Conference.

Beroggi, G. E. G., and W. A. Wallace. 1995. "Real-Time Decision Support for Emergency Management: An Integration of Advanced Computer and Communications Technology." *Journal of Contingencies and Crisis Management* 3 (1): 18–26.

Berry, D.; A. Usmani; J. Torero; A. Tate; S. McLaughlin; S. Potter, et al. 2005. "FireGrid: Integrated Emergency Response and Fire Safety Engineering for the Future Built Environment." Paper presented at the UK e-Science Programme All Hands Meeting, Nottingham, UK.

Berry, M. J. A., and G. Linoff. 2000. *Mastering Data Mining: The Art and Science of Customer Relationship Management.* New York: Wiley Computer.

Bessler, F. T.; D. A. Savic; and G. A. Walters. 2003. "Water Reservoir Control with Data Mining." *Journal of Water Resources Planning and Management* 129 (1): 26–34.

Bezir, N. C.; E. U. Kucuksille; M. Ozturk; and N. Ozek. 2007. "Modeling of Monthly Solar Radiation Using Data Mining." In Sixth International Conference of the Balkan Physical Union. American Institute of Physics Conference Proceedings, Vol. 899.

Birrer, F. A. J. 2005. "Data Mining to Combat Terrorism and the Roots of Privacy Concerns." *Ethics and Information Technology* 7 (4): 211–220.

Boyack, K. W. 2004. "Mapping Knowledge Domains: Characterizing PNAS." *Proceedings of the National Academy of Sciences of the United States of America,* 101 (Suppl. 1): 5192–5199.

Brewer, J. 2000. "Internet: Malacca Break Reveals New Web of Demands." Lloyd's List, December 5. http://www.lloydslistintelligence.com.

Breyer, J. P. 2000. "Tools to Identify Safety Issues for a Corridor Safety-Improvement Program." *Transportation Research Record* 1717: 19–27.

Brossette, S. E.; A. P. Sprague; J. M. Hardin; K. B. Waites; W. T. Jones; and S. A. Moser. 1998. "Association Rules and Data Mining in Hospital Infection Control and Public Health Surveillance." *Journal of the American Medical Informatics Association* 5 (4): 373–381.

Brossette; S. E.; A. P. Sprague; W. T. Jones; and S. A. Moser. 2000. "A Data Mining System for Infection Control Surveillance." *Methods of Information in Medicine* 39: 4–5, 303–310.

Brown, D. E., and L. F. Gunderson. 2001. "Using clustering to discover the preferences of computer criminals." *IEEE Transactions on Systems, Man, and Cybernetics.* 31 (4): 311–318.

Buckeridge, D. L.; H. Burkom; M. Campbell; W. R. Hogan; and A. W. Moore. 2005. "Algorithms for Rapid Outbreak Detection: A Research Synthesis." *Journal of Biomedical Informatics* 38 (2): 99–113.

Burl, M. C.; D. DeCoste; B. L. Enke; D. Mazzoni; W. J. Merline; and L. Scharenbroich. 2006. "Automated Knowledge Discovery from Simulators." In Proceedings of the Sixth Society for Industrial and Applied Mathematics International Conference on Data Mining, Bethesda, MD.

Cabena, P. 1998. *Discovering Data Mining: From Concept to Implementation.* Upper Saddle River, NJ: Prentice Hall.

Cameron, H. 2004. "CCTV and (In)dividuation." *Surveillance and Society* 2: 2–3.

Canter, D., and K. Fritzon. 1998. "Differentiating Arsonists: A Model of Firesetting Actions and Characteristics." *Legal and Criminological Psychology* 3 (1): 73–96.

Cao, L.; Y. Zhao; and C. Zhang. 2008. "Mining Impact-Targeted Activity Patterns in Imbalanced Data." *IEEE Transactions on Knowledge and Data Engineering* 20 (8): 1053–1066.

Carafano, J. 2003. "Preparing Responders to Respond: The Challenges to Emergency Preparedness in the 21st Century." Heritage Lectures (812). Washington, DC: Heritage Foundation.

Carley, K. M.; D. B. Fridsma; E. Casman; A. Yahja; N. Altman; L. C. Chen, et al. 2006. "BioWar: Scalable Agent-Based Model of Bioattacks." *IEEE Transactions on Systems, Man, and Cybernetics Part A: Systems and Humans* 36 (2): 252–265.

Chandra, B.; M. Gupta; and M. P. Gupta. 2007. "Adaptive Query Interface for Mining Crime Data." *Lecture Notes in Computer Science.* In Proceedings of the 5th International Conference on Databases in Networked Information Systems: 285–296.

Chang, Y. C., and N. B. Chang. 2002. "The Design of a Web-Based Decision Support System for the Sustainable Management of an Urban River System." *Water Science and Technology* 46: 6–7, 131–139.

Chen, H.; W. Chung; J. J. Xu; G. Wang; Y. Qin; and M. Chau. 2004. "Crime Data Mining: A General Framework and Some Examples." *Computer* 37 (4): 50–56.

Chen, H.; F. Y. Wang; and D. Zeng. 2004. "Intelligence and Security Informatics for Homeland Security: Information, Communication, and Transportation." *IEEE Transactions on Intelligent Transportation Systems* 5 (4): 329–341.

Chen, J. C., and N. B. Chang. 2007. "Mining the Fuzzy Control Rules of Aeration in a Submerged Biofilm Wastewater Treatment Process." *Engineering Applications of Artificial Intelligence* 20 (7): 959–969.

Chen, P. S. 2008. "Discovering Investigation Clues Through Mining Criminal Databases." *Studies in Computational Intelligence* 135: 173–198.

Chen, P. S.; K. C. Chang; T. P. Hsing; and S. Chou. 2006. "Mining Criminal Databases to Finding Investigation Clues—By Example of Stolen Automobiles Database." *Lecture Notes in Computer Science* 3917: 91–102.

Chen, R., and R. Sharman. 2005. "Design Principles of Coordinated Multi-Incident Emergency Response Systems." Paper presented at the IEEE International Conference on Intelligence and Security Informatics, Atlanta.

Chen, R.; R. Sharman; H. Rao; and S. Upadhyaya. 2007. "Design Principles for Critical Incident Response Systems." *Information Systems and e-Business Management* 5, 3.

Chen, S. T., and P. S. Yu. 2007. "Pruning of Support Vector Networks on Flood Forecasting." *Journal of Hydrology* 347: 1–2, 67–78.

Chen, Y.; J. Hu; Y. Zhang; and X. Li. 2006. "Traffic Data Analysis Using Kernel PCA and Self-Organizing Map." Paper presented at the Intelligent Vehicles Symposium, IEEE.

Cho, W. K. T., and B. J. Gaines. 2007. "Breaking the (Benford) Law: Statistical Fraud Detection in Campaign Finance." *American Statistician* 61 (3): 218–223.

Clippinger, J. H. 2007. *A Crowd of One: The Future of Individual Identity.* New York: PublicAffairs.

Coaffee, J. 2004. "Rings of Steel, Rings of Concrete and Rings of Confidence: Designing out Terrorism in Central London Pre and Post September 11th." *International Journal of Urban and Regional Research* 28, 1.

Coderre, G. D. 1999. *Fraud Detection: Using Data Analysis Techniques to Detect Fraud.* Vancouver: Global Audit.

Cole, M. 2004. "Signage and Surveillance: Interrogating the Textual Context of CCTV in the UK." *Surveillance and Society* 2, 2/3.

Cudahy, B. J. 2006. *Box Boats: How Container Ships Changed the World.* New York: Fordham University Press.

Dahbur, K., and T. Muscarello. 2003. "Classification System for Serial Criminal Patterns." *Artificial Intelligence and Law* 11 (4): 251–269.

Damle, C., and A. Yalcin. 2007. "Flood Prediction Using Time Series Data Mining." *Journal of Hydrology* 333: 2–4, 305–316.

Datta, R.; W. Ge; J. Li; and J. Z. Wang. 2007. "Toward Bridging the Annotation-Retrieval Gap in Image Search." *IEEE Multimedia* 14 (3): 24–35.

Datta, R.; D. Joshi; J. Li; and J. Z. Wang. 2008. "Image Retrieval: Ideas, Influences, and Trends of the New Age." *ACM Computing Surveys* 40, 2.

De Castro, O. C. F.; G. J. Lopez; B. E. G. Delgado; and J. A. Perez. 2007. "Application of data mining to the Cuban Pharmacovigilance System." *Revista Cubana de Farmacia* 41, 3.

de Nooy, W.; A. Mrvar; and V. Batagelj. 2005. *Exploratory Social Network Analysis with Pajek.* New York: Cambridge University Press.

Defense Advanced Research Projects Agency (DARPA). 2008. "Information Processing Techniques Office." www.darpa.mil (accessed September 11, 2008).

Delmater, R., and M. Hancock. 2001. *Data Mining Explained: A Manager's Guide to Customer-Centric Business Intelligence.* Boston: Digital Press.

Derr, K., and M. Manic. 2007. "Intelligent Control in Automation Based on Wireless Traffic Analysis." Paper presented at IEEE Conference for Emerging Technologies and Factory Automation, Patras, September 25–28.

Dick, S., and C. C. Bracho. 2007. "Learning to Detect the Onset of Disease in Cattle from Feedlot Watering Behavior." Paper presented at NAFIPS 2007 Annual Meeting of the North American Fuzzy Information Processing Society, San Diego.

Dickerson, J. E., and J. A. Dickerson. 2000. "Fuzzy Network Profiling for Intrusion Detection." Paper presented at the Nineteenth International Conference of the North American Fuzzy Information Processing Society (NAFIPS). Atlanta, July.

Donnelly, S. 2004. "You Say Yusuf, I Say Youssouf . . ." *Time,* September 25. www.time.com/time/nation/article/0,8599,702062,00.html.

Donoho, S. 2004. "Early Detection of Insider Trading in Option Markets." In *Proceedings of the Tenth ACM SIGKDD International Conference on Knowledge Discovery and Data Mining.* Seattle, August.

Douglas, C. C.; R. A. Lodder; R. E. Ewing; Y. Efendiev; J. Coen; M. Kritz, et al. 2006. "DDDAS Approaches to Wildland Fire Modeling and Contaminant Tracking." Paper presented at the Thirty-eighth Conference on Winter Simulation. Monterey, CA., December.

Drysdale, D. D.; A. J. R. Macmillan; and D. Shillito. 1992. "The King's Cross Fire: Experimental Validation of the Trench Effect." *Fire Safety Journal* 18, 1.

Dubrawski, A.; K. Elenberg; A. Moore; and M. Sabhnani. 2006. "Monitoring Food Safety by Detecting Patterns in Consumer Complaints." In *IAAI '06 Proceedings of the Eighteenth Conference on Innovative Applications of Artificial Intelligence,* vol. 2.

Duggan, M., and S. D. Levitt. 2002. "Winning Isn't Everything: Corruption in Sumo Wrestling." *American Economic Review* 92, 5.

Dunbar, J. 2007. "FBI Divulges New Surveillance Tactics." *Deseret News,* August 31.

Edelstein, H. 2007. "Data Mining Software Tools: The State of the Market." In *Data Mining for Business Intelligence: Concepts, Techniques, and Applications in Microsoft Office Excel with XLMiner,* ed. G. Shmueli, N. R. Patel and P. C. Bruce, xviii. Hoboken, NJ: John Wiley & Sons.

Eggen, D. 2005. "FBI Papers Indicate Intelligence Violations: Secret Surveillance Lacked Oversight." *Washington Post,* October 24.

Eining, M. M.; D. R. Jones; and J.K. Loebbecke. 1997. "Reliance on Decision Aids: An Examination of Auditors' Assessment of Management Fraud." *Auditing* 16 (2): 16–19.

Electronic Frontier Foundation. 2004. "CAPPS II: Government Surveillance via Passenger Profiling." http://w2.eff.org/Privacy/cappsii/ (accessed September 7, 2008).

Electronic Privacy Information Center. 2008. "Wiretapping." Epic.org. http://epic.org/privacy/wiretap/ (accessed September 7, 2008).

Ertoz, L., E. Eilertson, A. Lazarevic, P. N. Tan, V. Kumar, J. Srivastava, et al. 2004. "The MINDS—Minnesota Intrusion Detection System." In *Data Mining: Next Generation Challenges and Future Directions,* ed. Hillol Kargupta et al. Cambridge, MA: MIT Press.

Ertoz, L.; A. Lazarevic; E. Eilertson; P. N. Tan; P. Dokas; V. Kumar, et al. 2003. "Protecting Against Cyber Threats in Networked Information Systems." In *Proceedings,* vol. 5101, *Battlespace Digitization and Network-Centric Systems III,* ed. Raja Suresh, 51–56. SPIE, San Diego.

Fan, W.; P. S. Yu; and H. Wang. 2004. "Mining Extremely Skewed Trading Anomalies." *Lecture Notes in Computer Science* 2992: 801–810.

Fanning, K., and K. Cogger. 1998. "Neural Network Detection of Management Fraud Using Published Financial Data." *International Journal of Intelligent Systems in Accounting, Finance and Management* 7 (1): 21–41.

Fanning, K.; K. O. Cogger; and R. Srivastava. 1995. "Detection of Management Fraud: A Neural Network Approach." *International Journal of Intelligent Systems in Accounting, Finance and Management* 4 (2): 113–126.

Farmer, D., and C. C. Mann. 2003a. Part one: Surveillance Nation. *Technology Review,* 106, 3.

———. 2003b. Part two: Surveillance Nation. *Technology Review,* 106, 4.

Fitzpatrick, L. K.; J. A. Hardacker; W. Heirendt; T. Agerton; A. Streicher; H. Melnyk, et al. 2001. "A Preventable Outbreak of Tuberculosis Investigated Through an Intricate Social Network." *Clinical Infectious Diseases* 33 (11): 1801–1806.

Flynn, S. E. 2006. "Port Security Is Still a House of Cards. *Far Eastern Economic Review* (January/February).

Frawley, W. J.; G. Piatetsky-Shapiro; and C. J. Matheus. 1992. "Knowledge Discovery in Databases: An Overview." *AI Magazine* 13, 3.

Frederiksen, A. N., and M. Schmid. 2008. "Maintaining public railways with lower cost and improved safety." *Analog Dialogue* 42 (September).

Fugate, M., and J. R. Gattiker. 2003. "Computer Intrusion Detection with Classification and Anomaly Detection, Using SVMs." *International Journal of Pattern Recognition and Artificial Intelligence* 17 (3): 441–458.

Galdorisi, G., and R. Goshorn. 2006. "Maritime Domain Awareness: The Key to Maritime Security Operational Challenges and Technical Solutions." Paper presented at Eleventh ICCRTS Coalition Command and Control in the Networked Era, San Diego, September 26–28.

Galitsky, B., and B. Kovalerchuk. 2006. "Mining Emotional Profiles Using E-Mail Messages for Earlier Warnings of Potential Terrorist Activities." In *Proceedings, Data Mining, Intrusion Detection, Information Assurance, and Data Networks Security 2006,* vol. 6241, ed. Belur V. Dasarathy. Orlando: SPIE.

Garfinkel, S. 2001. *Database Nation: The Death of Privacy in the 21st Century.* Cambridge, MA: O'Reilly.

Garfinkel, S., and M. D. Smith. 2006. "Data Surveillance." *IEEE Security and Privacy* 4 (6): 15–17.

Geradts, Z., and J. Bijhold. 2002. "Data Mining in Forensic Image Databases." http://geradts.com/html/Documents/Articles/Datamining.pdf.

Germann, T. C.; K. Kadau; I. M. Longini Jr.; and C. A. Macken. 2006. "Mitigation Strategies for Pandemic Influenza in the United States." *Proceedings of the National Academy of Sciences of the United States of America* 103 (15): 5935–5940.

Ghafoor, A. 2004. "Web-Based Multimedia Databases: Prospects and Challenges." Paper presented at Conference on Information and Knowledge Management, Washington, DC, November 8–13.

Gong, X.; H. Gao; and F. Wang. 2005. "Global Monitoring and Security Assistance Based on Next Generation Internet for Public and Transportation Safety Enhancement." In *Intelligent Transportation Systems,* IEEE, Adelaide, Australia.

Gopinath, R. B., and M. R. Tamjis. 2008. "Water Quality in Healthcare." *International Journal of Environmental Technology and Management* 9 (1): 125–140.

Gras, M. L. 2004. "The Legal Regulation of CCTV in Europe." *Surveillance and Society* 2, 2/3.

Green, B. P., and J. H. Choi. 1997. "Assessing the Risk of Management Fraud Through Neural Network Technology." *Auditing* 16 (1): 25–28.

Grossman, R. 2001. *Data Mining for Scientific and Engineering Applications.* Boston: Kluwer Academic.

Grubesic, T., and M. Zook. 2007. "A Ticket to Ride: Evolving Landscapes of Air Travel Accessibility in the United States." *Journal of Transport Geography* 15 (6): 417–430.

Gunasekara, G. 2007. "The 'Final' Privacy Frontier? Regulating Trans-Border Data Flows." *International Journal of Law and Information Technology* 15 (3): 362–393.

Gunderson, L. F. 2002. "Using Data Mining and Judgment Analysis to Construct a Predictive Model of Crime." Paper presented at IEEE International Conference on Systems: Man and Cybernetics.

Guo, Y., and R. Grossman. 1999. *High Performance Data Mining: Scaling Algorithms, Applications, and Systems.* Boston: Kluwer Academic.

Hackworth, T., and P. Treleaven. 2007. "Modelling Terrorism." Paper presented at IEEE International Conference on Networking: Sensing and Control, London.

Haggerty, K., and R. Ericson. 2000. "The Surveillant Assemblage." *British Journal of Sociology* 51, 4.

Hale, A. 2000. "Railway Safety Management: The Challenge of the New Millennium." *Safety Science Monitor* 4, 1.

Han, J. G.; K. H. Ryu; K. H. Chi; and K. Y. Yeon. 2003. "Statistics Based Predictive Geo-Spatial Data Mining: Forest Fire Hazardous Area Mapping Application." *Lecture Notes in Computer Science* 2642: 370–381.

Hand, D. J.; H. Mannila; and P. Smyth. 2001. *Principles of Data Mining.* Cambridge, MA: MIT Press.

Hauben, M.; D. Madigan; C. M. Gerrits; L. Walsh; and E. P. Van Puijenbroek. 2005. "The Role of Data Mining in Pharmacovigilance." *Expert Opinion on Drug Safety* 4 (5): 929–948.

Hauben, M.; V. Patadia; C. Gerrits; L. Walsh; and L. Reich. 2005. "Data Mining in Pharmacovigilance: The Need for a Balanced Perspective." *Drug Safety* 28 (10): 835–842.

He, Q.; K. Chang; and E.-P. Lim. 2007. "Anticipatory Event Detection via Classification." *Information Systems and e-Business Management* 5, 3.

Hewett, R.; J. Leuchner; and M. Carvalho. 2001. "From Climate History to Prediction of Regional Water Flows with Machine Learning." Paper presented at IEEE International Conference on Systems, Man, and Cybernetics, Tucson, AZ.

Holmes, M. C.; D. D. Comstock-Davidson; and R. L. Hayen. 2007. "Data Mining and Expert Systems in Law Enforcement Agencies." *Issues in Information Systems* 8, 2.

Horan, T. A., and M. Marich. 2006. "Time-Critical Information Services: Analysis and Workshop Findings on Technology, Organizational, and Policy Dimensions to Emergency Response and Related E-Governmental Services." Paper presented at the International Conference on Digital Government Research.

Hormozi, A. M., and S. Giles. 2004. "Data Mining: A Competitive Weapon for Banking and Retail Industries." *Information Systems Management,* 21 (2): 62–71.

Horne, G. E. 2008. "Maritime Security Applications of Data Farming." Paper presented at the Sixth Security Workshop Globalization and Maritime Security.

Hornick, M. F.; E. Marcadé; and S. Venkayala. 2007. *Java Data Mining: Strategy, Standard, and Practice: A Practical Guide for Architecture, Design, and Implementation.* Boston: Morgan Kaufmann.

Hu, X.; X. Zhang; D. Wu; X. Zhou; and P. Rumm. 2006. "Integration of Instance-Based Learning and Text Mining for Identification of Potential Virus/Bacterium as Bio-Terrorism Weapons." *Lecture Notes in Computer Science* 3975: 548–553.

Huysmans, J.; B. Baesens; and J. Vanthienen. 2008. "A Data Miner's Approach to Country Corruption Analysis." *Studies in Computational Intelligence* 135: 227–247.

Huysmans, J.; D. Martens; B. Baesens; J. Vanthienen; and T. Van Gestel. 2006. "Country Corruption Analysis with Self Organizing Maps and Support Vector Machines." *Lecture Notes in Computer Science* 3917: 103–114.

Ishak, S. 2003. "Deriving Traffic-Performance Measures and Levels of Service from Second-Order Statistical Features of Spatiotemporal Traffic Contour Maps." *Transportation Research Record* 1858: 148–157.

Ishibashi, K.; T. Toyono; K. Toyama; M. Ishino; H. Ohshima; and I. Mizukoshi. 2005. "Detecting Mass-Mailing Worm Infected Hosts by Mining DNS Traffic Data." Paper presented at the SIGCOMM '05 Workshops, Philadelphia, August 22–26.

Jacob, B. A., and S. D. Levitt. 2003. "Catching Cheating Teachers: The Results of an Unusual Experiment in Implementing Theory." NBER Working Paper No. 9414, Philadelphia, January.

Jaenisch, H.; J. Handley; A. Lim; M. Filipovic; G. White; A. Hons, et al. 2004. "Data Modeling for Virtual Observatory Data Mining." In *Optimizing Scientific Return for Astronomy through Information Technologies,* ed. Peter J. Quinn and Alan Bridger, 198–220. SPIE, Glasgow, Scotland.

Jaenisch, H. M.; J. W. Handley; J. P. Faucheux; and B. Harris. 2003. "Data Modeling of Network Dynamics." In *Proceedings of the SPIE,* vol. 5200. San Diego.

Jensen, D., and J. Neville. 2002. "Data Mining in Networks." Paper presented at the Symposium on Dynamic Social Network Modeling and Analysis, Washington DC, November 7–9.

Jonas, J., and J. Harper. 2006. "Effective Counterterrorism and the Limited Role of Predictive Data Mining." The Cato Institute, www.cato.org/pub_display.php?pub_id=6784, (accessed October 31, 2001).

Kaufman, Z.; W. K. Wong; T. Peled-Leviatan; E. Cohen; C. Lavy; G. Aharonowitz, et al. 2007. "Evaluation of a Syndromic Surveillance System Using the WSARE Algorithm for Early Detection of an Unusual, Localized Summer Outbreak of Influenza B: Implications for Bioterrorism Surveillance." *Israel Medical Association Journal* 9 (1): 3–7.

Kennedy, R. L., et al. 1997. *Solving Data Mining Problems Through Pattern Recognition.* Upper Saddle River, NJ: Prentice Hall.

Kim, J. K., and R. Sharman. 2005. "An Investigation of Risk Management Issues in the Context of Emergency Response Systems." Paper presented at the Eleventh American's Conference on Information Systems, Omaha, NE.

———. 2006. "Framework for Analyzing Critical Incident Management Systems." Proceedings of the 39th Annual Hawaii International Conference on System Science, Kauai.

Kirkos, E.; C. Spathis; and Y. Manolopoulos. 2007. "Data Mining Techniques for the Detection of Fraudulent Financial Statements." *Expert Systems with Applications* 23: 32.

Klovdahl, A. S. 1985. "Social networks and the spread of infectious diseases: The AIDS Example." *Social Science and Medicine* 21 (11): 1203–1216.

Knoop, S. 2006. "Towards a National Healthcare Information Infrastructure." *Lecture Notes in Computer Science* 4075: 2.

Koch, D. B. 2004. "3D Visualization to Support Airport Security Operations." *Aerospace and Electronic Systems Magazine* 19 (6): 23–28.

Kolesar, P., and W. E. Walker. 1974. "An Algorithm for the Dynamic Relocation of Fire Companies." *Operations Research* 22, 2.

Korving, H., and E. C. Ottenhoff. 2008. "Analysis of the Causes of Pump Failure and Differences of Failure Characteristics." *Water Science and Technology* 57 (8): 1271–1276.

Kramer, M. 2008. "NYPD's 'Operation Sentinel' to Track Everything." CBS News, August 12. http://wcbstv.com/cbs2crew/operation.sentinel.nypd.2.793133.html (accessed September 11, 2008).

Krebs, V. E. 2002. "Uncloaking Terrorist Networks." *First Monday* (7) 4. http://firstmonday.org//htbin/cgiwrap/bin/ojs/index.php/fm/article/viewArticle/941/863.

Ku, C. H.; A. Iriberri; and G. Leroy. 2008. "Natural Language Processing and e-Government: Crime Information Extraction from Heterogeneous Data Sources." Paper presented at the Ninth Annual International Digital Government Research Conference, Montreal.

Kurlander, N. 2005. "The Crime Database Battle." *Security Journal* 42, 10.

Kutz, G. D., and J. J. Ryan. 2006. *Hurricanes Katrina and Rita Disaster Relief: Improper and Potentially Fraudulent Individual Assistance Payments Estimated to Be Between $600 Million and $1.4 Billion.* Washington DC: U.S. Government Accountability Office.

Laiho, J.; K. Raivio; P. Lehtimaki; K. Hatonen; and O. Simula. 2005. "Advanced Analysis Methods for 3G Cellular Networks." *IEEE Transactions on Wireless Communications,* 4 (3): 930–942.

Lamma, E.; P. Mello; A. Nanetti; F. Riguzzi; S. Storari; and G. Valastro. 2006. "Artificial Intelligence Techniques for Monitoring Dangerous Infections." *IEEE Transactions on Information Technology in Biomedicine* 10 (1): 143–155.

Langewiesche, W. 2004. *The Outlaw Sea: A World of Freedom, Chaos, and Crime.* New York: North Point Press.

Larson, R. C.; M. D. Metzger; and M. F. Cahn. 2006. "Responding to Emergencies: Lessons Learned and the Need for Analysis." *Interfaces* 36 (6): 486–501.

Lateef, F. 2007. "Of Pruritus and Terrorism." *Singapore Medical Journal* 48 (5): 380–382.

Lavrac, N.; M. Bohanec; A. Pur; B. Cestnik; M. Jermol; T. Urbancic, et al. 2005. "Resource Modeling and Analysis of Regional Public Health Care Data by Means of Knowledge Technologies." *Lecture Notes in Computer Science* 3581: 414–418.

Lavrac, N.; M. Bohanec; A. Pur; B. Cestnik; M. Debeljak; and A. Kobler. 2007. "Data Mining and Visualization for Decision Support and Modeling of Public Health-Care Resources." *Journal of Biomedical Informatics* 40 (4): 438–447.

Lee, I., and P. Phillips. 2008. "Urban Crime Analysis Through Areal Categorized Multivariate Associations Mining." *Applied Artificial Intelligence* 22 (5): 483–499.

Lee, W., and S. J. Stolfo. 1998. "Data Mining Approaches for Intrusion Detection." Paper presented at the USENIX Security Symposium, San Antonio, TX.

Levinson, M. 2006. *The Box: How the Shipping Container Made the World Smaller and the World Economy Bigger.* Princeton: Princeton University Press.

Li, C., and G. Biswas. 1995. "Knowledge-Based Scientific Discovery in Geological Databases." Paper presented at the First International Conference on Knowledge Discovery and Data Mining, Montreal.

Li, S. T.; F. C. Tsai; S. C. Kuo; and Y. C. Cheng. 2006. "A Knowledge Discovery Approach to Supporting Crime Prevention." In Proceedings of JCIS 2006. Kaohsiung, Taiwan.

Li, X., and N. Ye. 2002. "Grid- and Dummy-Cluster-Based Learning of Normal and Intrusive Clusters for Computer Intrusion Detection." *Quality and Reliability Engineering International* 18 (3): 231–242.

Li, Z.; A. Zhang; D. Li; and L. Wang. 2007. "Discovering Novel Multistage Attack Strategies." *Lecture Notes in Computer Science* 4632: 45–56.

Lin, J. W.; M. I. Hwang; and J. D. Becker. 2003. "A Fuzzy Neural Network for Assessing the Risk of Fraudulent Financial Reporting." *Managerial Auditing Journal* 18 (8): 657–665.

Lin, S., and D. E. Brown. 2003. "Criminal Incident Data Association Using the OLAP Technology." *Lecture Notes in Computer Science* 2665: 13–26.

Lisberg, A. 2007. "NYPD's 'Ring of Steel' Camera Rolls, Monitors License Plates." *Daily News,* October 1.

Litt, J.; D. L. Simon; C. Meyer; H. DePold; J. R. Curtiss; H. Winston, et al. 2000. "NASA Aviation Safety Program Aircraft Engine Health Management Data Mining Tools Roadmap." NASA Report, April.

Little, B. B.; W. L. Johnston; A. C. Lovell; R. M. Rejesus; and S. A. Steed. 2002. "Collusion in the U.S. Crop Insurance Program: Applied Data Mining." Paper presented at the ACM SIGKDD International Conference on Knowledge Discovery and Data Mining, Edmonton.

Liu, C.; D. Wang; B. Liu; D. Gao; C. Cui; and Y. Zhao. 2006. "An Astronomical Data Mining Application Framework for Virtual Observatory." In *Advanced Software and Control for Astronomy,* ed. Hilton Lewis and Alan Bridger, SPIE, Orlando.

Liu, Y.; A. Choudhary; J. Zhou; and A. Khokhar. 2006. "A scalable distributed stream mining system for highway traffic data." *Lecture Notes in Computer Science* 4213: 309–321.

Lopez-Ariste, A.; D. E. Rees; H. Socas-Navarro; and B. W. Lites. 2001. "Pattern Recognition Techniques and the Measurement of Solar Magnetic Fields." In *Astronomical Data Analysis,* ed. Jean-Luc Starck and Fionn D. Murtagh, 96–106. SPIE.

Lu, C. T.; L. N. Sripada; S. Shekhar; and R. Liu. 2005. "Transportation Data Visualisation and Mining for Emergency Management." *International Journal of Critical Infrastructures* 1 (2–3): 170–194.

Luo, Q. 2008. "Transportation Data Analyzing by Using Data Mining Method." Paper presented at 2008 International Symposiums on Information Processing (ISIP), Moscow, May 23–25.

Major, J., and J. Mangano. 1995. "Selecting Among Rules Induced from a Hurricane Database." *Journal of Intelligent Information Systems* 4 (1): 39–52.

Manganaris, S.; M. Christensen; D. Zerkle; and K. Hermiz. 2000. "Data Mining Analysis of RTID Alarms." *Computer Networks* 34 (4): 571–577.

Massengill, G.; M. L. Martin; G. Davies; R. L. Depue; C. A. Ellis; T. Ridge, et al. 2007. "Mass Shootings at Virginia Tech, April 16, 2007." Report of the Review Panel, presented to Governor Kaine of the Commonwealth of Virginia. August.

Mazzoni, D.; J. A. Logan; D. Diner; R. Kahn; L. Tong; and Q. Li. 2007. "A Data-Mining Approach to Associating MISR Smoke Plume Heights with MODIS Fire Measurements." *Remote Sensing of Environment* 107: 138–148.

McAndrew, D. 2000. "The structural analysis of criminal networks." In *The Social Psychology of Crime: Groups, Teams, and Networks,* ed. D. V. Canter and L. J. Alison, 53–94. Dartmouth University Press.

McCarthy, K.; B. Zabar; and G. Weiss. 2005. "Does Cost-Sensitive Learning Beat Sampling for Classifying Rare Classes?" Paper presented at the Conference on Knowledge Discovery in Data, Chicago.

McCue, C., and A. Parker. 2003. "Connecting the Dots: Data Mining and Predictive Analytics in Law Enforcement and Intelligence Analysis." *Police Chief* 70, 10.

McGinnis, M. D. 1991. "Richardson, Rationality, and Restrictive Models of Arms Races." *Journal of Conflict Resolution* 35, 3.

McIllwain, J. S. 1999. "Organized Crime: A Social Network Approach." *Crime, Law and Social Change* 32 (4): 301–323.

Memon, N.; D. L. Hicks; and N. Harkiolakis. 2008. "A Data Mining Approach to Intelligence Operations." In *Data Mining, Intrusion Detection, Information Assurance, and Data Networks Security 2008,* ed. Belur V. Dasarathy. SPIE, Orlando.

Memon, N.; D. L. Hicks; and H. L. Larsen. 2007. "Harvesting Terrorists Information from Web." Paper presented at Eleventh International Conference on Information Visualization, Zurich, July 4–6.

Mirhaji, P.; J. Zhang; A. Srinivasan; R. L. Richesson; and J. W. Smith. 2004. "Knowledge-Based Public Health Situation Awareness." In *Sensors, and Command, Control, Communications, and Intelligence (C3I) Technologies for Homeland Security and Homeland Defense III,* ed. Edward M. Carapezza, Proceedings of the SPIE, Volume 5403, 198–209. Baltimore, MD.

Monahan, T. (ed.). 2006. *Surveillance and Security: Technological Politics and Power in Everyday Life.* New York: Routledge.

Moore, M.; E. Chan; N. Lurie; A. G. Schaefer; D. M. Varda; and J. A. Zambrano. 2008. "Strategies to Improve Global Influenza Surveillance: A Decision Tool for Policymakers." *BMC Public Health* 8: 186–196.

Morais, J.; Y. Pires; C. Cardoso; and A. Klautau. 2007. "Data Mining Applied to the Electric Power Industry: Classification of Short-Circuit Faults in Transmission Lines." Paper presented at the Seventh International Conference on Intelligent Systems Design and Applications, Rio de Janeiro, October 20–24.

Mosca, R.; P. Giribone; and A. G. Bruzzone. 1995. "AI Techniques to Optimize the Distribution of Fire-Fighting Squads During an Accident." Paper presented at the Proceedings of Simulators International, Phoenix.

National Commission on Terrorist Attacks upon the United States. 2004. *The 9/11 Commission Report: Final Report of the National Commission on Terrorist Attacks upon the United States* (authorized ed.). New York: Norton.

National Incident Management System (NIMS). 2008. National Incident Management Systems Integration Division. www.fema.gov/emergency/nims/ (accessed September 30, 2008).

National Interagency Fire Center (NIFC). 2008. "Fire Information—National Fire News." www.nifc.gov/fire_info/nfn.htm (accessed September 30, 2008).

National Oceanic and Atmospheric Administration (NOAA). 2007. "NOAA Announces Next Solar Storm Cycle Will Likely Start Next March." www.noaanews.noaa.gov/stories2007/s2847.htm (accessed September 26, 2008).

———. 2008. "Sunspot Is Harbinger of New Solar Cycle, Increasing Risk for Electrical Systems." www.noaanews.noaa.gov/stories2008/20080104_sunspot.html (accessed September 26, 2008).

Norris, C., and G. Armstrong. 1999. *The Maximum Surveillance Society: The Rise of CCTV.* New York: Berg.

Norris, C.; M. McCahill; and D. Wood. 2004. "The Growth of CCTV: A Global Perspective on the International Diffusion of Video Surveillance in Publicly Accessible Space." *Surveillance and Society* 2, 2/3.

O'Harrow, R. 2005. *No Place to Hide.* New York: Free Press.

Oatley, G.; K. McGarry; and B. Ewart. 2006. "Offender Network Metrics." *WSEAS Transactions on Information Science and Applications* 3 (12): 2440–2447.

Oatley, G. C., and B. W. Ewart. 2003. "Crimes Analysis Software: 'Pins in Maps,' Clustering and Bayes Net Prediction." *Expert Systems with Applications* 25 (4): 569–588.

Office of the Inspector General (OIG). 2002. "The Federal Bureau of Investigation's Control over Weapons and Laptop Computers." U.S. Department of Justice, Washington, DC.

———. 2005. "The Federal Bureau of Investigation's Management of the Trilogy Information Technology Modernization Project." Audit Report No. 05–07, February. U.S. Department of Justice, Washington, DC.

———. 2006. "Federal Agencies' Efforts to Protect Sensitive Information: A Report to the Office of Management and Budget." PCIE/ECIE Report compiled by the PCIE Information Technology Roundtable and the Federal Audit Executive Council Information Technology Committee.

———. 2007a. "A Review of the Federal Bureau of Investigation's Use of National Security Letters." U.S. Department of Justice, Washington, DC.

———. 2007b. "Top Management and Performance Challenges in the Department of Justice: 2006." U.S. Department of Justice, Washington DC.

———. 2007c. "The Federal Bureau of Investigation's Control over Weapons and Laptop Computers Follow-Up Audit." U.S. Department of Justice, Washington DC.

———. 2008. "Office of the Inspector General Semiannual Report to Congress: October 1, 2007–March 31, 2008." U.S. Department of Justice, Washington DC.

Ong, G. G. 2002. "Pre-Empting Maritime Terrorism in Southeast Asia." Singapore: Institute of South East Asian Studies.

Ozgul, F.; Z. Erdem; and H. Aksoy. 2008. "Comparing Two Models for Terrorist Group Detection: GDM or OGDM?" *Lecture Notes in Computer Science* 5075: 149–160.

Ozkan, K. 2004. "Managing Data Mining at Digital Crime Investigation." *Forensic Science International* 146 (Suppl.): 37–38.

Packard, V. O. 1964. *The Naked Society.* New York: D. McKay.

Pagnoni, A.; S. Parisi; and S. Lombardo. 2001. "Analysis of Patient Flows via Data Mining." *Medinfo* 10 (Pt 2): 1379–1383.

Pan, J.; Q. Yang; Y. Yang; L. Li; F. T. Li; and G. W. Li. 2007. "Cost-Sensitive-Data Preprocessing for Mining Customer Relationship Management Databases." *IEEE Intelligent Systems* 22 (1): 281–287.

Pan, J. Y., and C. Faloutsos. 2002. " 'GeoPlo': Spatial Data Mining on Video Libraries." In *Proceedings of the Eleventh International Conference on Information and Knowledge Management*, McLean, VA.

Pande, A., and M. A. Abdel-Aty. 2006. "Application of Data Mining Techniques for Real-Time Crash Risk Assessment on Freeways." In *Proceedings of the Ninth International Conference on Applications of Advanced Technology in Transportation,* ed. Kelvin C. P. Wang, Brian L. Smith, Donald R. Uzarski, and S. C. Wong, Chicago.

Pang, M. B., and G. G. He. 2007. "Real-time Intelligent Recognition of Chaos in Traffic Flow Using Reduced Support Vector Machine." Paper presented at WiCom '07, International Conference on Wireless Communications, Networking and Mobile Computing, Shanghai, September 21–25.

Patcha, A., and J. M. Park. 2007. "An Overview of Anomaly Detection Techniques: Existing Solutions and Latest Technological Trends." *Computer Networks* 51 (12).

Pawling, A.; N. V. Chawla; and G. Madey. 2007. "Anomaly Detection in a Mobile Communication Network." *Computational and Mathematical Organization Theory* 13 (4): 407–422.

Peralta, J. O., and M. T. Casas de Peralta. 2002. "Security PIDS with Physical Sensors, Real-Time Pattern Recognition, and Continuous Patrol." *IEEE Transactions on Systems, Man and Cybernetics Part C: Applications and Reviews* 32 (4): 340–346.

Perry, W. G. 2006. "How ISO/IEC 17799 Can Be Used for Base Lining Information Assurance Among Entities Using Data Mining for Defense, Homeland Security, Commercial and Other Civilian/Commercial Domains." In *Data Mining, Intrusion Detection, Information Assurance, and Data Networks Security 2006,* ed. Belur V. Dasarathy. SPIE.

Persons, O. S. 1995. "Using Financial Statement Data to Identify Factors Associated with Fraudulent Financial Reporting." *Journal of Applied Business Research* 11 (3): 38–46.

Pollard, N. A.; R. V. Tuohy; T. M. Garwin; M. E. Powell; M. Royal; and G. T. Singley, III. 2004. "Project Responder: A National Technology Plan for Emergency Response." In *Sensors, and Command, Control, Communications, and Intelligence (C3I) Technologies for Homeland Security and Homeland Defense III,* ed. Edward M. Carapezza, 49–59. SPIE, Chicago.

Porter, D. 2003. "Insider Fraud: Spotting the Wolf in Sheep's Clothing." *Computer Fraud and Security* 2003 (4): 12–15.

Privacy Rights Clearinghouse. 2007. "Online Data Vendors: How Consumers Can Opt out of Directory Assistance and Non-Public Information." www.privacyrights.org/ar/infobrokers.htm (accessed April 24, 2007).

Pur, A.; M. Bohanec; B. Cestnik; N. Lavrac; M. Debeljak; and T. Kopac. 2005. "Data Mining for Decision Support: An Application in Public Health Care." *Lecture Notes in Computer Science* 3533: 459–469.

Raghu, T. S.; B. Jayaraman; and H. R. Rao. 2004. "Toward an Integration of Agent- and Activity-Centric Approaches in Organizational Process Modeling: Incorporating Incentive Mechanisms." *Information Systems Research* 15 (4): 316–335.

Ramadas, M.; S. Ostermann; and B. Tjaden. 2003. "Detecting Anomalous Network Traffic with Self-Organizing Maps." Paper presented at the Sixth International Symposium on Recent Advances in Intrusion Detection, Pittsburgh.

Reinhardt, W. 2008. "Landslide Vulnerability." *GIM International* 22 (9): 49.

Remagnino, P. 2002. *Video-Based Surveillance Systems: Computer Vision and Distributed Processing.* Boston: Kluwer Academic.

Ribaux, O., and P. Margot. 2001. "The Analysis of Serial Crime Through the Use of Different Sources of Data." 46: 93–100. In *Problems of Forensic Sciences,* vol. XLVI, Krakow, Poland.

Roach, J.; P. Ekblom; and E. Flynn. 2005. "The Conjunction of Terrorist Opportunity; a Framework for Diagnosing and Preventing Acts of Terrorism." *Security Journal* 18, 3.

Rosenberg, J. M. 1969. *The Death of Privacy.* New York: Random House.

Rosenberg, J. V., and T. Schuermann. 2006. "A General Approach to Integrated Risk Management with Skewed, Fat-Tailed Risks." *Journal of Financial Economics* 79, 3.

Rothenberg, R. B.; C. Sterk; K. E. Toomey; J. J. Potterat; D. Johnson; M. Schrader, et al. 1998. "Using Social Network and Ethnographic Tools to Evaluate Syphilis Transmission." *Sexually Transmitted Diseases* 25 (3): 154–160.

Ruegg, J.; V. November; and F. Klauser. 2004. "CCTV, Risk Management and Regulation Mechanisms in Publicly Used Places: A Discussion Based on Swiss Examples." *Surveillance and Society* 2, 2/3.

Rule, J. B. 1974. *Private Lives and Public Surveillance; Social Control in the Computer Age.* New York: Schocken Books.

———. 2007. *Privacy in Peril.* New York: Oxford University Press.

Safer, A. M. 2002. "The Application of Neural Networks to Predict Abnormal Stock Returns Using Insider Trading Data." *Applied Stochastic Models in Business and Industry* 18 (4): 381–389.

Salasznyk, P. P.; E. E. Lee; G. F. List; and W. A. Wallace. 2006. "A Systems View of Data Integration for Emergency Response." *International Journal of Emergency Management* 3 (4): 313–331.

Sanford, P.; D. J. Parish; and J. M. Sandford. 2007a. "Traffic Summary Analysis for Network Abuse Detection in High Date-Rate ISP Networks." Paper presented at the International Conference on Networking, Sainte-Luce, Martinique. April 22–28.

———. 2007b. "Understanding Increasing Traffic Levels for Internet Abuse Detection." *Security Journal* 20, 22.

Savic, D. A., and G. A. Walters. 1999. "Hydroinformatics, Data Mining and Maintenance of UK Water Networks." *Anti-Corrosion Methods and Materials* 46 (6): 415–425.

Seifert, J. W. 2004. "Data Mining: An Overview." CRS Report for Congress RL31798. Congressional Research Service, Washington, DC.

Shachtman, N. 2003. "Big Brother Gets a Brain—The Pentagon's Plan for Tracking Everything That Moves." *Village Voice*, July 8.

Shaikh, M. A.; J. Wang; H. Liu; and Y. Song. 2007. "Investigative Data Mining for Counterterrorism." *Lecture Notes in Computer Science* 4413: 31–41.

Shen, D.; Q. Yang; J.-T. Sun; and Z. Chen. 2006. "Thread Detection in Dynamic Text Message Streams." Paper presented at the Special Interest Group on Information Retrieval (SIGIR) 2006, Seattle, August 6–11.

Shen, S. Y., and M. J. Shaw. 2004. "Managing Coordination in Emergency Response Systems with Information Technologies." In *Proceedings of the Tenth Americas Conference on Information Systems.* New York, August.

———. 2005. "Using Information Technology for Effective Emergency Response." PowerPoint presentation. International Business Research Forum, www.fox.temple.edu/conferences/ibrf/2005/documents/ShawandShen.ppt.

Shmueli, G.; N. R. Patel; and P. C. Bruce. 2007. *Data Mining for Business Intelligence: Concepts, Techniques, and Applications in Microsoft Office Excel with XLMiner.* Hoboken, NJ: John Wiley & Sons.

Sickinger, L. 2005. "Effectiveness of Non-Lethal Capabilities in a Maritime Environment. Naval Postgraduate School." Naval Postgraduate School, Monterey, CA.

Sinai, J. 2004. "Future Trends in Worldwide Maritime Terrorism." *Quarterly Journal of the Consortium of Defense Academies and Security Studies Institutes* 3 (1).

Singel, R. 2007. "Point, click . . . eavesdrop: How the FBI Wiretap Net Operates." *Wired,* August 29.

Smith, B. L., and S. Babiceanu. 2004. "Investigation of Extraction, Transformation, and Loading Techniques for Traffic Data Warehouses." *Transportation Research Record* 1879: 9–16.

Smith, P. J.; M. Klopfenstein; J. Jezerinac; S. Augustine; R. Chapman; J. Obradovich, et al. 2000. "Detection and Quantification of Inefficiencies in the National Airspace System." In *Human Factors and Ergonomics Society Annual Meeting Proceedings,* Proceedings 6—Special Sessions.

Sokolova, M. V., and A. Fernandez-Caballero. 2009. "Modeling and Implementing an Agent-Based Environmental Health Impact Decision *Support System.*" *Expert Systems with Applications* 36 (2).

Solomon, J., and B. Gellman. 2007. "Frequent Errors in FBI's Secret Records Requests: Audit Finds Possible Rule Violations." *Washington Post,* March 9.

Solove, D. J. 2004. *The Digital Person: Technology and Privacy in the Information Age.* New York: New York University Press.

———. 2007. *The Future of Reputation: Gossip, Rumor, and Privacy on the Internet.* New Haven: Yale University Press.

Song, J.; H. Ohba; H. Takakura; Y. Okabe; K. Ohira; and Y. Kwon. 2007. "A Comprehensive Approach to Detect Unknown Attacks via Intrusion Detection Alerts." *Lecture Notes in Computer Science* 4846: 247–253.

Srivastava, J. 2002. "Data Mining for Information Vigilance." Paper presented at the IEEE International Conference on Data Mining, Workshop on Data Mining for Cyber Threat Analysis, Maebashi City, Japan.

Stanley, J. 2004. "The Surveillance-Industrial Complex: How the American Government Is Conscripting Businesses and Individuals in the Construction of a Surveillance Society." ACLU.

Stanley, J., and B. Steinhardt. 2003. "Bigger Monster, Weaker Chains: The Growth of an American Surveillance Society." ACLU, New York, January.

Sterling, T. R.; D. Thompson; R. L. Stanley; P. D. McElroy; A. Madison; K. Moore, et al. 2000. "A Multi-State Outbreak of Tuberculosis Among Members of a Highly Mobile Social Network: Implications for Tuberculosis Elimination." *International Journal of Tuberculosis and Lung Disease* 4 (11): 1066–1073.

Suskind, R. 2006. *The One Percent Doctrine: Deep Inside America's Pursuit of Its Enemies Since 9/11.* New York: Simon and Schuster.

Swartz, N. 2004. "Anti-Crime Database Losing Support." *Information Management Journal* 38 (3).

Sylvester-Evans, R., and D.D. Drysdale. 1998. "A Case Study: The Piper Alpha Explosion and Fire." *Philosophical Transactions of the Royal Society*, London.

Tachibana, Y., and M. Ohnari. 1999. "Prediction Model of Hourly Water Consumption in Water Purification Plant Through Categorical Approach." Paper presented at the IEEE International Conference on Systems, Man, and Cybernetics.

Tadesse, T.; J. F. Brown; and M. J. Hayes. 2005. "A New Approach for Predicting Drought-Related Vegetation Stress: Integrating Satellite, Climate, and Biophysical Data over the U.S. Central Plains." *ISPRS Journal of Photogrammetry and Remote Sensing* 59 (4): 244–253.

Taft, J. 2006. "The Intelligent Power Grid." IBM. http://cleanenergytransmission.org/uploads/en_intelligent_power_grid_final_a.pdf.

Tan, P.-N.; M. Steinbach; and V. Kumar. 2006. *Introduction to Data Mining*. Boston: Pearson Addison Wesley.

Tangredi, S. J. 2002. *Globalization and Maritime Power.* Washington, DC: National Defense University Press.

Technology and Privacy Advisory Committee (TAPAC). 2004. "Safeguarding Privacy in the Fight Against Terrorism." Executive Summary, March, Washington, DC.

Thongtae, P., and S. Srisuk. 2008. "An Analysis of Data Mining Applications in Crime Domain." Paper presented at the IEEE Eighth International Conference on Computer and Information Technology.

Thuillot, W.; J. Berthier; A. Sarkissian; A. M. Mickaelian; L. A. Sargsyan; J. Iglesias, et al. 2006. "Massive Physical and Dynamical Characterization of Asteroids." *Proceedings of the International Astronomical Union* 2 (14): 616.

Thuraisingham, B. 2004. "Data Mining for Security Applications." In *Intelligence and Security Informatics* (ISI). IEEE International Conference, Taipei, June.

Tiburcio, F. M. 2005. "Maritime Protection of Critical Infrastructure Assets in the Campeche Sound." Master's thesis, Naval Postgraduate School, Monterey, CA.

Tilak, S.; P. Hubbard; M. Miller; and T. Fountain. 2007. "The Ring Buffer Network Bus (RBNB) Dataturbine Streaming Data Middleware for Environmental Observing Systems." Paper presented at the Third IEEE International Conference on e-Science and Grid Computing (e-Science 2007), Bangalore, India, December 10–13.

Torero, J. L.; J. G. Quintiere; and T. Steinhaus. 2002. "Fire Safety in High-Rise Buildings: Lessons Learned from the WTC." Paper presented at the Jahresfachtagung der Vereingung zur Forderrung des Deutschen Brandschutzez, Dresden.

Trafalis, T. B., and A. White. 2003. "Data Mining Techniques for Pattern Recognition: Tornado Signatures in Doppler Weather Radar Data." *International Journal of Smart Engineering System Design* 5 (4): 347–359.

Trostle, J. A.; A. Hubbard; J. Scott; W. Cevallos; S. J. Bates; and J. N. S. Eisenberg. 2008. "Raising the Level of Analysis of Food-Borne Outbreaks: Food-Sharing Networks in Rural Coastal Ecuador." *Epidemiology* 19 (3): 384–390.

Tsai, F. S., and K. L. Chan. 2007. "Detecting Cyber Security Threats in Weblogs Using Probabilistic Models." *Lecture Notes in Computer Science* 4430: 46–57.

Turney, P. 2000. "Types of Cost in Inductive Concept Learning." Paper presented at the Workshop on Cost-Sensitive Learning at ICML-2000. Stanford: Stanford University.

Turoff, M. 2002. "Past and Future Emergency Response Information Systems." *Communications of the ACM* 45 (4): 29–32.

Turoff, M.; M. Chumer; B. Van de Walle; and X. Yao. 2004. "The Design of a Dynamic Emergency Response Management Information System (DERMIS)." *Journal of Information Technology Theory and Application* 5 (4): 1–36.

Uddameri, V. 2007. "Bayesian Analysis of Groundwater Quality in a Semi-Arid Coastal County of South Texas." *Environmental Geology* 51 (6): 941–951.

U.S. Department of Energy (DOE). 2003. " 'Grid 2030': A National Vision for Electricity's Second 100 Years." Office of Electric Transmission and Distribution, Washington, DC.

U.S. Department of Homeland Security (DHS). 2002. "The National Strategy for Homeland Security." Washington, DC. www.dhs.gov/xlibrary/assets/nat_strat_hls.pdf.

———. 2003. "National Strategy for the Physical Protection of Critical Infrastructures and Key Assets." Washington, DC.

———. 2004. "National Incident Management System." Washington, DC. www.dhs.gov/xlibrary/assets/NIMS-90-web.pdf.

———. 2005a. "National Plan to Achieve Maritime Domain Awareness." Washington, DC.

———. 2005b. "The National Strategy for Maritime Security." Washington, DC. www.dhs.gov/xlibrary/assets/HSPD13_MaritimeSecurityStrategy.pdf.

———. 2006. "Data Mining and Analysis Tools: Technology Review and Product Selection Guide. System Assessment and Validation for Emergency Responders (SAVER)." Washington, DC.

U.S. Fire Administration (ASFA). 2004. "Attacking the Violent Crime of Arson: A Report on America's Fire Investigation Units." U.S. Department of Homeland Security, Washington, DC.

Usmani, A. S.; Y. C. Chung; and J. L. Torero. 2003. "How Did the Towers Collapse—A New Theory." *Fire Safety Journal* 38.

Vaseashta, A.; M. Vaclavikova; S. Vaseashta; G. Gallios; P. Roy; and O. Pummakarnchana. 2007. "Nanostructures in Environmental Pollution Detection, Monitoring, and Remediation." *Science and Technology of Advanced Materials* 8 (1–2): 47–59.

Velastin, S. A., and P. Remagnino. 2006. *Intelligent Distributed Video Surveillance Systems.* London: Institution of Electrical Engineers.

Venkatasubramanian, M.; M. Kezunovic; and V. Vittal. 2008. "Detection, Prevention and Mitigation of Cascading Events: Power Systems Engineering Research Center." PSERC Report #08–05.

Virdhagriswaran, S., and G. Dakin. 2006. "Camouflaged fraud detection in domains with complex relationships." In *KDD'06 Proceedings of the Twelfth ACM SIGKDD International Conference on Knowledge Discovery and Data Mining,* New York.

Voisin, B.; A. Micol; S. O'Tuairisg; R. F. Butler; A. Golden; and A. Shearer. 2004. "Simulating Instruments for Mining Uncalibrated Archives." In *Proceedings, Vol. 5493, Optimizing Scientific Return for Astronomy through Information Technologies,* ed. Peter J. Quinn and Alan Bridger, 294–301. Bellingham, WA: SPIE.

Waissman, J.; R. Sarrate; T. Escobet; J. Aguilar; and B. Dahhou. 2000. "Wastewater Treatment Process Supervision by Means of a Fuzzy Automation Model." In Proceedings of the 2000 IEEE Symposium on Intelligent Control, 163–168. Rio Patras, Greece.

Wallace, R. 1993. "A Fractal Model of HIV Transmission on Complex Sociogeographic Networks: Towards Analysis of Large Data Sets." *Environment and Planning A* 25 (1): 137–148.

Wang, J.; Z. Ma; and L. Li. 2005. "Detection, Mining and Forecasting of Impact Load in Power Load Forecasting." *Applied Mathematics and Computation* 168 (1): 29–39.

Wang, J. H.; P. S. Deng; Y. S. Fan; L. J. Jaw; and Y. C. Liu. 2003. "Virus Detection Using Data Mining Techniques." In *Proceedings of the IEEE 37th Conference on Security Technology,* 71–76, Tao-Yuan, Taiwan.

Wang, X. L., and Z. J. Yin. 2007. "Artificial Immune Recognition System as a New Classifier for Reservoir Operating Rules Extraction." Paper presented at IIHMSP'07, Third International Conference on Intelligent Information Hiding and Multimedia Signal Processing, Kaohsiung, Taiwan, November 26–28.

Wang, Z. W.; M. Zhang; S. S. Ma; D. Y. Li; and D. Li. 2004. "Analysis of Monitoring Data of Landslide in Geheyan Reservoir Using Data Mining Algorithm." In *Proceedings of the Fourth International Conference on Dam Engineering,* ed. Martin Wieland, Qingwen Ren, and John S. Y. Tan. London: Taylor & Francis.

Wehenkel, L.; C. Lebrevelec; M. Trotignon; and J. Batut. 1999. "Probabilistic Design of Power-System Special Stability Controls." *Control Engineering Practice* 7 (2): 183–194.

Weigang, L.; Y. Yamashita; M. W. Koendjbiharie; R. C. De Moura Juca; and A. MacIver. 2004. "The Development and Implementation of the Operation System and Data Bank for the Intelligent Transportation System—Sitcuo." *Journal of Advanced Transportation* 38 (2): 163–186.

Weiss, S. M., and N. Indurkhya. 1998. *Predictive Data Mining: A Practical Guide.* San Francisco: Morgan Kaufmann.

Weiss, W. W.; S. Wo; J. W. Weiss; and J. Weber. 2001. "Data Mining at a Regulatory Agency to Forecast Waterflood Recovery." Paper presented at the SPE Rocky Mountain Petroleum Technology Conference, Keystone, CO, May 21–23.

Westin, A. F., and M. A. Baker. 1972. *Databanks in a Free Society; Computers, Record-Keeping, and Privacy.* New York: Quadrangle Books.

Whitehorn, M. 2006. "The Parable of the Beer and Diapers: Never Let the Facts Get in the Way of a Good Story." *A Register,* August 15. www.regdeveloper.co.uk/2006/08/15/beer_diapers/ (accessed August 26, 2007).

WinterCorp. 2007. "Winter TopTen Programs." www.wintercorp.com/VLDB/2005_TopTen_Survey/TopTenProgram.html (accessed April 20, 2007).

Witten, I. H., and E. Frank. 2005. *Data Mining: Practical Machine Learning Tools and Techniques* (2d ed.). Boston, MA: Morgan Kaufman.

Wong, Y. K., and W. L. Woon. 2008. "An Iterative Approach to Enhanced Traffic Signal Optimization." *Expert Systems with Applications* 34 (4): 2885–2890.

Wood, D. M. 2006. "A Report on the Surveillance Society." [UK] Information Commissioner's Office, London.

Xu, J. J., and Chen, H. 2005a. "Crimenet Explorer: A Framework For Criminal Network Knowledge Discovery." *ACM Transactions on Information Systems,* 23, 2 (April).

———. 2005b. "Criminal Network Analysis and Visualization: A Data Mining Perspective." *Communications of the ACM* 48 (6): 101–107.

Xu, K.; Z. L. Zhang; and S. Bhattacharyya. 2008. "Internet Traffic Behavior Profiling for Network Security Monitoring." *IEEE/ACM Transactions on Networking* 16 (6).

Xue, Y. 2006. "Space-Time Cooperative Framework for Defending Blackouts. Part I: From Isolated Defense Lines to Coordinated Defending." *Dianli xitong zidonghua* [Automation of Electric Power Systems] 30 (1): 8–16.

———. 2007. "Towards Space-Time Cooperative Defense Framework Against Blackouts in China." Paper presented at IEEE Power Engineering Society General Meeting, Tampa, FL, June 24–28.

Yamanishi, K.; J. I. Takeuchi; and Y. Maruyama. 2005. "Data Mining for Security." *NEC Journal of Advanced Technology* 2 (1): 63–69.

Yan, P.; R. P. Biuk-Aghai; S. Fong; and Y. W. Si. 2007. "Detection of Suspicious Patterns in Secure Physical Environments." 2007 IEEE Conference on e-Technology, June, Tokyo.

Yang, B.; Y. Tao; J. Sui; and F. Zhang. 2007. "Model Architecture of Intelligent Data Mining Oriented Urban Transportation Information." In *Proceedings,* vol. 6753, *Geoinformatics 2007: Geospatial Information Science,* ed. Jingming Chen, and Yingxia Pu. SPIE, Toronto.

Yang, C., and S. Letourneau. 2005. "Learning to Predict Train Wheel Failures." In *KDD'05 Proceedings of the Eleventh ACM SIGKDD International Conference on Knowledge Discovery in Data Mining.* New York: ACM.

Yang, C. C., and T. D. Ng. 2007. "Terrorism and Crime Related Weblog Social Network: Link, Content Analysis and Information Visualization." Paper presented at 2007 IEEE Intelligence and Security Informatics, New Brunswick, NJ, May 23–24.

Yang, L., and L. Jia. 2004. "A Fuzzy Modeling Method Using Rough Sets and Its Applications to Intelligent Transportation Systems." Paper presented at WCICA'04, Fifth World Congress on Intelligent Control and Automation, Hangzhou, China.

Yang, Q., and X. Wu. 2006. "10 Challenging Problems in Data Mining Research." *International Journal of Information Technology and Decision Making* 5 (4).

Yang, Y. B.; N. Li; and Y. Zhang. 2008. "Networked Data Mining Based on Social Network Visualizations." *Ruanjian xuebao* [Journal of Software] 19 (8): 1980–1994.
Yin, Z. J.; X. L. Wang; T. S. Hu; and Y. Q. Wu. 2006. "Water Supply Reservoir Operating Rules Extraction Based on Data Mining." *Xitong gongcheng lilun yu shijian* [System Engineering Theory and Practice] 26 (8): 129–135.
Yoshida, T. 2008. "Preparedness Response to Hazard and Toxic Incidents, and Food Terrorism." *Yakugaku zasshi* 128 (6): 851–857.
Yoshikura, H. 2005. "HIV Transmission Webs: HIV Infection Trends in Japan in 1989–2004." *Japanese Journal of Infectious Diseases* 58 (6): S19–S21.
Young Moon, C.; H. Seung Hee; C. Kyoung Won; L. Dong Ha; and J. Sun Ha. 2001. "Data Mining Approach to Policy Analysis in a Health Insurance Domain." *International Journal of Medical Informatics* 62 (2–3): 103–111.
Yu, D.; P. L. Gu; and C. G. Li. 2004. "Railway Freight Transportation Safety Data Management System (TMIS) Based on Data Exploration Technology." *Zhongguo tiedao kexue* [China Railway Science] 25 (2): 114–116.
Yuan, B.; X. Z. Wang; and T. Morris. 2000. "Software Analyser Design Using Data Mining Technology for Toxicity Prediction of Aqueous Effluents." *Waste Management* 20 (8): 677–686.
Yuan, X.; Z. Sun; Y. Varol; and G. Bebis. 2003. "A Distributed Visual Surveillance System." Paper presented at the IEEE International Conference on Advanced Video and Signal Based Surveillance, Miami.
Zadrozny, B., and C. Elkan. 2001. "Learning and Making Decisions When Costs and Probabilities Are Both Unknown." Paper presented at the Seventh ACM SIGKDD International Conference on Knowledge Discovery and Data Mining, San Francisco, August 26–29.
Zhang, C.; H. C. Zhou; and W. Li. 2008. "River Flow Forecasting Model Based on Fuzzy Inference and Associated Rules Analysis." *Dalian ligong daxue xuebao* [Journal of the Dalian University of Technology] 48 (2): 263–269.
Zhang, Z.; J. J. Salerno; and P. S. Yu. 2003. "Applying Data Mining in Investigating Money Laundering Crimes." Paper presented at the Ninth ACM SIGKDD International Conference on Knowledge Discovery and Data Mining, Washington, DC, August 24–27.
Zhao, J.; C. T. Lu; and Y. Kou. 2003. "Detecting Region Outliers in Meteorological Data." Paper presented at the Eleventh ACM International Symposium on Advances in Geographic Information Systems, New Orleans, November 7–8.
Zhou, Y., and H. Lu. 2005. "Mobile Transportation Information Service System: The Architecture and Its Technical Solution." Paper presented at Intelligent Transportation Systems, Proceedings IEEE, Hong Kong, March.
Zimmerman, R., and T. A. Horan. 2004. *Digital Infrastructures: Enabling Civil and Environmental Systems Through Information Technology.* New York: Routledge.

CHAPTER 9

# Can Catastrophes Lead to Good Policy?

JEFFREY D. STERN

*One can only imagine what it must have been like at that moment frozen in time, the immediate and unfolding horror as a beautiful fall day on the Mediterranean coast was transformed into the dark, ash-filled sky. We have all seen the pictures of the stone bodies caught in that moment, mouths agape in their anguish. A moment from 79 C.E., when the world stopped and Mount Vesuvius erased Pompeii in a shower of volcanic ash and lava.*

From the destruction of Sodom and Gomorrah to the flood of Noah, the first book of the Bible recounts stories of catastrophe. The eruption of Mount Vesuvius was one of the earliest human calamities to be recorded and rediscovered by archaeologists. Human history is marked by these moments of disaster and rebirth, and in the aftermath of every catastrophe a common refrain is articulated: never again, in instances where human intervention may prevent or mitigate future disasters. Yet within a short time people return to their routines and put aside the lessons learned from the disaster's impact.

Perhaps this reaction is an aspect of our social psychology, as survivors move on and the challenges of the brief hazard fade into the past, but a majority of recent postdisaster routines can be attributed to our political environment, in the United States and around the world. This chapter explores the challenge and the question of whether catastrophe ever leads to good policy and whether we, in our humanity, ever learn the lessons of the past or are condemned to repeat them.

The history of the United States is marked with many poignant instances of catastrophes: the Great Chicago Fire, the San Francisco earthquake of 1906, the 1920s dust bowl, the Alaskan earthquake of 1964, and more recent examples such as the nuclear reactor meltdown at Three Mile Island, Hurricanes Andrew and Katrina, the September 11, 2001, terror attacks, and the 2010 BP *Deepwater Horizon* explosion and oil spill. Each serves as an example of how disasters affect policy decisions enacted in its aftermath.

Rational choice dictates that postcatastrophic policy changes be optimized to prevent repetition of the same calamity. However, the reality is that usually we end up with suboptimal policy changes in the aftermath of an event. This is a function of the political system in the United States, a system purposely designed for incrementalism and evolutionary change rather than wholesale reengineering and revolutionary change (Lindblom 1959). Multiple stakeholders and interest groups exist in a perpetual cycle of engagement and activism in the United States, and when a disaster strikes, they are often the ones who temper the outcome of policy changes.

However, a narrow policy window opens in the aftermath of most disasters, when calls for reform arise and laws and regulations are enacted to prevent such incidents from occurring in the future. These changes are inevitably tempered by other forces, competing interests, other needs, and institutional inertia, resulting in a sort of punctuated equilibrium, an effort to implement change, and then a return to stability (Baumgartner and Jones 1993; Birkland 2006).

These forces contend in what could be termed the disaster policy domain, where an intersection takes place of the event, the attribution of cause, the agendas of the stakeholders who are affected by the event, and the policy changes that flow from it in the aftermath.

## THE EVOLUTION OF THE DISASTER POLICY DOMAIN

Four intersecting shifts in the past century have contributed to how the current disaster policy domain functions: growth in the role of government at all levels, an increase in technological hazards and disasters, development of an increasingly litigious and blame-oriented society, and an increasing media-driven politicization of disaster events.

The first social shift is the growth in government as a provider of services across society during disasters. Government at the federal, state, and local levels has grown in both size and scope of responsibility since the turn of the twentieth century, mirroring the growth in the wealth and population of the United States. The national system of disaster response in the country has similarly evolved as government at all levels became increasingly involved in public safety and disaster mitigation, prevention, response, and recovery.

At the turn of the nineteenth century, the federal government had little involvement in most disasters. After the San Francisco earthquake of 1906, local elected officials and businessmen in the ashes of the destroyed city made the decision to raze and rebuild on the very territory where the city had come crumbling and burning down. Safety and standards were traded for speed in an effort to ensure that investment and growth would continue for the city, and it would be decades before serious efforts were made to ensure that more stringent building codes were put in place.

Aside from the occasional dispatch of federal troops to preserve order, the federal role was largely reduced during catastrophes. Society accepted the occasional "act of God," and policy changes in the aftermath were few and incremental. Until the 1960s, the federal government largely stayed out of the domestic disaster business. A 1964 earthquake in Alaska changed that. The quake, which measured 9.2 on the Richter scale (at the time, the second most powerful in recorded history), destroyed much of Alaska's infrastructure. President Lyndon Johnson and his staff, Congress, and Alaskan leaders led the combined effort to rebuild the state's infrastructure, lest the whole state need to be evacuated, jeopardizing its newly designated statehood (Carafano and Weitz 2008). It was the beginning of the shift toward increased executive branch and legislative branch roles in postdisaster response and recovery.

The cold war also led to the growth in civil defense on the domestic front, and the growth of local and state governments, which were increasingly providing public safety and social services. This increased the role for government and built more interdependent relationships between government and citizens during catastrophes. Under the U.S. Constitution, domestic disaster response falls under the purview of the state governors, and they were increasingly tasked with carrying out those responsibilities.

Most policy actions that followed events were minor and often local in nature. Nevertheless, safety regulations were promulgated; building and construction standards were strengthened. More sweeping changes began to be made in the second half of the twentieth century in response to disasters, especially as more events resulted from manmade causes.

The emergence of technologies was the second social shift affecting the disaster policy domain. The incredible human achievements that coincided with the growth of the industrial economy have brought their own hazards. In the twentieth century the country's wealth and population soared. Cities grew into metropolitan regions, fueled by factory jobs, waves of immigration, and increases in the standard of living at a speed and scale unmatched in prior human history. Transportation networks spread across the continent: ship, rail, road, and air. With that technological growth came technological hazards. Toxic industrial chemicals were developed, transported, and used widely to provide resources for this new economy. Energy sources from newly discovered fossil fuel sources, and nuclear technologies led to the establishment of power plants, fuel depots, and transportation stations. Society began to depend on supply chains that provide the flow of commerce, energy, fuel, and food to support the booming population; society moved from individual to local to national to global interdependence.

For the first time in human history, the hazards associated with these changes resulted in technological disasters. Unlike the acts of God from natural causes, these various technologies and systems became the object of blame in the aftermath of an incident, subject to policy changes as lessons were learned, and increasingly regulated by government.

The ability to assign blame to forces under human control ran parallel to the development of an increasingly blame-driven society and the growth of litigation and regulation in the aftermath. This represents the third key social shift in the past century to affect the disaster policy domain. As various hazards were identified, laws were enacted and regulations produced. Incidents such as the chemical explosion on the Cuyahoga River in 1969 led to the Clean Water Act of 1972, the nuclear reactor explosion at Three Mile Island in 1979 led to changes in the Nuclear Regulatory Commission and increased regulation, as well as slowing the development of nuclear power as an energy source in the United States. Tremendous growth in civil litigation also resulted from these incidents. Other incidents also contributed to changes in postdisaster policies. Secular disasters were attributable to man, and man could be blamed when the cause and effect were assigned.

Private industry and government policies shifted in response to this increased litigation. A Google search for "disaster litigation history" provides a telling illustration of the growth in litigation in the past 150 years (see Figure 9.1).

Figure 9.1 **Disaster Litigation History: 1860 to Present**

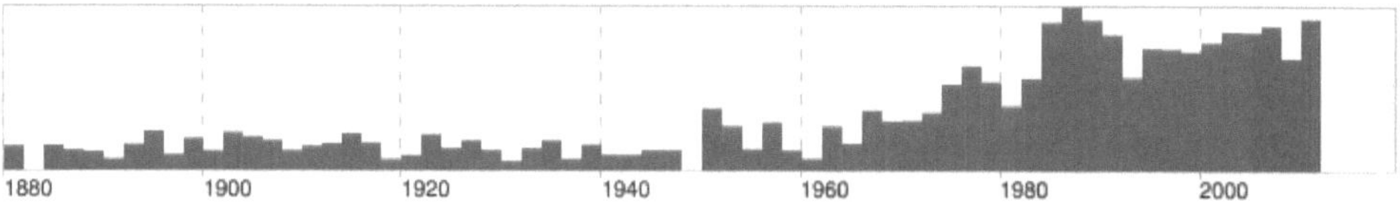


*Source:* Google Timeline, search with the words disaster, litigation, and history. http://www.google.com/search?q=disaster+litigation+history&hl=en&source=lnt&tbs=tl:1&sa=X&ei=eFxHTMqQKMGBlAfWpYHuAw&ved=0CA8QpwU.

Federal agencies such as the Environmental Protection Agency (1970) and the Federal Emergency Management Agency (1979) were created to ensure that proper regulatory regimes were developed to limit the negative effects of industrial technologies and that federal government could carry out its increased role in coordinating disaster responses. Internationally, other societies experienced these shifts as well. Incidents such as the Union Carbide leak of toxic gas and other chemicals in Bhopal, India, that killed between 4,000 and 15,000 people in 1984 played similar roles in the countries in which they occurred.

The fourth great shift has occurred in the past thirty years with the expansion of the media industry into a 24-hour news cycle and even more recently into the immediate reporting that takes place in online media. As columnist and author Thomas Friedman has described, we are now in a period when everyone with a computer blog or Facebook account and everyone with a cell-phone camera is an on-the-spot reporter (Friedman 2007). Events are depicted in real time via uploads to YouTube from cell phones with embedded video cameras. Interpretation of events is no longer controlled by an editorial staff, as it was from the outbreak of the Spanish-American War in 1898 until the beginning of the twenty-first century.

The explosion of the media cycle has occurred in symbiosis with the politicization of disaster response over the past thirty years. This emerging phenomenon was highlighted after in the aftermath of Hurricane Andrew in 1992, as President George H.W. Bush's administration struggled to provide support to local communities and effectively manage the public outcry in the wake of the media response to the event (Wamsley and Schroeder 1996). Despite prepositioning of equipment and resources before the storm struck Florida, the area was not ready for the powerful storm. Gary Wamsley, Professor Emeritus of public administration, dubbed the media frenzy that occurred in the immediate aftermath "the CNN syndrome," epitomized when an exasperated local emergency manager asked while on air, "Where the hell is the cavalry on this one?" (Wamsley and Schroeder 1996). From this time forward, disasters in the United States of various proportions have drawn presidential attention, if not direct presidential leadership, and response is exercised in a political context. At the same time, through various disasters, policy changes have been enacted to change the response system. Yet few presidents have escaped the challenge posed in the modern era, when public expectations of government response have been difficult, if impossible, to meet as disasters unfold in real time. (September 11, 2001, serves as noteworthy milestone not only for the horrific act it represents but also as the first time that millions of people around the world witnessed live on TV the mass murder of 3,000 people, including the act of destruction as the second aircraft struck Tower 2 of the World Trade Center at 9:03 A.M. Eastern Standard Time. Part of the government's response was also broadcast live, as the explosions over Baghdad were transmitted all over the world.) In a society that has shifted from self-reliance to dependence on the supply chains and systems of private and public services that provide sustenance for the routines of everyday life, meeting public expectations is an almost impossible task.

After the outcry for help from local officials during Hurricane Andrew, President George H.W. Bush dispatched his transportation secretary, Andrew Card, to oversee and coordinate the federal response; this political leadership team side-stepped FEMA's bureaucratic processes that were already in place and may have actually delayed an effective response but was an essential part of the president's effort to demonstrate command of the situation. Despite these efforts at political management of the aftermath of the hurricane, the storm is widely thought to have served as a tipping point, helping Bill Clinton to win the election as president.

Several policy changes were called for in the wake of Hurricane Andrew, notably reforms at FEMA recommended in the National Academy of Public Administration report (National Academy of Public Administration 1993). Congress failed to pass legislation enacting these reforms, but some of them were adopted by the Clinton administration. One notable recommendation that was not adopted was reduction in the number of political appointees to FEMA.

To illustrate how politicized disasters have become, Figure 9.2 shows FEMA's distribution of disaster aid from 2000 to 2008. A quick glance might cause one to wonder whether anywhere in the United States is safe, but the table makes much more sense when it is overlaid on congressional districts and with the understanding that the map is a reflection of increased disbursement of federal money and resources that has come to dominate response and recovery. It, therefore, reflects both the increased federal role in disasters from the second social shift as well as the politicization of the disaster policy domain.

Andrew Card later served as chief of staff to President George W. Bush in 2004, when a series of four hurricanes crisscrossed Florida in that presidential election year. The president's brother, Jeb Bush, was governor of Florida at the time, and the storms attracted a high level of political attention, including trips from the president in the aftermath. In this instance, the politics of the disaster were properly managed, and this President Bush was reelected. One year later, the same disaster response system failed, operationally and politically, as Hurricane Katrina rolled across South Florida, up the Gulf of Mexico, and into New Orleans.

Katrina was not the first catastrophe faced by President George W. Bush. The September 11, 2001, terrorist attacks launched by al Qaeda and the administration's response to them resulted in the largest policy changes in U.S. national security policy since the creation of the Department of Defense (in 1949), the Central Intelligence Agency, and the National Security Council (both in 1947). The policy window created by 9/11 led to the creation of the Department of Homeland Security, the passage of the Patriot Act, and to the preemptive war strategies promulgated through the Global War on Terror, primarily in Afghanistan and

Figure 9.2 **Presidential Disaster Declarations, January 10, 2000 to January 8, 2008**

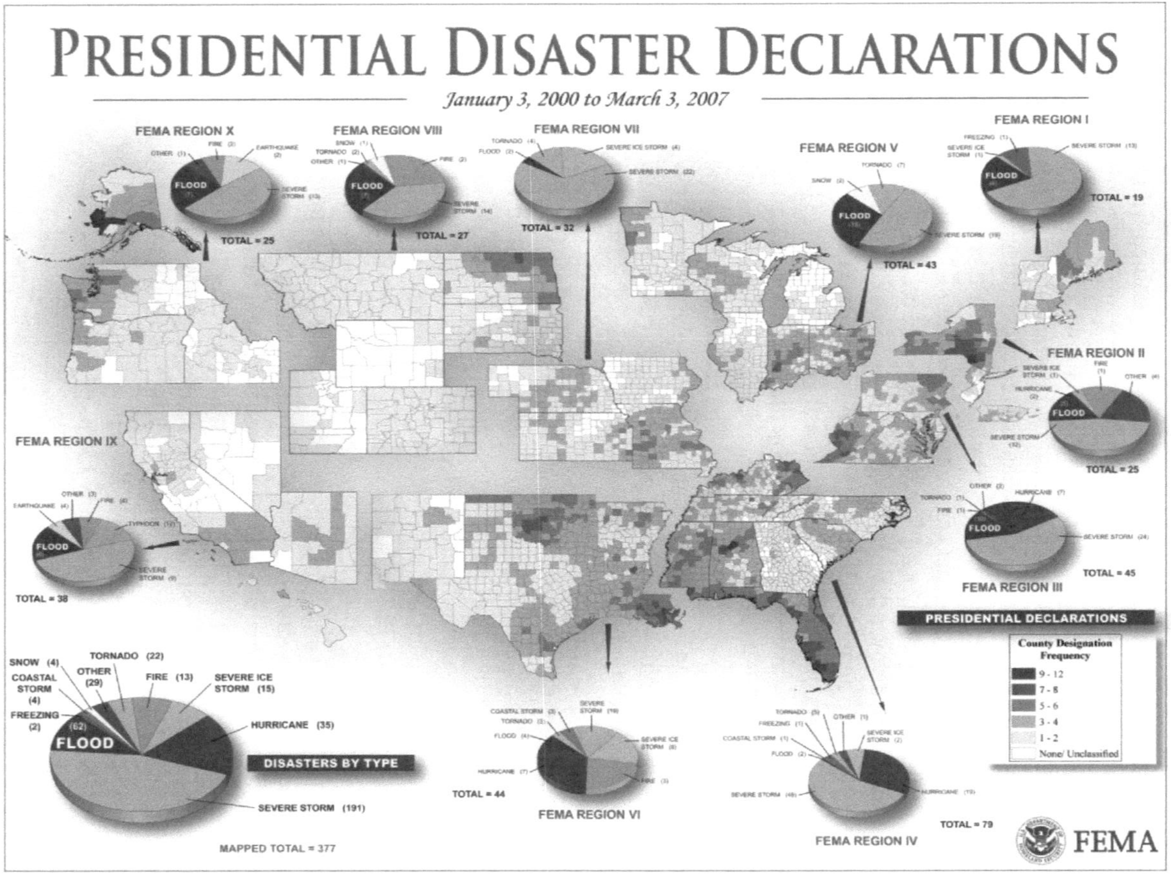


*Source:* FEMA, http://www.fema.gov/pdf/hazard/map/declarationsmap2000_07.pdf.

Iraq but also elsewhere. The long-term outcome of these efforts remains uncertain, and the administration of Barack Obama retained some of the policies while changing others. Ongoing efforts to counter Islamic terrorism have met with short-term success, but terror attacks continue worldwide at an accelerated rate, and huge challenges to U.S. hegemony have developed as diplomatic relations have become strained with long-standing allies. The asymmetric threat posed by the possibility that a group of terrorists could obtain a weapon of mass destruction remains one without the coherent strategy that existed during the cold war. The policies of containment and mutually assured destruction of that era prevented the world from nuclear holocaust. Since political and economic forces caused the collapse of the Soviet Union, the strategy has been less clear. The lack of a grand strategy to combat terrorism that can transcend presidential politics and administrations over time is something the United States continues to seek. Al Qaeda's grand strategy to recreate the caliphate over the next hundred years serves as a call to action, and, as has been said, we have all the clocks, but they have all the time.

In his first years in office, President Barack Obama faced the fiscal meltdown of the economy (which began before he even took the oath of office), a crisis in Haiti with the death of more than 300,000 people in the aftermath of the earthquake of 2010, and a catastrophic oil spill in the Gulf of Mexico. The administration's handling of these events dissipated confidence in its ability to manage catastrophes. Perhaps the greatest lesson here is that disasters are bipartisan in their political damage—a lesson that should rationally lead to efforts to swing the pendulum back to a depoliticized approach, however wistful that desire may be in reality.

The interplay of federal, state, and local government and the growth of private industry responsible for many technology-related hazards leave us with a multilayered system of policy promulgation that complicates implementation of positive changes that are attempted in the aftermath. In the United States, local officials remain responsible for things like building codes, for the allocation of emergency response resources, and for preparedness programs. Public safety is a local issue, and the players involved at any local level are the same as in 1906. Confronted with a patchwork of 87,000 local government entities across the United States, from county, city, and town governments, to school boards, to utility districts, to fire districts, to others, efforts to build a unified and rational policy in the aftermath are fragmented by the nature of our governance. A lesson learned in one community must be learned in 87,000 others. We must accept this sloppy mosaic that is the result of our democratic system and understand that it can result in a levee failure in New Orleans during Katrina. But that failure cannot be wholly attributed to one fixed cause among the myriad cities, parishes, levee boards, and other governing authorities, engineering firms, or Army Corps of Engineers—each with its own varying roles, responsibilities, finances, accountability, and capabilities. All own some of the thousand points of minor failure that lead to the greater catastrophe. Our challenge is to improve things at a systemic level, across all these levels. It is a challenge of federalism and freedom of choice in society as much as a challenge of finding the "right" policy.

As Charles Perrow (1984) cited in his seminal work, *Normal Accidents,* highly interdependent complex and coupled technological systems will fail at multiple points. Even the safety systems will fail with unintended consequences. These failures are a predictable, normal result of complex technologies, and his theory could be equally applied to complex and interdependent policymaking and implementing systems in the policy domain generally and in the disaster policy domain specifically. The levees designed for safety from routine floodwaters thus failed in the midst of a hurricane, leading to a catastrophe that killed 1,800 people and eroded confidence in American governance along with the foundations of the houses in the Ninth Ward. This failure resulted from the complexities of our structures of governance and accountability as much as from flaws in engineering design or construction techniques. Certainly, these failures continue, despite our best efforts to enact change in the aftermath.

Claire Rubin and colleagues created a Disaster Time Line series for the United States that illustrates the interrelated disasters and policy changes that resulted from their occurrence between 1988 and 2008 (Ruben, Renda-Tanali, and Cumming 2009a). They illustrate 23 catastrophic events that led to 11 major reports, 23 separate federal statutes, 17 presidential executive orders, 29 new or rewritten national strategic guidance documents, and 17 government reorganizations, including the creation of the Department of Homeland Security in 2003. Their analogous Terrorism Time Line focuses just on the events of September 11, 2001, and the October 2001 anthrax attacks, as well as Hurricanes Katrina, Rita and Wilma in 2005. These two periods and five events led to 23 policy reports, 24 federal statutes, 36 presidential executive orders (excluding secret, classified executive orders), 38 national strategic documents (again, excluding secret documents created for the Global War on Terror), and 14 government reorganizations (Ruben, Renda-Tanali, and Cumming 2009b). There does not seem to be any shortage of policymaking in the aftermath. As Figure 9.3 shows, this policymaking is further complicated by the transfer of oversight to the entire federal homeland security and emergency management complex, which is carried out through a complex web of almost 90 congressional oversight committees, many with overlapping jurisdictions and competing agendas. Successfully navigating any policy changes through the disaster policy domain is very complex.

The international disaster policy domain also leads to suboptimal unintended outcomes as they interact with political and social events. The League of Nations is an example of this dynamic: It was designed to prevent all wars, but quickly failed. Its successor, the United Nations, may have helped to prevent nuclear war between superpowers, yet the world continues to have more low-intensity conflicts on an ongoing basis than ever before. It remains unclear whether the UN will be able to act as a relief valve when the world faces the added threat of chemical or biological attacks by a small state or even a nonstate actor, notwithstanding its large standing military force, the world's second largest, after the U.S. military. The threat of such weapons is no longer the stuff of science fiction (Graham and Talent 2008). We have only begun to grasp the impact of asymmetric warfare in this new technological world, but we have yet to grasp the impact of asymmetric politics that play out on the global stage and within which policies must be enacted, in which one person with a camera can incite a revolution. The pictures of abuses at Abu Ghraib prison come to mind as asymmetric victories for an enemy from self-inflicted wounds that do irreparable harm.

Figure 9.3 **Congressional Oversight of DHS**

In the 110th Congress, 108 committees and subcommittees oversee the Department of Homeland Security.

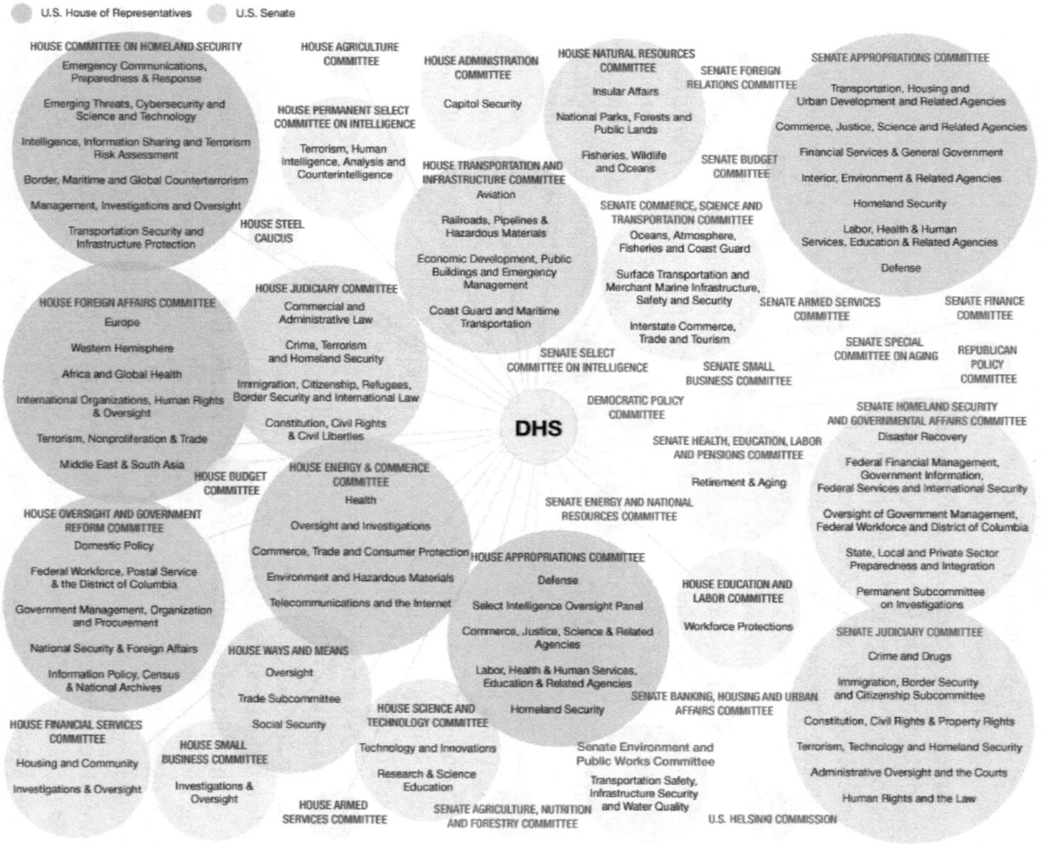


*Source:* Adrienne Wollman, National Public Radio. http://www.npr.org/templates/story/story.php?storyId=128650264.

Internationally, the system of government varies widely—from authoritarian regimes of socialized nations, centralized to decentralized, from governed to ungoverned, from rich to poor and from well-resourced to poorly resourced—and so does the ability to cope with calamity. In China, after the 2008 Sichuan earthquake, which claimed 69,000 lives and left 4.8 million homeless, largely from the collapse of poorly constructed buildings, the central government elected to move the city of Wenchuan to a different location. Haiti, already the poorest country in the Western Hemisphere, was left with no defined leadership by the earthquake of 2010. Despite the efforts of numerous humanitarian organizations, the opportunity to use the catastrophe and the world's attention to force drastic underlying changes to improve Haitian society was wasted.

## NEWTON'S THIRD LAW: ENACTING POLICY CHANGE IN THE AFTERMATH

Newton's Third Law of Motion, commonly phrased as "for every action, there is an equal and opposite reaction" serves as good jumping-off point to explain policy change in the aftermath of a calamity. Catastrophes provide a unique policy window: a period of opportunity within the disaster policy domain when the executive and legislative agendas are aligned with public desire, resulting in a flurry of reaction to whatever calamity has become the issue of focus (Sabatier 2007). Such actions typically result in one of several kinds of policy changes: policies designed to repair damage from the disaster, policies designed to prevent similar disasters from occurring again in the future, policies designed to add government regulation to the policy area of the disaster, and, significantly, policies directed at reorganization of the government institutions that were designed to prevent or respond to whatever occurred.

Hurricane Katrina dealt a severe blow to the legacy of President George W. Bush, even though it resulted from failure at all levels of government (local, state, and federal) and from the absorption of the emergency management system into the newly created homeland security system. The failure of the federal government and the challenges within FEMA and the Department of Homeland Security (DHS) serve as an excellent example of a policy change (post-9/11) that was imperfectly implemented, resulting in an unintended consequence, in this case the weakening of the federal emergency management system. The fact that a hurricane striking New Orleans was one of the existing National Planning Scenarios at the DHS highlights the challenges of implementing effective policy outcomes from policy changes. In the wake of Katrina, the emergency management system reasserted itself and reform implemented at the DHS through the Post Katrina Emergency Management Reform Act: Punctuated equilibrium and institutional forces illustrate Newton's Third Law at work. The policy window opens, changes are enacted, the reorganized system stabilizes itself based on the new arrangements, but with much of the old institutional way of working. Action and reaction, until the next event, when the forces and inherent tensions within the system clash.

In the quest for more rational policy outcomes after disasters, we often fail to account for the stakeholders with different policy agendas and perspectives who exist within the policy domain around the catastrophe. Classic public administration and political science have long-standing literature on these players and the way in which they seek to set and influence the policy agenda. During these focusing events that were created by the given disaster, these players rally to influence the interpretation of the event, to advocate for changes (or prevent changes) that will support their agendas (or negatively affect them). In the immediate period of response to a disaster, competition among the stakeholders begins to develop the story that will be told about the event, about the players, and where blame resides (Birkland 2006). As indicated, this effort to frame the interpretation of events has become more difficult and accelerated and unfocused with the explosion of the 24-hour media cycle and the immediacy of the online new media environment in the past decade. Incident managers, policymakers, public and private organizations that are involved in the incident, even nonprofit organizations, now have large staffs of public affairs and public communications specialists whose sole role is to furnish the talking points and storylines to fight for the interpretation of unfolding events. These public affairs and media relations staff have little role in actual response and exist solely to provide information and their version of the storyline for the viral media. For the incident managers, the role of the staff is to provide accurate information as quickly as possible and to amass information on how the rest of the media-political-policy sphere is interpreting the events, giving insight into gaps and vulnerabilities in the response. For the political and organizational leadership, the goal is to provide the image of a strong, unified, robust, and effective response. For the private sector, the goal is to ensure positive images of whatever product or service is involved or to deflect blame if their product or service is involved in the incident. For the nonprofits, the goal is often to highlight their role in responding to the disaster while engaging in fundraising to support ongoing relief operations. This is the outcome of the third social shift discussed earlier, the media-driven politicization of disasters, in which the perception of response outweighs, in many cases, the actual performance of the responders.

Throughout all these disasters, the stakeholders in the disaster policy domain remain in action. A key example is the policy debate around the National Flood Insurance Program, a core component of hurricane protection along the Gulf Coast and around the country. This program is designed to develop an insurance pool for catastrophic flooding yet runs a multibillion-dollar deficit. The policy changes necessary to correct this problem involve an interplay between the federal government, the states, the private insurance industry (which does not wish to take on the burden of paying for extreme catastrophes), and private landowners, who are forced to pay the insurance premiums if their homes are deemed to stand in historic flood zones as mapped by FEMA. The program remains an imperfect design, yet forces that would be necessary for major insurance reform remain locked in political struggle, and Congress is forced to extend the program incrementally on an ad hoc basis without fixing the overall problem of insurance reform for disasters.

Similarly, the BP oil spill resulted in rapid organizational reforms to the Minerals Management Service at the Department of the Interior, increased calls for regulation of offshore oil drilling, and a temporary moratorium on drilling, as the safety and environmental policy domains clashed with the energy policy domain, with each domain representing the interests and agendas of dozens or hundreds of interest groups and stakeholders. Coalitions of advocates for or against certain reforms lobby in their own best interests (Sabatier 2007). Policymaking has been described as a stream. This policy stream meanders on a downward trajectory toward the next set of rapids, which will change and influence its direction, but it never stop it in one final place, always shaped by the agendas of the interest groups that represent the terrain through which it flows.

## THE CHALLENGE FOR THE FUTURE

Some successful policy changes have been made in the aftermath of previous disasters: stronger building codes, safety standards, engineering standards, and industrial safety regulations. Each catastrophe brings lessons—some new; some revisited.

The future poses more challenges for the disaster policy domain: Catastrophes grow more complex as technologies advance and society becomes more interdependent. The various advocacy groups with their own interests are increasingly opportunistic in their ability either to exploit or mute changes in the aftermath of an incident, often clouding efforts at making rational, well-metered changes. The policy windows that open will more often need to be explosively fractured by dynamic events to create the momentum for punctuated change, and those changes will always result in imperfect implementation and institutional resistance. The increasing challenges of international incidents in this new interdependent world will further exacerbate the difficulty in enacting positive policy changes. BP's role as a global company is a case in point, limiting U.S. government options, and even causing minor turmoil in public opinion between the citizens of the United States and the citizens of Great Britain. The clash of domains is not between good and evil but between competing agendas that both lay claim to the greater good. This is a classic ethical dilemma, requiring Solomon-like wisdom for policymakers to develop the right trade-offs. In the case of oil, we need to support the economy and national security against environmental conservation and fisheries. The challenge is to find a way to support both business people and fishermen in a political environment in which compromise is not an acceptable option.

Future threats, such as bioengineered weapons of mass destruction or cyberattacks, exceed the ability of any single government to design strategies for preventing catastrophe. The uncertainty around the impact of potential changes in climate or demographic shifts around the world ensures that many of the challenges in the disaster policy realm fall into the tragedy of the commons, the economists' no-man's-land where no one is accountable for the problem, but everyone contributes to the cause.

Efforts to rationalize the disaster policy system will always be tempered by various stakeholders. As in any emerging field, critical incident analysts should understand the political players and agendas involved in the particular disaster policy domain, their interplay in the aftermath, and further study how they influence efforts toward more rational policy outcomes. Analysts must understand that there are political winners and losers in any policy change, yet only through our political processes can we develop better policy outcomes. In the analysis, these stakeholders could be mapped and potential policy choices developed that provide economic incentives for more rational policy outcomes. Often, it is only through economic incentives and regulatory penalties that advances in the mitigation of hazards can be made in a democratic and heterogeneous political society. We do not have the option of being autocratic and moving cities wholesale, without undermining the fundamental nature of our political system. China can decide to move the cities near the site of the Sichuan earthquake, but it is up to the citizens of New Orleans to decide the fate of the Ninth Ward. To change that policy would be to risk our democracy and the constitutional freedoms that we enjoy—even if those freedoms result in suboptimal policy outcomes.

Moreover, we must reinforce the lesson that the field of critical incident analysis is inherently interdisciplinary. It is more than engineering or science or psychiatry or sociology or history; it draws from public administration, political science, and the law, among others. A key function of the analyst must be to understand the political and organizational context within which the specific disaster policy domain is shaped—a context characterized by cultural influences, institutions, interest groups, political design, agendas, power and influence, authority and governance. But the methods to shape policy choices in the disaster policy domain and the capacity to implement those changes so that they result in rational and positive policy outcomes are equal in importance must be explored by practitioners of critical incident analysis. The ability of society to learn from the catastrophes of the past and increase our individual and social resilience to deal with the inevitable uncertainties of the future rests on understanding this political context within which policy is shaped, as well as the administrative institutions that are designed to carry it out.

The title of this chapter is a question that we have attempted to answer by looking at some historical events within the disaster policy domain. Can catastrophes lead to good policy? Yes, they can. But those policies are developed in a political context that is part of the disaster policy domain. In that domain, policymaking develops in an unfolding interplay between various agendas and competing interests. It increasingly unfolds in an environment characterized by high public expectations, increasingly complex and unintended consequences of technological hazards, in a very litigious and blame-driven environment, and in a media environment that creates high political stakes. Society expects these events to be preventable, and in the aftermath of large disasters there will be increasing calls for policy reforms to fix the perceived problem from a regulatory or organizational reform perspective. But the challenges inherent in implementing these policies among the diverse spectrum of players ensure that we will probably always have suboptimal policy outcomes.

We do have good policies. They have helped create a safer society. Safety standards, fire codes, and seismic building engineering serve as examples. But these policy and regulatory changes were enacted when the cause and effect were much simpler to diagnose, and when these systems fail, they tend to fail catastrophically. Sometimes policy has unintended outcomes (e.g., the League of Nations); sometimes technology leaps ahead without full human understanding of its side effects (e.g., nuclear energy and nuclear waste, industrial chemicals and pollution, the use of fossil fuels and the climate change debate). As the disaster policy domain becomes more complex, so will the challenge of designing and implementing good policy. In addition, further social shifts will continue to influence the disaster policy domain. For example, for the first time in human history, more people live in cities than in the countryside; the world's population continues to grow in spite of finite resources; and newly emergent countries like China, India, and Brazil will attempt to exert increased influence over world events. Technology changes will continue to shape how we communicate, how we use our resources, and what energy sources we will require. The potential for unexpected and negative consequences of these shifts is real and not very predictable until we look back in the aftermath. The confluence of technological hazards may now drive natural disasters and vice versa. Perhaps Hurricane Katrina serves as our first warning of this confluence, as the manmade levees gave way under the natural winds and rain of the storm—each magnifying the catastrophe caused by the other. It is a warning to be heeded by all who hope to shape policy in the aftermath of catastrophe.

*One can imagine the beauty of the mountains and hillsides in the surrounding country. For many years, families have lived in the fertile and rich hillside, cooled by the breezes from the sea. Grapes grown on the vines and fresh olives can be bought in the marketplace. It is the splendor of an enviable lifestyle that comes from a moderate climate, a rich cultural history, and the fertile lands, made more fertile by the minerals embedded in the soil, minerals produced by the great mountain from time to time over history. Today, 3 million people go about their daily lives, living in the shadow and the reach of Mount Vesuvius.*

## REFERENCES

Baumgartner, F., and B. Jones. 1993. *Agendas and Instability in American Politics*. Chicago: University of Chicago Press.

Birkland, T. 2006. *Lessons of Disaster: Policy Change After Catastrophic Events.*. Washington, DC: Georgetown University Press.

Carafano, J. J., and R. Weitz. 2008. *Mismanaging Mayhem: How Washington Responds to Crisis*. Westport, CT: Praeger Security International.

Friedman, T. 2007. "The Whole World Is Watching." *New York Times*, June 27.

Graham, B., and J. Talent. 2008. *World at Risk: The Report of the Commission on the Prevention of Weapons of Mass Destruction and Terrorism*. New York: Vintage Books.

Lindblom, C. 1959. "The Science of Muddling Through." *Public Administration Review* 19 (2): 79–88.

National Academy of Public Administration (NAPA). 1993. *Coping with Catastrophe: Building an Emergency Management System to Meet People's Needs in Natural and Manmade Disasters*. Washington, DC.

Perrow, C. 1984. *Normal Accidents: Living with High Risk Technologies*. Princeton: Princeton University Press.

Rubin, C.; I. Renda-Tanali; and W. Cumming. 2009a. "Disaster Time Line: Major Focusing Events and U.S. Outcomes 1988–2008." Version 5.0. Claire B. Rubin and Associates, Arlington, Virginia.

———. 2009b. "Terrorism Time Line: Major Focusing Events and U.S. Outcomes 2001–2008." Version 7.0. Claire B. Rubin and Associates, Arlington, Virginia.

Sabatier, P. (ed.). 2007. *Theories of the Policy Process*, 2d ed., Boulder, CO: Westview Press.

Wamsley, G., and A. D. Schroeder. 1996. "Escalating in a Quagmire: The Changing Dynamics of the Emergency Management Policy Subsystem." *Public Administration Review* 56 (3): 235–244.

CHAPTER 10

# Living with Disasters

## A Planning Approach to Critical Incidents

DIVYA CHANDRASEKHAR

As a field, planning has traditionally been concerned with managing the physical development of human settlements. However, the field also recognizes that the built environment is usually a manifestation of the social, cultural, economic, political, and ecological context of a place. For example, housing quality might differ according to the race and ethnicity of residents (Rosenbaum 1996); local politics influences access to infrastructure (Lake and Baum 2001; Swyngedouw 1997) and the location of new residential and commercial activity (Lewis 1996); local cultural practices determine the use of forests, parks, and other public spaces (Croll and Parkin 1992; Low and Lawrence-Zúñiga 2006); and hazard risk is different within different economic classes, as is their adaptation to it (Brouwer, Akter, Brander, and Haque 2007). In recognition of these interrelationships, planners also deal with issues such as economic development, housing finance, policy development, environmental management, and local community organizing. Thus, as a field, planning addresses both the structural and nonstructural aspects of human settlements. A second significant characteristic of the profession is that it is concerned with making decisions about the future in the present. Planners usually do this by creating visions or scenarios for the future and identifying proactive steps that can be taken to help achieve this. Finally, planning is as much a process as an outcome, and in recent years the profession has tended to focus as much on the procedural dimensions of practice as the design outcomes.

Knowing how the profession operates is vital to understanding how planners address critical incidents. Planners must examine both the physical and nonphysical impacts of a critical incident, work to create a holistic strategy to recover from it, and facilitate processes to mitigate impacts of future incidents. Thus planning brings two distinct perspectives to critical incident analysis. First, it connects the recovery of individuals to the recovery of the various physical, social, cultural, economic, political, and ecological systems in which these individuals live and operate. Second, planning gives recovery a long-term and proactive perspective—one with particular relevance for the redevelopment and reconstruction phases of recovery (see below). Through planning, settlement communities can evaluate long-term implications of immediate decisions and take appropriate actions in the present to improve their own resilience to future critical incidents.

This chapter describes how planning addresses natural disasters. The first section describes the postdisaster recovery period and examines the role for planning in it. The second section discusses the impacts of disasters on human settlements. The final section describes some salient planning activities undertaken for postdisaster recovery. It is beyond the scope of this chapter to describe comprehensively all the disaster impacts and recovery actions of planners. Books and manuscripts are devoted to exploring, examining, explaining the impact of disasters on society and the economy, and formulating new planning strategies to deal with it. The aim of this chapter is therefore only to give the reader a broad (and brief) understanding of what planners do during postdisaster recovery and of their justification behind these actions.

## DISASTER RECOVERY AND RELEVANCE FOR PLANNING

When a disaster strikes a human settlement, many different types of activities are set in motion. Residents are evacuated to safety and provided with food, temporary shelter, and other basic life services; damages are assessed and debris removed; basic infrastructure facilities are restored to serve returning populations; and decisions are made on what to rebuild, where, when, and how, and followed up with implementation on the ground. The nature of these activities differs significantly depending on the hazard, their immediate goals, the actors they engage, the resources they require, and the mechanism by which they come about. Moreover, while some of these actions have clear start and end characteristics (e.g., evacuation), others have only identifiable start or end points (e.g., temporary shelter), and some other still have neither (e.g., permanent reconstruction). Differentiating between the various stages of postdisaster recovery is useful for timely and effective action as well as for codifying results of research (Neal 1995).

For decades, researchers have been attempting to define the various stages of postdisaster recovery. While most researchers agree that postdisaster recovery occurs in various phases, there is no clear consensus on the exact number of such phases, their degree of mutual exclusivity, or their sequence of occurrence. Haas et al. (1977) suggested that recovery occurs in four consecutive phases, with each successive phase much lengthier than the preceding one, thus implying a logical and logarithmic relationship between them. The four phases identified by Haas et al. were: an *emergency period,* referring to the initial hours or days of relief and rescue activities; a *restoration period,* when evacuees return, major urban services and transportation are restored, and rubble debris is cleared; a *replacement reconstruction period,* during which capital stock is rebuilt to predisaster levels and social and economic activities return to previous levels, and a *commemorative, betterment, and development reconstruction period,* during which "major projects are undertaken and future growth and development begin to take hold" (Haas et al. 1977, xxvi–xxviii). However, a 1985 study of fourteen disaster-affected communities in the United States by Rubin et al. called this classification too "neat" to describe the reality on the ground of recovery. The study found that the recovery phases as described by Haas et al. did not necessarily occur in a logical sequence. That is, they often overlapped and, in some cases, happened at different times for different parts of the community, depending on access to resources (Rubin, Saperstein, and Barbee 1985). In a similar vein, Johnson (1999) also found that these phases did not follow a uniformly sequential pattern and suggested instead an alternative set of phases to describe recovery: an *immediate response,* which involves rescuing people and property and providing immediate aid; *restoration,* which involves "bringing about some sense of normalcy" such as utility restoration and short-term housing; *short-term or interim recovery,* which is a "period of intense policy making and interim solutions, restoring predisaster levels of functioning to impacted groups or areas"; and *long-term recovery* or *permanent reconstruction,* which involves "repair, rehabilitation and redevelopment, and dealing with unresolved problems" (Johnson 1999, 78). The exact sequence and timing of these phases, according to Johnson, would be different for different places and for disaster events.

Despite the lack of consensus on the exact timing and sequence of phases, most research tends to agree that, over time, the focus of recovery moves from immediate response to restoration and, finally, reconstruction. Researchers also appear to agree that there is a qualitative difference between immediate response, which involves activities like emergency shelter, evacuation, and food provision, and later phases of recovery, which involve permanent reconstruction, redevelopment, and rehabilitation initiatives. Therefore planning, which is concerned with building communities, has more relevance in later phases of recovery than earlier ones. This does not mean that planners are not engaged in recovery from the very start: Most planners, for example, begin damage assessments almost immediately after a disaster and are usually involved in locating temporary housing. But the long-term recovery phase has the most impact and, in turn, affects planning. For example, predisaster building codes, zoning regulations, plans, planning processes, and institutional mechanisms can dictate who does the reconstruction and how they do it. At the same time, it is also the time when new elements can be introduced to the substantive and procedural aspects of planning that affect the long-term sustainability of a settlement community, especially those in hazard-prone areas.

## IMPACT OF DISASTERS

Disasters have structural and nonstructural impacts on settlements. Structural impacts are more physically evident in the damage to the housing stock, commercial areas, public facilities, and physical infrastructure such as roads, the water supply, and sanitation systems. Nonstructural impact refers to effects on population composition and settlement patterns, local economy, and sociopolitical organization of a place, which, despite having some physical manifestation, are usually less visible than structural impacts.

Structural and nonstructural impacts do not occur in mutual exclusivity. Physical damage to downtown buildings, for example, might have a significant impact on the local economy, and the loss of schools or hospitals might act as deterrents to return for families with children or the elderly. Moreover, both kinds of impacts are largely dependent on predisaster societal conditions because disaster events might be seen as precipitants for crisis and disaster behavior (i.e., as "social catalysts") rather than their cause (Wenger 1978). This perspective is built on the core premise that "communities and regions are not homogeneous, unified systems but rather mosaics of overlapping subsystems cross-cut by social and economic inequalities," which experience disasters in different ways (Elliott and Pais 2006), and that disaster impacts are "not random, but determined by everyday patterns of social interaction and organization" (Morrow 1999).

Planning, therefore, must not only address the existing inequities highlighted by the structural and nonstructural impacts of a disaster during recovery but also utilize this opportunity to build more equitable and sustainable settlement systems. The following section describes three major activity areas of recovery in which planners play a significant role: settlement reconstruction, economic redevelopment, and local capacity building. More emphasis is given to settlement reconstruction than the other two areas because it is the most obvious and common concern of the planners among the three. Moreover, there is a high degree of interaction between these activities, and, in reality, they are undertaken simultaneously.

## STRUCTURAL IMPACT OF DISASTERS

Structural damage can differ based on the type of hazard experienced (earthquakes, flood, hurricanes, tornadoes) as well as the nature of development (urban or rural). Planners usually assess structural impact using measures such as physical damage (usually measured on an ordinal scale—fully destroyed to no damage) and damage cost (sometimes also assessed as reconstruction cost). However, these measures are composite by nature and do not adequately express the disparate nature of this impact. For example, Peacock and Girard (1997) show that minority groups (especially blacks) are disproportionately located in poor-quality housing segregated into low-value neighborhoods, making them more vulnerable to damage from hurricanes. Further, they point out that this choice of low-quality housing is not an outcome of cultural preferences but of race and ethnicity, which are still important determinants of economic resources, such as income and credit, that are critical in obtaining better-quality housing (Peacock and Girard 1997). Similarly, mobile homes are more likely to be destroyed than conventional homes in disasters such as tornadoes and hurricanes, as seen in 1992 when Hurricane Andrew destroyed almost all the 6,600 mobile homes in the disaster-hit areas (Morrow 1999). After the 2004 Indian Ocean tsunami in India, homes of lower-caste *dalits* and small-scale fisherfolk experienced more damage than higher-income and higher-caste fisherfolk who tended to have homes made out of brick and mortar (Chandrasekhar 2010). Squatter settlements, in which a significant majority of the urban poor in developing countries reside, tend to be more vulnerable to disasters because of their location in high-hazard-risk areas and due to the high density of these settlements, both of which are outcomes of real estate markets in urban settlements (Montz 2000; Morrow 1999; Napier and Rubin 2002). Homelessness can increase after disasters, and the previously homeless are even less likely to get homes afterward (Cherry and Cherry 1996).

Other structural impacts include the destruction of physical infrastructure of a place or its public facilities. Between 1990 and 1998, the annual cost of floods in rural Asia was approximately $15 billion, and by some estimates loss to infrastructure accounts for about 65 percent of this loss (Freeman 1999; Munich Re 1998). Although few studies have looked into indirect or secondary costs of this loss (Freeman 1999), an obvious implication is that populations that are more dependent on this kind of infrastructure for their daily life and livelihood are probably more adversely affected. Moreover, since access to infrastructure is often seen as a measure of poverty, the loss of it might affect low-income populations more than others. For instance, loss of agricultural infrastructure can affect marginal farmers more than large farmers of the same settlement, just as the loss of transportation infrastructure can affect workers who travel long distances more than those who do not. Where basic services are destroyed, lower-income populations will be less likely to return or wait until they are completely restored. In post-Katrina New Orleans, neighborhoods where municipal services were restored quickly could begin home reconstruction sooner than neighborhoods where restoration was delayed (Bates and Green 2009).

Similarly, the destruction of local schools, hospitals, and other community facilities like day-care centers can also affect some socioeconomic groups, such as the poor or the elderly, who depend on these services more than other groups. A study conducted in 2007 found that there were fewer households with children, both married couples and female-headed households, in New Orleans after Hurricane Katrina than before the storm, which may have been related to the slow rebuilding of the city's school system (Frey, Singer, and Park 2007). For more than two years after Hurricane Katrina, New Orleans residents were forced to undertake long commutes to out-of-neighborhood schools and faced reductions in police protection, fire patrols, trash collection, and residential postal deliveries (Green, Bates, and Andrew 2007).

## NONSTRUCTURAL IMPACT OF DISASTERS

### Demography and Settlement Pattern

Nonstructural impacts of disasters can manifest as changes in settlement pattern or demography. In some cases, disasters can lead to the large-scale displacement and change in the demographic composition of a settlement. Nearly a third of the black population of Cameron Parish was killed during Hurricane Audrey in 1957, as opposed to only 2.5 percent of its white residents, resulting in large-scale out-migration of black survivors, who had lost their homes or personal ties to the community (Bates, Fogleman, Parenton, Pittman, and Tracy 1963). Residents of post-Katrina New Orleans were generally found to be older, more educated, and less poor than the prestorm residents, and the city had fewer renters as compared to homeowners before the storm (Frey, Singer, and Park 2007). In other cases, such as Friuli (Italy) after the 1976 earthquake, disasters might quicken preexisting trends of migration (Geipel 1991). Where a large portion of the population is displaced, new housing arrangements that emerge during settlement reconstruction can change the patterns of social interaction and cause more community conflicts (Bolin 1982). But nonstructural impacts are not always negative. After a massive landslide in 1970 in Yungay, Peru, which killed many of its prominent community members, new residents moved into the community and took over their political and economic roles (Oliver-Smith 1977). The community of Huaraz, which was also affected by the same disaster and lost approximately 30 percent of its population, experienced a growth rate that was almost three times the regional average, owing to the influx of reconstruction aid (Doughty 1986). In some south Indian villages where previously lower-caste *dalits* and higher-caste fisherfolk lived in adjacent hamlets within the same settlements, after the 2004 Indian Ocean tsunami, *dalit* families relocated to separate settlements away from the fisherfolk (Chandrasekhar 2010).

The nature and speed of recovery can also influence postdisaster demographic composition. For example, low-income populations in South Carolina were left behind during recovery from Hurricane Hugo in 1989, because reconstruction of multifamily housing had been largely overlooked in recovery assistance. This was also the case after the Loma Prieta earthquake in 1989, when multifamily housing was the last to be rebuilt even though it accounted for over 60 percent of all destroyed structures (Comerio 1998). In New Orleans, the Housing Authority demolished at least four public housing projects after Hurricane Katrina without providing adequate and affordable replacements. This reduced the number of public housing units in the city from 5,100 to 2,000 and disproportionately affected black female-headed households, which formed a majority of ex-occupants of public housing in the city (Luft and Griffin 2008). Renters affected in the same disaster found themselves unable to return to the city as quickly as homeowners because of the lack of affordable rental housing in the city after the disaster, while low-income homeowners who lacked flood insurance, savings, or access to credit found it more difficult to rebuild their homes than their well-insured, single-family counterparts (Green et al. 2007). To exacerbate the problem, the existing federal aid system in the United States is heavily oriented toward well-insured single-family homeowners who intend to rebuild on the same site (Comerio 1998). Timeliness is a major factor in repopulation: the longer the delay, the less the inclination to return, especially for those who are not able to bear the cost of relocating twice in succession (first from and then back to their original locations).

## Economy

The economic impact of disasters can be understood in two ways: first, in terms of economic loss, and, second, changes in the economic systems at large. Until recently, postdisaster economic loss estimation was done by calculating the extent of physical damage caused (Rose and Lim 2002). While this approach took into account the *stock* value lost in the disaster (such as property damage), it does not account for losses to *flows* (e.g., business disruption), which are the services or outputs of stocks over time. For example, damages to a convenience store must include not only the losses to the building itself (i.e., the stock) but also loss of business activity (i.e., the flow). The inclusion of flows in economic loss estimation is useful (and important) because it provides a more comprehensive understanding of economic impact even in cases where there is no physical damage and because it can help clarify the secondary effects of this loss (Rose and Lim 2002).

Loss estimation usually comprises two aspects. The first is "direct effect" or "direct loss," which represents property damage, loss in production due to this damage, and loss in production due to loss of utility and infrastructure services (Tierney, Chang, Eguchi, Rose, and Shinozuka 1999; Rose and Lim 2002). The last component refers to instances in which businesses not directly affected in a disaster are forced to shut down for a period because infrastructure, such as roads, water supply, power, or transport, has been damaged or destroyed. For example, loss of boats and nets during the 2004 Indian Ocean tsunami posed a serious economic threat to affected village communities that depend almost entirely on fishing for their livelihood. For those engaged in the import and export of fish, economic loss in the same disaster arose from the destruction of wholesale markets and the closure of port facilities and other transport infrastructure (Chandrasekhar 2010). In the lower Ninth Ward of New Orleans nearly 60 percent of the commercial buildings sustained significant damage to their structure, thus posing a major challenge to business recovery in the region (ACORN Housing/University Partnership 2007).

Loss can also be represented by estimating "indirect effects," which refers to the ripple effect of direct losses. These effects can be experienced in terms of loss of earnings due to impact on an interdependent business or industry, job loss, and to other social impacts, such as large-scale displacements (Kreimer and Arnold 2000). For example, if a manufacturing facility closes down, it can both hinder its suppliers and its customers who depend on it for their own operations as well as render its employees jobless. In post-Katrina New Orleans, a local labor shortage due to mass displacement of residents was a serious concern for the construction industry, and small business recovery was heavily affected by the slow reopening of local markets upon which they depended (Green et al. 2007; New Orleans Community Support Foundation 2007). Similarly in South India, the reduced fishing activity in the months following the 2004 tsunami led some lower-caste *dalit* fish vendors, laborers, and transporters to take up other professions, such as driving a taxi or selling food to substitute for their lost income (Chandrasekhar 2010).

Another way of looking at economic impact is by looking at changes in the economic system at large. For example, many of the *dalit* vendors described above did not return to the fishing business, because it took over three months for the business to resume and they did not have economic support to help them through this disruption. As a result, the *dalit* community is less involved in fishing than it was before the disaster. Similarly, farmers whose lands were inundated with seawater during this same disaster were faced with the necessity of desalinating their land—a process that takes many crop cycles—or growing cash crops such as Casuarina, which requires a longer time to grow (Chandrasekhar 2010). In Dominica, the share of bananas in agricultural output increased after Hurricanes David and Frederick (both 1979) because its cultivation provided a fast and low-investment means of restoring agricultural livelihood; bananas continue to be a dominant crop in the island nation (Benson and Clay 2004). The number of small shops owned and operated by women increased significantly in the village communities of South India because of the microfinancing programs initiated by nongovernmental organizations (NGOs), indicating that not all long-term economic changes are necessarily negative.

As was the case with other impacts, prevalent economic and social conditions greatly affect how direct, indirect, and systemic changes play out on the ground. Moore's study (1958) of two tornado-hit towns showed that the blacks were much more economically vulnerable because they had lower incomes, a low ability to bear economic loss, and rarely had pensions or insurance coverage. A study of human response in Hurricane Katrina showed that racial differences were highly salient to job security and that black workers from New Orleans were four times as likely as their white counterparts to lose their jobs after the storm (Elliott and Pais 2006). In particular, Moore (1958) pointed to their poor access to channels of communication, lack of information about the sources of aid available, unfamiliarity with bureaucratic organizations, and racial discrimination as specific reasons African Americans were less able to meet their emergency needs effectively (Moore 1958). Small businesses and businesses in the trade sector suffered more losses than those in the service sector during the Loma Prieta earthquake, and businesses that were already financially marginal before the disaster had more difficulty recovering from the 1984 earthquake in Coalinga, CA (Durkin 1984; Kroll, Landis, Shen, and Stryker 1991). In some South Indian villages, female marginal farmers who could not reclaim their land after seawater flooding during the 2004 tsunami turned to other low-paying jobs, such as selling food, to sustain them for many months after disasters (Chandrasekhar 2010).

## Sociopolitical Organization

Research on how communities are affected in disasters has been a topic of discussion since early 1900s. For the first half of the twentieth century, disasters were associated with the phenomena of social disorganization—a view that was refuted in a spate of studies conducted in the 1950s and 1960s. Fritz (1961) and Baker and Chapman (1962) found that the incidence of rampant looting, hysterical behavior, and panic-led action following disasters was actually atypical. Fritz claimed the observable breakdown of coordination and control was found to arise from the informal, spontaneous "convergent actions" of persons living outside the disaster areas, rather than the victim population. Disaster-related convergence, therefore, referred to the collective behavior involving the mass movement of people, goods, and other resources into disaster-stricken areas, and a "convergence theory" was proposed to account for this altruistic response and the collective behavior (Tierney 2001). Convergence was seen as both beneficial and problematic, because it brought much-needed volunteers and resources to disaster-stricken areas while also creating substantial management challenges (Tierney 2001). The convergence theory also acted as the foundation for the concept of "therapeutic communities," in which "information about the needs of community members is widely shared; there is rapid consensus on actions to be taken to meet those needs, and highly motivated work for common purposes. These behavior patterns persist until the more urgent needs are met" (Barton 1969, 207). Dynes (1970) attempted to resolve the paradox of disorganization and convergence by presenting them as dual aspects of a larger restructuring process during which, first, there is disintegration of activities and social groups from which eventually new patterns of interaction emerge.

More recently, however, researchers have begun to question whether the postdisaster communities are indeed as integrative as previously portrayed. Abrams et al. (2004) point out that the previously studied aspects of convergence ignore questions of culture, community, and sense-making, except as they are relevant to the question of managing victimized populations. In their study on post-9/11 New York, they argue that local residents actively created, negotiated, and contested the meaning of "community" in the city after the disaster. According to them, this negotiation "was rooted in uncertainty about identity, interaction, and the boundaries of community in the wake of the attacks" which begs the question of whether postdisaster community "restructuring" is experienced by all its residents in the same manner (Abrams et al. 2004, 189). Tierney also points out that, in the long term, the events of September 11, 2001, exposed differential vulnerabilities and community fault lines in New York society and gave rise to "competing and often conflicting disaster framing processes" (2001, 3). The recovery period after damaging disasters is often marked by conflicts over the recovery process—for example, debates over whether a disaster-stricken community should be restored consistent with predisaster development patterns or rebuilt in ways consistent with new community visions (Tierney 2001). Albright notes that despite the recognition of a common disadvantage by disaster victims and their reliance on one another, inequalities exist in the manner in which they are disadvantaged, and that, as resources grew scarce in post-9/11 New York, "racial, class, and spatial divisions became more salient again, testing the cohesiveness of a weary and frayed community" (2006, 311).

Postdisaster recovery also presents opportunities for new organizations to emerge, often based on common perceptions of disaster victimization or racial and ethnic identity. Conflicts among groups of residents in disaster areas can lead to the creation of new collectivities (formal, informal, or as coalitions) in opposition or support of some government action (Nigg and Tierney 1993) or in order to gain legitimacy in a resource distribution or recovery decision process (Chandrasekhar 2010). In post-tsunami Banda Aceh (Indonesia) and post-Katrina New Orleans (United States), numerous organizations emerged for purposes such as citizen representation, network building, recovery management, and monitoring recovery progress (Chandrasekhar and Olshansky 2007). In India after the same disaster, previously disenfranchised groups revived their traditional governance structures to gain more visibility and to actively pursue their own resource agendas with external aid agencies and the state government during long-term recovery. In some villages, new labor unions emerged to represent the fishing labor as separate from boat owners (a historic move in the context of these close-knit communities), and in other villages, animosity between lower-caste *dalits* and higher-caste fisherfolk increased during the resettlement process (Chandrasekhar 2010).

These impacts of disasters on the sociopolitical organization of human settlements present a challenge and an opportunity for planners. First, the increasing divisiveness within postdisaster communities makes consensus building on a common future course of action harder to achieve. Second, the increased awareness and organizing allows more voices to be recognized and heard, thereby making decisions more equitable while also setting precedents for future planning decision processes.

## PLANNING APPROACHES TO RECOVERY

It is obvious from the discussion above that geographic location, race, class, and gender can play an important role in reconstruction after disasters and that the recovery process for different locations, different communities, and different social groups takes a different path. It is also obvious that in communities of high hazard risk, there is a need to mitigate the risk of disasters. There is, therefore, a need for "site-specific, adaptive planning strategies, and citizen participation"(Rubin et al. 1985) in disaster recovery. The goal here is not only to recover from disaster events but to do so in a way that improves on predisaster conditions and, at the same time, builds more resilient and sustainable communities. While there is an ongoing debate on the exact meaning and scope of terms such as resilience and its specific measures (Manyena 2006; McEntire, Fuller, Johnston, and Weber 2002; Mileti 1999), there is a common acknowledgment that future disaster impacts can be minimized through active intervention in the present and that such interventions must address both the outcomes and the process of postdisaster recovery.

Planning actions in postdisaster recovery target both outcomes and processes. Outcome-oriented actions usually take three forms: disaster mitigation projects to reduce the vulnerability and risk of local populations; long-term disaster rehabilitation measures (structural and nonstructural) undertaken after disasters; institutionalization of disaster-mitigation planning in development planning. These activities often overlap and are almost always concurrent. For instance, it is not always easy to differentiate between disaster recovery and disaster mitigation because recovery actions commonly include disaster mitigation and some recovery actions can be undertaken proactively as mitigation. Experience suggests that conversations on disaster mitigation often start during recovery but that mitigation can outlast recovery by being integrated into local development principles and mechanisms. Thus disaster mitigation can be thought of as over and above recovery and a step toward building sustainable communities.

Disaster mitigation programs are not limited to a particular disaster phase or time. They can be undertaken before and after disasters, though there is usually more awareness of the need for mitigation after a disaster than before it. Planning tools used for mitigation differ based on the local context and need; examples include development regulations, disaster-resistant building standards, proactive property buyouts, structural flood and hurricane protection solutions such as building levees or sea walls, wetland and other wildlife habitat restoration, and promoting sustainable agricultural practices. In some cases, these form part of a mitigation plan created under an overarching agenda of environmental management. Tulsa, Oklahoma, which experiences annual flooding, created a Flood Mitigation

Plan as part of its floodplain and storm water management program in the late 1990s; it included strategies such as land acquisition, preemptive relocation, and flood-proofing to reduce future risks. In post-tsunami South Indian villages, mitigation plans made after the disaster also comprise components of emergency management such as evacuation routes and identification of emergency shelters. Sometimes, mitigation is undertaken through separate initiatives, such as in California, where building codes were revised to include seismic safety after a major earthquake in 1971, or in India, where mangrove restoration was undertaken after the 2004 tsunami to mitigate impacts of future disasters.

Despite these efforts at actively pursuing mitigation policies and actions, it is not always possible to eliminate all impacts of disasters (Tobin 1999). Disaster recovery is a complex process with multiple actors, multiple agendas, high levels of uncertainty, and a strong sense of urgency for action. At the same time, any recovery action also has to take into account the disparate nature of disaster impacts and of recovery capabilities as well as of local aspirations for the future of the place. Where there are preexisting development plans, communities often use the resource opportunities presented by recovery to implement ideas and strategies in them. Recovery programs initiated by planners attempt to address both structural and nonstructural impacts of disasters. Common structural recovery strategies include comprehensive *in situ* settlement reconstruction programs (including housing, infrastructure, and public facilities), full or partial relocation programs (including buyouts), land-readjustment schemes, and affordable housing programs. Common nonstructural strategies include programs for livelihood restoration, local economic development, agricultural land reclamation, and creating tax incentive districts. Planners can also be involved in creating nonstructural strategies, such as setting up resource centers or recovery management organizations and creating mechanisms to distribute financial resources. Like mitigation, recovery strategies are also implemented concurrently in most places, either as different components of a single recovery plan or as separate projects conducted on an ad-hoc basis. For instance, the Unified New Orleans Plan (UNOP) created as a recovery plan after Hurricane Katrina in 2005 included programs such as those for partial relocation and land readjustment, home-buyer assistance, affordable and rental housing, neighborhood resource centers, corridor revitalization programs, small business incubator and assistance programs, workforce training programs, rejuvenating neighborhood health-care centers, creating neighborhood centers, reopening and renovating schools, historic preservation, renovating public parks, brownfield development, and public safety (New Orleans Community Support Foundation 2007). By contrast, in India, few communities had a comprehensive recovery plan, and housing reconstruction, infrastructure restoration, issues of health and education, livelihood restoration, and natural resource management were addressed in separate projects undertaken by different NGOs at different times within the same community (Chandrasekhar 2010).

The latest focus in planning with regard to disaster management is on integrating mitigation into "regular" development planning processes. In the United States, the conversation has revolved around including a component of mitigation in regular comprehensive plans (Schwab 1998). Schwab suggests that these components be integrated into a mitigation plan, which can then be used to take action. Recovery plans, according to this model, would be part of a mitigation plan. This system seems workable in places such as New Orleans, where there is an official comprehensive planning process, but is less likely to work in places like India, where development is pursued on a project basis (Chandrasekhar 2010). A better understanding of planning under "normal" circumstances and for long-term recovery (which is the least-studied recovery phase) will contribute significantly to this effort, and research on these topics is still under way.

Planning also addresses the process aspect of recovery by enabling citizen participation in decision making. Citizen participation is crucial to the creation and successful implementation of policies and plans, such as disaster mitigation plans, which usually lack strong and supportive interest groups (Burby 2003). Active local involvement allows for local learning, and case studies have shown that such citizen participation in planning and construction can lead to successful resettlement after disasters (Berke, Kartez, and Wenger 1993; Oliver-Smith 1991). The actual mechanisms used for citizen input in recovery decision making are highly dependent on the local sociocultural and policy context. Sometimes planners use preestablished mechanisms for citizen input, such as planning commission hearings or neighborhood meetings, while in other cases they create new mechanisms, particularly for recovery planning, such as the Sub-District Coordination Mechanisms (SDCMs) in post-tsunami Banda Aceh or the Community Support Organization in New Orleans. Not all participatory planning efforts are successful, however, and inequitable decision making has always been the Achilles' heel of planners. Existing participatory planning methods have often been criticized for being overly technical and for ignoring underlying power structures and trust issues in local communities. Participatory planning during recovery poses an even bigger challenge because of the sense of urgency that pervades it. Reduced time for deliberation can affect the comprehensiveness of stakeholder identification, which can in turn affect the level of their involvement in the decision-making process. Interest groups that are not very prominent or those that are not immediately visible in the postdisaster situation (such as displaced residents) are in danger of being left out because of the time that it would take to locate these stakeholders. Time constraints can impede efforts to address imbalances in institutional or social power, class conflicts, racism, or mistrust underlining the participatory planning process. Finally, participating stakeholders need time to identify the information and skills that they can bring to the table; familiarize themselves with the other stakeholders, their needs, skills, and cultural practices; and build consensus or resolve conflicts on decisions—each of which is a time-consuming proposition even under normal circumstances (Maginn 2007).

## CONCLUSION

Planning is not a precise science, and planning in the postdisaster context is even less so. There can be no perfect planning solutions to postdisaster recovery in a world that is constantly changing and from a profession that varies from location to location. The best planners can do is to remember to be contextual, aware, balanced, and flexible in their approach to recovery and to use each recovery experience as an opportunity to learn something new about disaster recovery and about planning itself.

## ACKNOWLEDGMENTS

The author thanks Tim Green, Kanako Iuchi, and Dr. Rob Olshanksy of the University of Illinois for their input and help in writing this chapter.

## REFERENCES

Abrams, C. B.; K. Albright; and A. Panofsky. 2004. "Contesting the New York Community: From Liminality to the 'New Normal' in the Wake of September 11." *City and Community* 3 (3): 189–220.

ACORN Housing/University Partnership. 2007. "A People's Plan for Overcoming the Hurricane Katrina Blues." New Orleans.

Albright, K. 2006. "The Meaning(s) of Community in Post-Disaster New York." *Sociological Forum* 21 (2): 307–314.

Baker, G. W., and D. W. Chapman. 1962. *Man and Society in Disasters.* New York: Basic Books.

Barton, A. H. 1969. *Communities in Disaster.* Garden City, NY: Doubleday.

Bates, F.; C. Fogleman; V. Parenton; R. Pittman; and G. Tracy. 1963. "The Social and Psychological Consequences of a Natural Disaster: A Longitudinal Study of Hurricane Audrey." Disaster Research Group Disaster Study No. 18. National Academy of Sciences-National Research Council Publication 1081. Washington, DC.

Bates, L. K., and R. A. Green. 2009. "Housing Recovery in the Ninth Ward: Disparities in Policy, Process, and Prospects." In *Race, Place, and Environmental Justice After Hurricane Katrina: Struggles to Reclaim, Rebuild, and Revitalize New Orleans and the Gulf Coast,* ed. R. D. Bullard and B. Wright, 229–245. Boulder, CO: Westview Press.

Benson, C., and E. J. Clay. 2004. "Understanding the Economic and Financial Impacts of Natural Disasters." Disaster Risk Management Series No. 4. World Bank, Washington, DC.

Berke, P. R.; J. Kartez; and D. Wenger. 1993. "Recovery After Disaster: Achieving Sustainable Development, Mitigation and Equity." *Disasters* 17 (2): 93–109.

Bolin, R. 1982. *Long-Term Family Recovery from Disaster.* Boulder, CO: Institute of Behavioral Science, University of Colorado.

Brouwer, R.; S. Akter; L. Brander; and E. Haque. 2007. "Socioeconomic Vulnerability and Adaptation to Environmental Risk: A Case Study of Climate Change and Flooding in Bangladesh." *Risk Analysis* 27 (2): 313–326.

Burby, R. J. 2003. "Making Plans That Matter: Citizen Involvement and Government Action." *Journal of the American Planning Association* 69 (1): 33–50.

Chandrasekhar, D. 2010. "Understanding Stakeholder Participation in Post-Disaster Recovery Process." PhD dissertation, University of Illinois at Urbana-Champaign.

Chandrasekhar, D., and R. B. Olshansky. 2007. "Managing Development after Catastrophic Disaster: A Study of Organizations That Coordinated Post-Disaster Recovery in Aceh and Louisiana." Paper presented at the Annual Conference of the Association of Collegiate Schools of Planning, Milwaukee.

Cherry, A. L., and M. E. Cherry. 1996. "Research as Social Action in the Aftermath of Hurricane Andrew." In *Research on Social Work and Disasters,* ed. C. Streeter and S. Murty. New York: Haworth.

Comerio, M. 1998. *Disaster Hits Home: New Policy for Urban Housing Recovery.* Berkeley: University of California Press.

Croll, E. J., and D. J. Parkin. 1992. *Bush Base, Forest Farm: Culture, Environment and Development.* London: Routledge.

Doughty, P. L. 1986. "Decades of Disaster: Promise and Performance in the Callejon de Huaylas, Peru." In *Natural Disasters and Cultural Responses: Studies in Third World Societies,* 35–80. Williamsburg, VA: College of William and Mary.

Durkin, M. 1984. "The Economic Recovery of Small Businesses After Earthquakes: The Coalinga Experience." Paper presented at the International Conference on Natural Hazards Mitigation Research, New Delhi.

Dynes, R. R. 1970. *Organized Behavior in Disaster.* Lexington, MA: Heath Lexington Books.

Elliott, J., and J. Pais. 2006. "Race, Class, and Hurricane Katrina: Social Differences in Human Responses to Disaster." *Social Science Research* 35: 295–321.

Freeman, P. K. 1999. *Infrastructure, Natural Disasters, and Poverty.* Laxemberg, Austria: International Institute for Applied Systems Analysis (IIASA).

Frey, W. H.; A. Singer; and D. Park. 2007. "Resettling New Orleans: The First Full Picture from the Census." Brookings Institution, Washington, DC.

Fritz, C. E. 1961. "Disasters." In *Contemporary Social Problems,* ed. R. K. Merton, 651–694. New York: Harcourt, Brace and World.

Geipel, R. 1991. *Long-Term Consequences of Disasters: The Reconstruction of Friuli, Italy in Its International Context, 1976–1988.* New York: Springer.

Green, R.; L. K. Bates; and S. Andrew. 2007. "Impediments to Recovery in New Orleans' Upper and Lower Ninth Ward: One Year After Hurricane Katrina." *Disasters* 31 (4): 311–335.

Haas, E. J.; R. Kates; and M. Bowden (eds.). 1977. *Reconstruction Following Disaster.* Cambridge: MIT Press.

Johnson, L. 1999. "Empowering Local Governments in Disaster Recovery Management: Lessons from Watsonville and Oakland in Recovering from the 1989 Loma Prieta Earthquake and Other Recent Disasters." In *Lessons Learned over Time,* vol. 1, 41–84. Oakland, CA: Earthquake Engineering Research Institute.

Kreimer, A., and M. Arnold (eds.). 2000. *Managing Disaster Risk in Emerging Economies.* Washington, DC: World Bank.

Kroll, C. A.; J. D. Landis; Q. Shen; and S. Stryker. 1991. "Economic Impacts of the Loma Prieta Earthquake: A Focus on Small Business." Working Paper 91–187. U.C. Transportation Center and the Center for Real Estate and Urban Economics, University of California at Berkeley.

Lake, D. A., and M. A. Baum. 2001. "The Invisible Hand of Democracy." *Comparative Political Studies* 34 (6): 587–621.

Lewis, P. G. 1996. *Shaping Suburbia: How Political Institutions Organize Urban Development.* Pittsburgh: University of Pittsburgh Press.

Low, S. M., and D. Lawrence-Zúñiga. 2006. *The Anthropology of Space and Place: Locating Culture,* 5th ed. Oxford: Wiley-Blackwell.

Luft, R. E., and S. Griffin. 2008. "A Status Report on Housing in New Orleans After Katrina: An Intersectional Analysis." In *Katrina and the Women of New Orleans,* ed. B. Willinger, 50–53. New Orleans: Tulane University.

Maginn, P. 2007. "Towards More Effective Community Participation in Urban Regeneration: The Potential of Collaborative Planning and Applied Ethnography." *Qualitative Research* 7 (1): 25–43.

Manyena, S. B. 2006. "The Concept of Resilience Revisited." *Disasters* 30 (4): 433–450.

McEntire, D. A.; C. Fuller; C. W. Johnston; and R. Weber. 2002. "A Comparison of Disaster Paradigms: The Search for a Holistic Policy Guide." *Public Administration Review* 62 (3): 267–281.

Mileti, D. 1999. *Disasters by Design: A Reassessment of Natural Hazards in the United States.* Washington, DC: Joseph Henry Press.

Montz, B. 2000. "The Generation of Flood Hazards and Disasters by Urban Development on Floodplains." In *Floods,* ed. D. Parker, 116–127. London: Routledge.

Moore, H. 1958. *Tornadoes Over Texas: A Study of Waco and San Angelo in Disaster.* Austin: University of Texas Press.

Morrow, B. H. 1999. "Identifying and Mapping Community Vulnerability." *Disasters* 23 (1): 1–18.

Munich Re. 1998. *Topics: Annual Review of Natural Catastrophes 1997.* Munich.
Napier, M., and M. Rubin. 2002. "Managing Environmental and Disaster Risks Affecting Informal Settlements: Lessons from Southern Africa." Paper presented at the International conference and meeting of CIB Task Group 40 on informal settlements, Surabaya, Indonesia, October 10–13.
Neal, D. M. 1995. "Transition from Response to Recovery: A Look at the Lancaster, Texas, Tornado." Quick Response Report 79, December. Natural Hazards Research and Application Center, University of Colorado, Boulder. www.colorado.edu/hazards/research/qr/qr79.html (accessed September 14, 2009).
New Orleans Community Support Foundation. 2007. "Unified New Orleans Plan: Citywide Strategic Recovery and Rebuilding Plan." New Orleans: Greater New Orleans Foundation.
Nigg, J. M., and K. J. Tierney. 1993. "Disasters and Social Change: Consequences for Community Construct and Affect." DRC Preliminary Paper. Disaster Research Center, University of Delaware, Newark.
Oliver-Smith, A. 1977. "Disaster Rehabilitation and Social Change in Yungay, Peru." *Human Organization* 36 (1): 5–13.
———. 1991. "Successes and Failures in Post-Disaster Resettlement." *Disasters* 15 (1): 12–23.
Peacock, W., and C. Girard. 1997. "Ethnic and Racial Inequalities in Hurricane Damage and Insurance Settlements." In *Hurricane Andrew: Ethnicity, Gender and the Sociology of Disasters,* ed. W. Peacock, B. Morrow, and H. Gladwin, 171–190. New York: Routledge.
Rose, A., and D. Lim. 2002. "Business Interruption Losses from Natural Hazards: Conceptual and Methodological Issues in the Case of the Northridge Earthquake." *Environmental Hazards* 4: 1–14.
Rosenbaum, E. 1996. "Racial/Ethnic Differences in Home Ownership and Housing Quality." *Social Problems* 43 (4): 403–426.
Rubin, C. B.; M. D. Saperstein; and D. G. Barbee. 1985. *Community Recovery from a Major Natural Disaster.* Institute of Behavioral Science, University of Colorado, Boulder.
Schwab, J. 1998. "Planning for Post-Disaster Recovery and Reconstruction." Planning Advisory Service Report 483/484. American Planning Association, Chicago.
Swyngedouw, E. 1997. "Power, Nature, and the City. The Conquest of Water and the Political Ecology of Urbanization in Guayaquil, Ecuador: 1880–1990." *Environment and Planning A* 29 (2): 311–332.
Tierney, K. J. 2001. "Strength of a City: A Disaster Research Perspective on the World Trade Center Attack." Disaster Research Center, University of Delaware, Newark.
Tierney, K. J.; S. E. Chang; R. T. Eguchi; A. Rose; and M. Shinozuka. 1999. "Improving Earthquake Loss Estimation: Review, Assessment and Extension of Loss Estimation Methodologies." In *Research Progress and Accomplishments, 1997–1999,* 13–28. Buffalo: State University of New York at Buffalo, Multidisciplinary Center for Earthquake Engineering Research.
Tobin, G. A. 1999. "Sustainability and Community Resilience: The Holy Grail of Hazards Planning?" *Global Environmental Change Part B: Environmental Hazards* 1 (1): 13–25.
Wenger, D. 1978. "Community Response to Disaster: Functional and Structural Alterations." In *Disasters: Theory and Research,* ed. E.L. Quarantelli, 17–47. Beverly Hills, CA: Sage.

CHAPTER 11

# Disaster Investigations

## A 9/11 Case Study

GLENN P. CORBETT

Although the term *disaster* is commonly used by the public to describe a variety of events, attempts to clearly define it prove somewhat elusive. Merriam-Webster (2010) broadly defines disaster in terms of speed, size, and impact as "a sudden calamitous event bringing great damage, loss, or destruction." Some have defined disasters quantitatively, such as Sheehan and Hewitt (1969), who proposed that a "natural" disaster is an event that satisfies at least one of the following criteria: 100 injuries, 100 fatalities, or $1 million in damage. International organizations such as the World Bank define a disaster in terms of scope and the need for external assistance, enumerated as "a serious disruption of the functioning of society, causing widespread human, material, or environmental losses which exceed the ability of affected society to cope using only its own resources. Disasters are often classified according to their cause (natural or man-made)" (2010).

In the United States, federal disaster declarations by the president are based upon federal law (the Stafford Act), in which a "finding must be made that an incident is of such severity and magnitude that effective response is beyond state and local capabilities and that Federal assistance is necessary" (FEMA 2010). The Federal Emergency Management Agency (FEMA) is charged with evaluating requests from state governors, including factors such as the amount and type of damage, impact on the infrastructure, imminent threats to public health and safety, and the impact on essential government services.

As noted in the World Bank definition above, disasters can be framed in the context of their origin: a "natural" disaster or a "man-made" disaster. Natural occurrences that can directly result in a disaster include floods, tornadoes/hurricanes, and earthquakes. These events are typically easily observed and generally understood. Man-made disasters, however, can be the result of intentional acts, such as acts of terrorism, or the result of unintentional errors such as the poor design, construction, or maintenance of a structure.

Failures like these, in some cases, can be difficult to understand and study because of their complexities.

It is important to note the interaction of nature and humans. Many disasters are the result of synergy between man and nature, beginning as a natural occurrence such as a snowstorm, and becoming truly disastrous only as the result of the failure of a man-made structure, such as a building's roof. Many "engineering" failures, such as bursting dams or falling bridges, fit into this category.

Disasters, therefore, can be described and defined in many ways. A single standard accepted definition does not exist. As can be seen, a variety of descriptors are used: size/magnitude, damage, deaths/injuries, impact, monetary losses, and the overwhelming of local resources. Definitions are important; determining whether an event is a disaster depends heavily upon the perspective of the evaluator. In the case of nonprofit organizations and governments, using defined thresholds to differentiate disasters from nondisasters provides them with clear criteria as to their obligations and need to respond.

When it comes to investigating disasters, there is even less standardization. Disaster investigations are often entangled in issues such as competition between government entities for jurisdiction and control (between different levels of governments and within levels of government), presence/lack of technical capabilities (including concerns about lack of expertise at various government entities), the desire to attribute "blame," civil and criminal liability concerns, the creation of both "official" and "nonofficial" inquiries, and the ability to technically establish the origin and cause of the disaster.

## HISTORICAL PERSPECTIVE ON DISASTER INVESTIGATIONS

By their very nature, disasters attract the focused attention of a concerned public. Disasters shock the conscience. The public seeks answers to questions: Why and how did it happen? Could it have been prevented? Who is to blame for the disaster? Public officials are the targets of these questions and are expected to take action to satisfy public pressure for answers and accountability.

Historically, there have been a wide variety of approaches to investigating disasters. The following set of examples provides a snapshot of different disasters, all of them "engineering failures" (disasters in which a man-made structure or object failed), and their subsequent investigations.

### February 4, 1850

A boiler at 3 Hague Street, New York City, exploded, killing 67 people. The coroner's jury scene investigation included "practical boilermakers" and machinists, who examined pieces of boiler scattered on the site. "Minute examinations" were made by those who were "subpoenaed witnesses." Following the coroner's jury decision, responsibility was attributed to the building owner, the boiler repair company, and the boiler manufacturer. A resolution was appended to the verdict that called on the state and national legislatures and municipal governments to take immediate steps to prevent a "similar calamity" (Little 1860).

**January 10, 1860**

In Lawrence, Massachusetts, a mill collapsed and a fire broke out, killing 145 people. The coroner's jury inquiry found substandard construction and excessive loads. The coroner's jury placed "extreme censure" on the building's design engineer (*Boston Journal* 1860).

**December 30, 1903**

A theater fire in Chicago, Illinois, killed 602 people. The Iroquois Theater, which had recently been opened, caught fire when a lamp on stage ignited combustible drapes. The lack of automatic fire sprinklers over the stage, a defective asbestos curtain, and a poorly designed means of egress led to the country's most deadly theater fire. Numerous inquiries resulted, including one by a private sector insurance engineer (Freeman 1906).

**June 15, 1904**

In the East River, New York City, the excursion steamboat the *General Slocum* caught fire, killing 1,021 people. A fire broke out below deck but, instead of landing the boat, the captain piloted it past nearby piers and into the wind, finally beaching the ship in the waters surrounding North Brother Island. Most victims drowned after jumping ship. A federal grand jury was impaneled. The ship's captain was convicted of criminal negligence and sentenced to ten years in prison. President Theodore Roosevelt appointed a special commission to investigate the disaster, and New York City's coroner also conducted an inquiry (Kornblum 2002).

**March 25, 1911**

A fire at the Triangle Shirtwaist Company factory in New York City left 146 people dead. A Factory Investigation Commission was established by the state of New York, looking broadly at workplace safety (beyond the fire itself). The commission submitted seventeen workplace safety bills to the state legislature, of which thirteen became law (McLaury n.d.). A number of other investigations were conducted, including ones by the city's coroner, the city's buildings and fire departments, and private organizations such as New York City's Chamber of Commerce and the Architectural League (Von Drehle 2003).

**January 15, 1919**

A 2.3-million-gallon molasses tank, erected by the Purity Distilling Company in December 1915, ruptured in Boston, Massachusetts, killing 21 people. Although the tank was filled and emptied several times, it eventually failed to maintain containment, spilling all its contents and flooding the surrounding North Boston neighborhood. A grand jury was impaneled, and other investigations were conducted, including one by the city's corporation counsel. Improper engineering was determined to be the cause of the collapse; engineering certification/licensure requirements were passed in many states as a result of this incident (Puleo 2003).

**March 18, 1937**

A natural gas explosion took place at a school in New London, Texas, killing 298 students and teachers. A natural gas leak that had occurred under the school building had ignited, leading to a massive explosion. The natural gas did not have a detectable odor. This disaster led the state of Texas (followed by other states soon after) to require the placement of a detectable malodorant (mercaptan) in natural gas (Jackson 1938).

**November 28, 1942**

A cocktail lounge caught fire in Boston, Massachusetts, killing 492 people. Fire ignited combustible decorations in the Cocoanut Grove nightclub, where an inadequate means of egress led to the high death toll. Investigations were conducted, including one by the fire department. A criminal trial led to conviction of the club owner for involuntary manslaughter. Massachusetts and other states enacted regulations limiting the combustibility of decorations as well as adding requirements for additional standard outward swinging doors next to revolving doors (Grant 2007).

**July 17, 1981**

A walkway in Kansas City, Missouri, collapsed, killing 114 people. The Hyatt Regency hotel, opened a year earlier, included concrete and steel walkways suspended in the hotel's atrium. On the day of the disaster, hundreds of people were standing on the walkways watching a dance contest below. The walkways at the second and fourth floors collapsed. A technical investigation found that the original walkway design and subsequent changes made to this design were structurally inadequate. The design engineers and the engineering firm lost their engineering licenses. The National Bureau of Standards, a federal agency (now known as the National Institute of Standards and Technology, described on page 208), conducted a technical investigation of this disaster (Marshall 1982).

**July 17, 1996**

An airplane crashed off Long Island, New York, killing 230 people. Transworld Airlines Flight 800 suffered an onboard explosion soon after takeoff from John F. Kennedy Airport. The aircraft crashed into the waters of the Atlantic Ocean. The National Transportation Safety Board (NTSB) claimed jurisdiction, joined by the Federal Bureau of Investigation (FBI) in the belief that it was a terrorist attack. They initiated parallel investigations, creating friction. The investigation was made even more difficult by having to recover all evidence from the ocean floor. It was later concluded that the disaster was due to a vapor explosion inside a nearly empty fuel tank (Negroni 2000).

## ESTABLISHING BLAME AND ESTABLISHING THE CAUSE

Although engineering-disaster investigations can take a variety of forms, as seen earlier, they can be divided into two broad categories based on the objective of the investigation: establishing blame and establishing the cause for the disaster. Some disaster investigations fall into both categories.

The following case study of the 9/11 World Trade Center disaster investigation, in which the author was a participant, was a technical investigation whose objective was establishing the cause. However, the "blame" objective played a role in the investigation. The account is written in the first person, representing the observations, impressions, and, most important, opinions, of the author. This account focuses primarily on the events leading up to the investigation and the mechanics of the actual investigation itself and, to a minor extent, on the actual engineering issues.

## A 9/11 CASE STUDY

Several weeks after the September 11, 2001, attacks on the World Trade Center (WTC), Associate Professor Charles Jennings and I (both of John Jay College) were approached by a newspaper reporter asking a simple question: How would high-rise firefighting change in the future, given the large loss of firefighters? Our reply was that we did not really know since we did not know what actually happened in the firefighting efforts. We advocated for an investigation (Calderone 2001).

We were immediately contacted by Sally Regenhard, mother of Christian Regenhard, a probationary firefighter killed at the WTC. We met with her and found that she also wanted to know what happened and wanted a comprehensive investigation into the disaster.

The result of our meeting was the creation of an advocacy group, the Skyscraper Safety Campaign (SSC). The primary goal of the organization was the creation of a fully resourced technical investigation into not only the failures of the towers but also the emergency response and evacuation. First, however, the SSC had to deal with an ongoing 9/11 inquiry sponsored by FEMA and conducted by the American Society of Civil Engineers (ASCE): the *World Trade Center Building Performance Study* (FEMA 2002).

The *World Trade Center Building Performance Study* was prepared by an ASCE group, known as a Building Performance Assessment Team (BPAT). A similar study was conducted on the Pentagon attack by a parallel BPAT. The BPAT studies were focused exclusively on the structural/fire protection issues of the buildings themselves and did not directly address concerns of emergency response and evacuation.

It became readily apparent to me that the BPAT approach to the WTC disaster had major flaws, primarily the fact that the "study" was focused too narrowly. It needed to be broadened to include the emergency response and evacuation issues—they are all interrelated and of significant concern. In addition, the BPAT was staffed with volunteers with limited financial resources and no authority to investigate and collect evidence. Since the team was made of volunteers (albeit well-qualified volunteers) with regular employment elsewhere, they had limited time to spend on the WTC inquiry.

One pressing issue was the fact that the steel of the twin towers—the keys to understanding the impact of the aircraft, the effects of the subsequent fire exposure, and the collapse mechanisms—was being removed from the site and sent to scrapyards in the immediate months after the disaster. Without this physical evidence, it would be impossible to definitively establish how the structures reacted up to the point of collapse. Even then, I knew that heavy emphasis would be placed on computer modeling instead of actual physical evidence.

Attempts to preserve the steel were met with silence on the part of the city of New York. When the BPAT team leader asked the city to save the steel, his request was essentially ignored (Glanz and Chang 2001). Sally Regenhard sent an overnight request to FEMA director Joe Allbaugh, making a similar request (FEMA was also involved in the recovery effort, which included steel removal). Her request was also ignored.

Limited inspections of the steel were performed at the scrapyards by structural engineering volunteers as well as others, such as Dr. Abolhassan Astaneh-Asl, a structural engineering professor from the University of California at Berkeley who was at the site conducting his own investigation with a National Science Foundation grant. There was no comprehensive evaluation of the steel moving through the scrapyards, however. It was a sporadic effort. During a conference call, when I questioned BPAT members whether *each piece of steel* was evaluated for its proximity to the impact/collapse zone, they admitted that only the visible pieces on the top surface of the enormous piles of steel were actually reviewed.

The SSC had its first press conference near New York's City Hall on December 14, 2001 (Glanz 2001). Joined by Deputy Fire Chief Vincent Dunn (ret.) and other 9/11 family members, Sally Regenhard called for a comprehensive investigation into the WTC disaster. I argued for an investigation on the order of the investigations conducted by the NTSB; we needed a national capability to conduct building failure investigations, just as we investigate plane crashes. The twin towers of the WTC were the largest buildings in recorded human history to collapse, yet the federal response was to dispatch BPAT volunteer engineers with virtually no money or authority to conduct a proper investigation.

The call for a wider, deeper, and more comprehensive investigation continued into early 2002. By February, it caught the attention of the House Committee on Science, Space, and Technology. The chairman of the committee, Representative Sherwood Boehlert (R-NY), called for a hearing on March 6, 2002 (U.S. Congress 2002). It would prove to be a very significant event, leading to major changes in the manner in which building disasters are investigated.

## CONGRESSIONAL HEARINGS

The House Committee on Science, Space, and Technology hearing on March 6, 2002, included a single panel of five witnesses. In addition to me, the panel included Robert F. Shea, acting administrator, Federal Insurance and Mitigation Administration, FEMA; Dr. W. Gene Corley, senior vice president, CTL Engineering and leader of the FEMA BPAT; Dr. Abolhassan Astaneh-Asl, professor, Department of Civil and Environmental Engineering, University of California, Berkeley; and Dr. Arden L. Bement Jr., director, National Institute of Standards and Technology, U.S. Department of Commerce.

The lengthy March hearing focused on the shortcomings of the BPAT approach to investigating the WTC disaster. It identified many of the issues described above—lack of authority, volunteer personnel, minimal financial resources, and the unabated recycling of the critical structural steel remnants. It also highlighted other problems that had arisen, including the complete secrecy of the scope and details of the BPAT as the result of confidentiality agreement requirements of FEMA and the ASCE. This issue created a situation in which other investigators, such as Dr. Astaneh, were not informed about what the BPAT was actually doing, making it impossible to know whether they were duplicating efforts or missing issues of concern. The hearing, however, was dominated by discussion over the uncoordinated efforts among all the inquiries that were in progress.

The problem of coordination was highlighted when Representative Anthony Weiner (D-NY) asked a simple question: Who is in charge of the investigation? Three hands on the panel were raised, illustrating the uncoordinated overlap between a group of government and private sector inquiries.

Chairman Boehlert summed up the problems in his closing remarks:

> Well, look, I think we have learned a lot of things this afternoon—some things that raised concerns and that underscore some of the points I made right at the beginning.
>
> First, right now, there are no clear lines of authority as to who in the Federal Government is to conduct an investigation of a building failure. No one is in charge. No one is sure what powers the Federal Government can exercise. No one is sure of the scope of an investigation, and that has to be fixed right away. And I see a lot of nodding of heads from the Panel, and I appreciate that.
>
> We need an enhanced disaster investigation protocol in place—thank you, Dr. Corbett—so that from minute one, someone will assume a recognized leadership role. Someone has to be in charge or no one is. And having no one in charge is unacceptable.
>
> Some of it can be fixed by the Federal agencies, and I understand that handing the ball from FEMA to NIST, which I think is logical and makes sense, but some of it has to be fixed by those of us up here in the Congress with legislation. And we are going to be looking for very specific proposals.

The hearing in March was followed by another hearing on May 1, 2002 (U.S. Congress 2005a). The witness panel consisted of the same individuals who had appeared at the previous hearing, with the exception of Dr. Astaneh. A primary focus of the hearing was discussion of the results of the BPAT study that was to be released that day. However, given the severe criticism of the BPAT study and the obvious need for change, the hearing was actually a starting point for legislatively addressing the problems that had been highlighted in the preceding months.

The BPAT study, while informative, did not provide definitive conclusions about what had occurred on 9/11. In essence, it highlighted specific areas that it believed needed to be investigated in much greater detail. As noted above, the BPAT study did not address the concerns of emergency response and building evacuations.

The National Institute of Standards and Technology (NIST) became the obvious successor to the WTC investigation. In the months between the hearings, the Bush administration had requested $16 million for NIST to conduct an investigation. (Ironically, this amount was twice as large as the budget of the 9/11 Commission, which would conduct its own inquiry during the same period as NIST [Kean and Hamilton 2006].) The hearing debated the scope of the anticipated investigation and specific questions that needed to be answered.

I pushed for the issue of most importance to me: the need to translate the findings of the investigation into actual positive changes in building and fire codes as well as emergency response procedures. The WTC disaster exposed myriad problems (many of them well known before 9/11), ranging from inadequate high-rise stairwell widths, portable radio signals attenuated by structural steel building frames that made communications impossible, poor fireproofing insulation standards for steel, to an outdated 1920s-era fire resistance test standard. From my perspective, NIST's conclusions, whatever they were, had to result in actual safety improvements.

Of particular note in this hearing was the issue of NIST's investigative authority, particularly the need for subpoena power. The BPAT study had shown how the lack of such power was a significant impediment, particularly getting access to critical documents such as blueprints. The House Committee on Science, Space, and Technology focused on this issue, particularly in the context of having NIST assume the lead role in the investigation. To some extent, this became a political issue because the director of NIST, Dr. Bement, hesitated to endorse the concept without consulting the administration. It was obvious to many of the questioning members of Congress that such authority was critical.

In retrospect, this hearing was the genesis of the legislative action that would follow. The federal government needed to have the ability to investigate significant building failures, just as it investigates plane crashes. The following July, legislation was introduced to create this capability.

## THE NATIONAL CONSTRUCTION SAFETY TEAM ACT

After the May hearing, the House Committee on Science, Space, and Technology quickly drafted a bill, the National Construction Safety Team Act (NCST), which placed the responsibility for investigating building disasters squarely on NIST. Passed by the House of Representatives on July 22, it was then sent to the Senate for consideration, and by the first anniversary of 9/11, it was being discussed by the Senate Committee on Commerce, Science, and Transportation.

In an effort to speed passage by the Senate, Sally Regenhard and SSC co-chair Monica Gabrielle visited the chairman of the committee, Senator John McCain (R-AZ), while Congress was in session in New York City (in commemoration of the 9/11 attacks). Within a few days, the bill was passed by the committee and the Senate itself and then sent to President Bush, who signed it into law on October 1, 2002. The speed with which the bill passed into law indicates an appreciation of its obvious importance (National Construction Safety Team Act 2002).

The NCST included all the desired objectives highlighted during the congressional hearings: The technical investigations were to be comprehensive in nature, including issues of building failure, emergency response, and emergency evacuation. It gave NIST subpoena power and the ability to secure evidence and to hold public hearings in which witnesses could be questioned. Most important, it required that NIST prepare recommendations for changes to national building/fire codes and policies and procedures. It also included a provision for the creation of a federal advisory committee, which would provide investigations with input and "outside" technical expertise.

## THE NIST INVESTIGATION

I was appointed as a charter member of the ten-member federal advisory committee to NIST and the NCST. Although each member was appointed as a subject matter expert on a particular area of a disaster investigation (I was appointed to represent "emergency responder" issues), all members were permitted to comment on other issues raised by the investigation.

NIST held several public hearings to discuss their proposed investigation plan. The goals, objectives, and guiding principles of the 24-month investigation, from the "final plan," were as follows (NIST 2002).

**Goals:**

- To investigate the building construction, the materials used, and the technical conditions that contributed to the outcome of the World Trade Center (WTC) disaster.
- To serve as the basis for:
  - Improvements in the way buildings are designed, constructed, maintained, and used;
  - Improved tools, guidance for industry and safety officials;
  - Revisions to codes, standards, and practices; and
  - Improved public safety.

**Objectives:**

The primary objectives of the NIST-led technical investigation of the WTC disaster are to:

1. Determine why and how WTC 1 and 2 collapsed following the initial impacts of the aircraft and why and how WTC 7 collapsed;
2. Determine why the injuries and fatalities were so high or low depending on location, including all technical aspects of fire protection, occupant behavior, evacuation, and emergency response;
3. Determine what procedures and practices were used in the design, construction, operation, and maintenance of WTC 1, 2, and 7; and
4. Identify, as specifically as possible, areas in building and fire codes, standards, and practices that are still in use and warrant revision.

**Guiding Principles:**

- Active, comprehensive, independent, and objective technical investigation that is fully informed of the concerns and issues of all interested parties and within the limits of available resources.
- Open and inclusive process in planning and conducting the investigation, and in publishing and disseminating findings and recommendations.
- Contribute to improving standards, codes, and practices to reduce future risks by focusing on:
  - Fact-finding and analysis of the facts, and
  - Validating and verifying existing knowledge.
- Nontechnical issues are outside scope of investigation: no findings of fault or negligence by any individual or organization.
- Maintain ongoing liaison with professional community, public, and local authorities.
- Project teams that include NIST staff and external world-class technical experts.

As can be seen, these goals, objectives, and guiding principles established an ambitious investigation plan. Many aspects of the investigation plan, particularly the guiding principles, were based on protocols developed at the NTSB and the U.S. Chemical Safety and Hazard Investigation Board (CSB, which primarily investigates explosions in industrial facilities). Note the specific NIST guiding principle indicating that this was solely a technical investigation and that there would be no "findings of fault or negligence by any individual or organization."

The investigation covered a broad range of issues and topics. Given the size and complexity of the disaster, the WTC NIST investigation became a lightning rod for many high-rise fire safety problems and concerns. The investigation itself was divided into eight "project areas": analysis of building and fire codes and practices; baseline structural performance and aircraft impact damage analysis; mechanical and metallurgical analysis of structural steel; investigation of active fire protection systems; reconstruction of thermal and tenability environment; structural fire response and collapse analysis; occupant behavior, egress, and emergency communications; and fire service technologies and guidelines. It is important to note the interrelationship of each of these project areas—data gained in one project area informed one or more of the other projects. For example, the "Reconstruction of Thermal and Tenability Environment" (fire modeling) project was used for input data for the "Structural Fire Response and Collapse Analysis" project. This issue placed great importance on coordination, including the availability of data at crucial junctures to keep the investigation moving.

Although NIST employs a significant number of researchers and has a substantial testing facility, a significant portion of the investigation was contracted out to engineering firms. For example, much of the data collection in the "Analysis of Building and Fire Codes and Practices" was subcontracted to a fire protection engineering firm. Fire testing of the floor trusses—newly built floor beams constructed to the specifications of the actual trusses used in the WTC—was conducted in a test furnace at a private laboratory. Although this "outsourcing" concerned some observers, the awarding of a monetary contract to the WTC's original structural engineer of record to provide data, including a structural analysis of the design, raised many eyebrows. This real potential for conflict of interest forced NIST to hire a third-party engineering firm to review the data submitted (NIST 2003).

This is not to say that NIST failed to employ many of its own resources. NIST assigned numerous employees and used several of its testing facilities. Of particular note was the enormous amount of computer modeling employed, specifically the modeling of the plane impacts, fire spread, and collapse sequences. These were perhaps the most complicated models of such subject matter ever created.

Over the next three years, the advisory committee held meetings to discuss the progress of the investigation, in closed sessions, as well as open sessions that allowed the public to comment. As an advisory board member, I was asked to sign a confidentiality agreement, which I thought odd, given the claim of an "open process" in the investigation's guiding principles (and the well-known concerns over the lack of transparency of the BPAT study).

As the investigation progressed, the conflict between the technical aspects of the investigation and the issue of "avoiding blame" became apparent. Each of the WTC towers had three stairwells, despite the prevailing code requirement (at the time of construction) of four, based upon the high number of occupants in the restaurant and observation deck at the top of each tower (Dwyer 2005). In addition, fire testing of the floor trusses encapsulated in half an inch of "fireproofing" insulation, as specified in the original building design, showed that this half-inch would not have met the prevailing code requirements at the time of construction.

On February 20, 2003, in the midst of the WTC investigation, a fire broke out in the Station nightclub in West Warwick, Rhode Island, killing 100 people. A few days after the fire, NIST decided to initiate another NCST investigation, this time on the Station fire. The NCST advisory provided input on this investigation as well. Of particular note was the difficulty that NIST encountered with the Rhode Island attorney general, who refused NIST access to the physical evidence of the fire, including foam rubber wall finish which was central to the spread of the fire.

In both investigations, I became concerned over NIST's lack of aggressiveness in obtaining information and evidence. I was particularly concerned about NIST's reluctance to use its subpoena power and to conduct public hearings of witnesses under oath. For example, although the Rhode Island attorney general had refused to give NIST access to evidence, there was no reason why NIST could not hold a public hearing with witnesses under oath (Corbett 2006).

I was also concerned that the most important ultimate goal of the WTC investigation—the preparation of recommendations for improvements to our codes, policies, and procedures—was not being met. While I had hoped that NIST would prepare the recommendations as part of the investigation, NIST asked the National Institute for Building Sciences (NIBS) to perform that function. NIBS was contracted not only to create the specific code change language but to shepherd the changes through the code modification process (NIST 2008).

The final reports on the Towers 1 and 2 of the WTC were released to the public in September 2005 and contained thirty recommendations. The final report on Tower 7 was delayed for an extended period and was released in November 2008.

## REGULATORY AND TECHNOLOGICAL IMPROVEMENTS

On October 26, 2005, the House Committee on Science, Space, and Technology held a hearing in conjunction with the release of the NIST report on Towers 1 and 2. I testified at the hearing, expressing my frustration over several of the issues identified above, including the lack of aggressiveness in acquiring information and evidence and the lack of specificity in many code-related recommendations. I also called attention to NIST's reluctance to implement the NSCT for other disasters, noting that it did not deploy an NCST investigative team to the Gulf Coast after Hurricane Katrina struck the region (to this day, NIST has not deployed NCST capabilities to investigate any other building failure). While recognizing the talents of many NIST researchers, I recommended that the NCST be modified, turning over responsibility for its implementation to the CSB (given CSB's investigative focus) (U.S. Congress 2005b).

NIBS eventually developed a series of specific code-change proposals that were presented to our primary model code-writing groups, the National Fire Protection Association and the International Code Council. Some of these changes have made their way into these code groups' model building and fire codes, including requirements for high rises to have additional stairwells, redundancies in water supplies for sprinkler systems, larger fire command centers, and improved emergency responder radio communications coverage. Some other recommendations have yet to be implemented in the model codes (NIST 2008).

## CONCLUSION

The NIST investigation of the 9/11 disaster at the World Trade Center was the largest technical disaster investigation ever conducted. It covered a multitude of issues, many of them well known for many years in the fire protection community. It highlighted a variety of investigative issues, including the negative influence on the investigation of attempting to avoid "blame," the loss of the steel (from the points of aircraft impact) to recyclers (only a few dozen pieces were actually recovered from both towers, making it impossible to challenge the validity of the computer model results with physical evidence), a substantial reliance on outside contractors to perform important investigative functions, the need for aggressiveness to obtain information and evidence, and the need for specific recommendations for building and fire code improvements as part of the investigation process, to ensure more timely adoption.

Individuals conducting future disaster investigations are well advised to review the history of the NIST WTC investigation to gain important insight into the complexities and difficulties of such an effort. The references cited in this chapter provide a roadmap to conducting such a review.

## REFERENCES

*Boston Journal*. 1860. "The Lawrence Disaster: Capt. Bigelow's Reply to the Verdict of the Jury." *New York Times,* February 7.

Calderone, J. 2001. "2 Seek Probe of Collapse." *New York Daily News,* October 30. http://articles.nydailynews.com/2001–10–30/news/18361752_1_fire-safety-fire-prevention-fire-engineering/ (accessed December 27, 2010).

Corbett, G. 2006. "The Station Fire—A.G.'s Office Blocked Experts from Getting Facts." *Providence Journal,* October 16. www.projo.com/opinion/contributors/content/projo_20061016_ctcorb.31d9d1b.html (accessed December 27, 2010).

Dwyer, J. 2005. "Staircases in Twin Towers Are Faulted." *New York Times,* April 6. http://query.nytimes.com/gst/fullpage.html?res=9F04E2DE163EF935A35757C0A9639C8B63&scp=6&sq=dwyer%20world%20trade%20center%20stairwell&st=cse&pagewanted=1/ (accessed December 27, 2010).

Federal Emergency Management Agency (FEMA). 2002. "World Trade Center Building Performance Study." Washington, DC. www.fema.gov/pdf/library/fema403_cover-toc.pdf (accessed December 27, 2010).

———. 2010. "Evaluating a Request for a Major Disaster Declaration." Washington, DC. www.fema.gov/rebuild/recover/evaluation.shtm (accessed December 27, 2010).

Freeman, J. 1906. *On the Safeguarding of Life in Theaters.* New York: American Society of Mechanical Engineers.

Glanz, J. 2001. "Demand Rises for Widening Investigation into Collapse." *New York Times,* December 15. www.nytimes.com/2001/12/15/nyregion/15TOWE.html?scp=1&sq=regenhard%20glanz&st=cse/ (accessed December 27, 2010).

Glanz, J., and Kenneth Chang. 2001. "Engineers Seek to Test Steel Before It Is Melted for Reuse." *New York Times,* September 29. http://query.nytimes.com/gst/fullpage.html?res=9E03E7DC173DF93AA1575AC0A9679C8B63&scp=2&sq=corley%20new%20york%20times%20steel%20world%20trade&st=cse/ (accessed December 27, 2010).

Grant, C. 2007. "Fire Safety Regulations." *NFPA Journal* (November/December). www.nfpa.org/itemdetailjournal.asp?categoryID=1517&itemID=36758&cookie%5Ftest=1/ (accessed December 27, 2010).

Jackson, R. L. 1938. *Living Lessons from the New London Explosion.* Nashville: Parthenon Press.

Kean, T., and L. Hamilton. 2006. *Without Precedent: The Inside Story of the 9/11 Commission.* New York: Knopf.

Kornblum, W. 2002. *At Sea in the City: New York from the Water's Edge.* Chapel Hill, NC: Algonquin Books. www.newyorkhistory.info/Hell-Gate/General-Slocum.html.

Little, G. 1860. *The Fireman's Own Book: Containing Accounts of Fires Throughout the United States.* New York: George P. Little.

Marshall, R. 1982. "Investigation of the Kansas City Hyatt Regency Walkways Collapse, Gaithersburg, Maryland." U.S. Department of Commerce, National Bureau of Standards, Washington, DC.
McLaury, Judson. n.d. "The New York Factory Investigating Commission." In *Government Regulation of Workers' Safety and Health, 1877–1917*. U.S. Department of Labor, Washington, DC. www.dol.gov/oasam/programs/history/mono-regsafepart07.htm (accessed December 27, 2010).
Merriam-Webster. 2010. "Disaster" [definition]. Merriam-Webster.com. http://mw1.merriam-webster.com/dictionary/disaster (accessed December 27, 2010).
National Construction Safety Team Act. 2002. 107th Congress, 2d Sess., January 23. www.nist.gov/public_affairs/factsheet/upload/hr46871.pdf (accessed December 27, 2010).
National Institute of Standards and Technology (NIST). 2002. "Final Plan: Federal Building and Fire Safety Investigation of the World Trade Center Disaster." NIST Special Publication 1000–1. http://wtc.nist.gov/media/WTCplan_new.htm (accessed December 27, 2010).
———. 2003. "Contracts: Awards." http://wtc.nist.gov/solicitations/LERA_Award.htm (accessed December 27, 2010).
———. 2008. "NIST's Recommendations Following the Federal Building and Fire Investigation of the World Trade Center Disaster." October. http://wtc.nist.gov/recommendations/recommendations.htm (accessed December 27, 2010).
Negroni, C. 2000. *Deadly Departure: Why the Experts Failed to Prevent the TWA Flight 800 Disaster and How It Could Happen Again*. New York: William Morrow.
Puleo, S. 2003. *Dark Tide: The Great Boston Molasses Flood of 1919*. Boston: Beacon Press.
Sheehan, L., and K. Hewitt. 1969. "A Pilot Survey of Global Disasters of the Past Twenty Years." Natural Hazards Research Working Paper 11. University of Colorado, Institute of Behavioral Science, Boulder.
U.S. Congress. 2002. "Learning from 9/11: Understanding the Collapse of the World Trade Center." Hearing before the Committee on Science, 107th Congress, 2d Sess., March 6, Serial No. 107–46. http://commdocs.house.gov/committees/science/hsy77747.000/hsy77747_0f.htm (accessed December 27, 2010).
———. 2005a. "The Investigation of the World Trade Center Collapse: Findings, Recommendations, and Next Steps." Hearing before the Committee on Science, 107th Congress, 2d Sess., May 1, Serial No. 107–61. http://commdocs.house.gov/committees/science/hsy78961.000/hsy78961_0f.htm (accessed December 27, 2010).
———. 2005b. Statement of Professor Glenn P. Corbett. Hearing before the Committee on Science, House of Representatives, 109th Congress, 1st Sess., October 26. Serial No. 109–28. www.access.gpo.gov/congress/house/pdf/109hrg/24133.pdf (accessed December 27, 2010).
Von Drehle, D. 2003. *Triangle: The Fire That Changed America*. New York: Atlantic Monthly Press.
World Bank. 2010. "Disaster" [definition]. In Glossary. http://info.worldbank.org/etools/docs/library/230308/Course%20Resources/Glossary.doc (accessed December 27, 2010).

# PART III

# Issues in Critical Incident Analysis Management

CHAPTER 12

# The Critical Role of Street-Level Bureaucrats in Disaster and Crisis Response

ALEXANDER C. HENDERSON

Front-line emergency responders occupy a fundamental role in organizations that provide services in response to a disaster or crisis yet are often neglected in scholarly examinations of these events. Research examining responses to these critical incidents has addressed intergovernmental relations during the four phases of disasters (i.e., mitigation, preparedness, response, and recovery) (Comfort 1985, 2007; Waugh 1994), leadership and management capabilities necessary for successfully handling these incidents (Boin and 't Hart 2003; Farazmand 2007; Kapucu and Van Wart 2008; Waugh and Streib 2006), detailed studies of the often complex network of organizations that respond to large incidents (Comfort et al. 2001; Farazmand 2007), and communication and coordination during responses (Comfort and Haase 2006; Kapucu 2006). As expected, many of these studies are presented as case studies (Comfort 2007; Farazmand 2007; Kapucu 2006). This research has often focused on the systems level, including the effects of critical incidents on political, economic, and social systems at the organizational level on issues such as capacity, adaptability, and resiliency, or multilevel studies examining interactions and network responses. While this critically important research is necessary and exceptionally important, a microperspective examining the behavior of front-line responders, both professional and volunteer, is largely missing from the discussion and will serve to enhance knowledge on improving responses to these incidents.

Responders on the front lines are essentially the last stop in the implementation of response policies and plans and thus substantially affect the lives of those affected by these incidents. The jobs performed by street-level bureaucrats (SLBs)—defined as those "who interact directly with citizens in the course of their jobs, and who have substantial discretion in the execution of their work" (Lipsky 1980, 3)—during the response phase of critical incidents vary widely. These individuals conduct search and rescue operations, provide for safety and security, assist with the evacuation of residents, distribute emergency goods, coordinate resources, and work to restore normalcy to social, political, and economic systems. These SLBs include federal, state, and local government personnel, with occupations ranging from emergency responder to city planner, and include both part-time employees hired to temporarily increase agency staffing during critically necessary periods and those who respond in a volunteer capacity to crises and disasters.

Although previous research on SLBs commonly associated with response to disasters and crises (e.g., police officers, firefighters, and emergency medical services personnel) has been conducted, in many cases it is only loosely applicable to these large-scale incidents. First responders charged with providing "routine" emergency services (e.g., responding to residential building fires, car accidents, and emergency medical incidents) are in most cases the first individuals to respond to larger-scale crises. However, there are critical differences in the primary tasks of these different types of first responders as they respond to routine emergencies and crises. As Howitt and Leonard (2009) note, important situational, environmental, and scale problems exist between these types of incidents. The key functions that these individual serve, specific tasks they must complete, and service outcomes can differ from the routine emergency incidents that make up their normal workday.

Examples of the behavior of front-line workers in disasters and crises are easy to imagine: firefighters perform search and rescue operations in collapsed buildings, police officers maintain order and enforce laws in chaotic situations, and emergency medical services workers triage, treat, and transport patients to hospitals. Other jobs, though less visible, are also critically important. Scott provides anecdotal evidence of the strict application of rules by employees of the Federal Emergency Management Agency (FEMA) during the response to Hurricane Katrina:

> The bureaucracy of FEMA also seemed to interfere with the smooth delivery of supplies. Despite Brown's urging FEMA staff to "push the envelope" and cut red tape in prestorm meetings, there were numerous tales of trucks being halted or diverted or refused entry by agency workers. The problem, Brown explained in his House testimony, lay with disaster assistance employees, a "cadre" of part-time workers the agency hired to provide "surge-capacity in times of disaster." (2009, 58)

The aggregated behavior of individual actors at the front lines of emergency response in essence becomes the response of organizations to critical incidents. As Comfort et al. (2001) note, organizational performance in the mitigation of an emergency incident is a product of both individual behavior and environmental characteristics of the emergency incident. Thus, a better understanding of individual behavior can contribute to improving organizational performance during the response phase of disasters and crises and leads to tangible effects on both citizens and society. The ramifications of both individual and organizational failure in the response to emergency incidents are high. Bigley and Roberts note that "operational failures resulting in inappropriate, incomplete, laggardly, or otherwise mindless organizational responses to unexpected and demanding environmental contingencies . . . are ever more likely to be immediately and critically disabling" (2001, 1281). "Normal" political, social, and economic functioning are, to some extent, disrupted by emergency incidents. These disruptions can be even more significant in situations where the response organizations perform poorly.

This chapter discusses the nature of street-level work, including an examination of discretion at the front lines of crisis and disaster response, autonomy and management in street-level work, and sources of influence that shape the behavior of those providing front-line work.

Two of these sources of influence are highlighted, including occupational culture and client characteristics. These two factors are particularly salient, as they both represent routinization of key work factors in an effort to process massive amounts of information to which they are exposed (van Ryn and Burke 2000).

## LITERATURE REVIEW

### Complexity, Uncertainty, and Performance in Critical Incidents

Although already complex in nature, individual behavior in emergency mitigation is complicated by the uncertain environment created by the very nature of an emergency or crisis. As Moynihan notes, crises and emergencies are "characterized by high consequentiality, limited time, high political salience, uncertainty, and ambiguity" (2008, 351). Boin and 't Hart add that these events "disrupt a wide range of social, political, and organizational processes. Crises are dynamic and chaotic processes, not discrete events sequenced neatly on a linear time scale" (2003, 545). High task complexity, task uncertainty, and environmental uncertainty all provide additional challenges in responding to crises and disasters (Boin and 't Hart 2003; Comfort 1985, 2007; Comfort et al. 2001; Farazmand 2007). In many cases, the volatile nature of the trigger event (Kapucu and Van Wart 2008), the physical and social repercussions of that event, and the resulting disruption to normalcy all contribute to substantial uncertainty. Complexity and uncertainty can, in turn, lead to declines in organizational performance (Comfort et al. 2001, 144).

Scholarly work on improving organizational performance in the management of emergency incidents has focused on the role of new technologies in the form of hardware and electronics. Indeed, the introduction of these technological improvements has provided a "means to overcome the long-observed decrease in organizational performance in complex environments" (Comfort et al. 2001, 144). However, this vein of research tends to focus on the hardware and ignore the importance of individual actors and their behavior in an individual and group context. Others address the nature of individual and group action in a complex and uncertain environment and most often note humanness as a limitation. La Porte (1975), quoted in Comfort et al. (2001), notes that "performance in complex environments has been viewed as necessarily limited by human information processing capacity" (144). Though this "humanness" can be viewed as an obstacle, it is nonetheless one that must be both acknowledged and examined.

Past research points to the importance of front-line workers in responding to critical incidents. Schneider (1992) contrasts the traditional bureaucratic norms associated with a structured, orderly response developed under an assumption of relatively normalcy with a more chaotic, unpredictable response and the resulting difficulties for individuals at all levels of the response framework:

> Some emergent norms are highly consistent with previously existing behavior patterns and routines (e.g., orderly evacuations, voluntary relief efforts). In such cases, the government's response can work as intended. In other situations, the emergent norms conflict with the usual social environment. This conflict may take many different forms, from vocal public dissatisfaction to violence and social unrest. This creates patterns of interaction that are unanticipated by the government response system. Hence, no specific contingencies exist for dealing with them. When this occurs, the gap between government planning and human behavior is particularly wide, and the response process will probably be viewed as a failure. (Schneider 1992, 138)

McLoughlin, discussing the intergovernmental system of emergency response, notes that successful emergency services organizations require cooperative and interdependent action of individuals at all levels of the organization (1985, 165). Though focused primarily at the managerial level, this certainly extends to the front lines as well. Likewise, Waugh and Streib (2006) indicate that interpersonal skills, rather than technical skills, have been recognized as important in mitigation of emergency incidents. This is certainly true in cases in which SLBs from multiple organizations must work together closely in providing services. Given the importance of the human factors in emergency response, a discussion of key role characteristics of a street-level bureaucrat is necessary.

## Street-Level Bureaucrats

Those studying the topic have reached a consensus on the definition of a "street-level bureaucrat." Lipsky identifies street-level bureaucrats as those "who interact directly with citizens in the course of their jobs, and who have substantial discretion in the execution of their work" (1980, 3). Similarly, Vinzant and Crothers define street-level bureaucrats as "those who are directly responsible for service delivery to the public and who exercise a significant level of discretion in carrying out their responsibilities" (1998, 12). Riccucci defines street-level bureaucrats as workers "who have direct, face-to-face contact with . . . clients" (2005, 9). These definitions make clear that street-level workers are the face of public service, tasked with direct provision of public services. Street-level bureaucrats are, in other words, an "analytically unique category" of public servant that can be viewed as different and distinct from other types of public servants and can be "understood in terms generic to the role rather than in organizationally specific, ad hoc terms" (Prottas 1978, 288).

Ideally, the function of street-level bureaucrats within the larger context of governance is, according to Vinzant and Crothers, "quite simple: such workers execute the rules, programs, and policies established by their agencies in accordance with the law" (1998,

10). Bovens and Zouridis note that "policy comes alive in the daily practice of street-level bureaucracy" (2002, 175). Street-level bureaucrats are responsible, at the most basic level, for policy implementation. As Prottas notes, "these bureaucrats do the 'actual' work of the agency—they interview clients, arrest suspects, treat patients, and so forth—and the way they do their work must therefore affect the way the public bureaucracy fulfills its public responsibility" (1978, 286). The accumulated actions of street-level bureaucrats constitute a critical area of interest, not just for individual clients but for analysis and evaluation of public programs. As Riccucci notes, "it is at the street level where policy delivery may be most critical for social programs, because the actions of front-line workers have substantial and sometimes unexpected consequences for the *actual* direction and outcome of benefit programs" (2005, 5). In short, street-level bureaucrats distribute the goods and services of the state (Prottas, 1978).

Riccucci (2005) situates the actions of street-level bureaucrats in the larger context of the provision of public services in a democratic system. As agents of the state, street-level bureaucrats are charged with fulfilling a public need or achieving a desired outcome through distribution of goods and services. However, the process of crafting legislation, administrative rules and regulations, and additional policies and procedures can shape the actions of bureaucrats in a manner that deviates from the intent of democratically elected officials. In some cases, though the behavior of front-line workers may be satisfactory to managers, "the workers are not doing what the elected officials—in particular, governors, and state legislators—and high-level bureaucrats expected them to do. That is to say, street-level bureaucrats are not implementing the policies that the 'state' intended to be delivered" (Riccucci 2005, 75). This is important in that it points to the role of street-level bureaucrats in sustaining the democratic ideals of the country.

The jobs of street-level bureaucrats are anything but simple (Bovens and Zouridis 2002; Hill 1974). Indeed, "not only is there great variation and unpredictability in the problems these workers face, but the problems themselves are often multifaceted, intractable, and emotionally laden" (Vinzant and Crothers 1998, 4). These individuals are, in the words of Maynard-Moody and Musheno the "coal miners of policy: they do the hard, dirty and dangerous work of the state" (2003, 157). The vital work of street-level bureaucrats raises the question of how these individuals make decisions and behave in an uncertain and stressful environment.

In their service encounters with the public, street-level bureaucrats act as boundary-spanners (Hupe and Hill 2007; Prottas 1978; Wenger and Wilkins 2008). As they distribute the public goods and provide services, they interact directly with citizens, a position that confers the street-level bureaucrat with some power (Prottas 1978; Thompson 1967). By virtue of the position, street-level bureaucrats work in:

> two interdependent worlds. This simultaneity is the key to the bureaucrats' power. Slotting labels a client and defines a bureaucratic process. As such, it combines information internal to an organization with information external to it. The bureaucracy and its non-street-level employees have access to the internal facts—rules, formal categories, procedures, and so forth. The clients have access to the external

> information—facts about themselves, their demands, and so forth. Only the street-level bureaucrat has routine access to both. (Prottas 1978, 293)

Thompson adds that this power emanates from the need to meet contingences in a "heterogeneous and shifting environment" (1967, 111), thus requiring the use of discretion in the course of the service interaction. However, this confers power only in cases in which these individuals are operating in a complex situation and the magnitude of their actions is great. A simple, unchanging environment—or low task complexity—would entail little use of discretion or would not have consequences of interest to the organization. At the same time, environments characterized by significant change or substantial consequences confer greater power on the street-level bureaucrat (Thompson 1967). Therefore, an in-depth look at discretion is warranted.

### Discretion on the Front Lines of Public Service

Street-level bureaucrats exercise a significant amount of discretion in carrying out their duties on the front lines of public service (Lipsky 1980). For the purpose of this discussion, discretion is defined as "the making of a choice, within certain constraints, regarding the task process or outcomes" (Lipsky 1980; Scott 2009; Vinzant and Crothers 1998).

According to Lipsky (1980), discretion is a necessity on the front lines for three reasons. First, street-level bureaucrats work in situations that are often "too complicated to be reduced to programmatic formats" (1980, 15). In the words of Frederickson, "every application of a law involves further elaboration of that law" (in Maynard-Moody and Musheno 2003, 10). The complexities of street-level work do not allow for the promulgation of rules and regulations to cover every possible work situation, thus requiring some amount of discretion. As Bovens and Zouridis note, it is at the street level where, "despite detailed rules and regulations, reality is shown to be far more complex and varied than legislators had ever dreamed" (2002, 175). Even in situations where a substantial body of rules and regulations are created to guide behavior, street-level bureaucrats may not be able to internalize and follow every guideline. As Prottas notes,

> To serve multiple and conflicting and ambiguous goals, agencies promulgate an even greater multitude of even more ambiguous and conflicting regulations. A small number of these rules are given de facto priority and are obeyed. The rest are all to be obeyed equally yet manifestly cannot be. They are too numerous and too inconsistent in content. Therefore they cannot provide clear priorities for street-level behavior nor a clear basis for evaluating that behavior. (1978, 295)

Second, street-level work often results in situations that "require responses to the human dimensions of situations" (Lipsky 1980, 15). Front-line workers must continually evaluate their actions for appropriateness as client needs and situational factors change. Indeed, the act of responding to clients can, by itself, pose substantive difficulties. Prottas notes that when we try to find a clear definition of street-level responsibilities, the impediments from above are matched by impediments from below. The behavior of clients is not rule-bound and their demands are wholly unbureaucratic in form. This injects an irreducible element of uncertainty into the job of the street-level bureaucrat and the agency's expectations of him/her cannot help by reflect that uncertainty (1978, 295–296).

Finally, discretion in front-line work "promotes workers' self-regard and encourages clients to believe that workers hold the key to their well-being" (Lipsky 1980, 15). Street-level bureaucrats may be more likely to consider the impact of their actions if the ramifications of their actions are proximal and if clients believe in the efficacy of the individuals with whom they are interacting. Indeed, discretion is the hallmark of the professionalization of an occupation and "presupposes a degree of trust in [the street-level bureaucrat] to produce desired responses, and to deal with situations that may be exceptional in a sensible and creative way" (Hupe and Hill 2007, 282). The resulting discretion can be viewed both positively and negatively. "At best, street-level bureaucrats invent benign modes of mass processing that more or less permit them to deal with the public fairly, appropriately, and successfully. At worst they give in to favoritism, stereotyping, and routinizing—all of which serve private or agency purposes" (Lipsky 1980, xii).

Several types of discretion exist in the world of street-level bureaucrats. First, discretion can be built in to the rules, policies, and procedures created to regulate action on the front lines. This "within-rule" discretion is bounded by the officially promulgated rules and comprises two or more outcomes that presumably have the same intended outcome. Second, as noted earlier, discretion can be found in those areas not regulated by the rules, policies, and procedures. Street-level work is inherently complex, and not every aspect of the job can be legislated. Finally, due to the rule-saturated environment of street-level work, rule complexity, and situational stress, front-line workers may purposefully not follow all rules created to guide their actions. Deviation from complex rules sets has been previously documented in emergency services organizations (Davis 1975; Salerno, Wrenn, and Slovis 1991). These deviations from rules open an additional area of decision making to influence from outside sources. As Hupe and Hill note, "discretion and rules are interrelated: As rules specify the duties and obligations of officials, discretion allows them freedom of action" (2007, 280–281).

These types of discretion are constrained (Lipsky 1980; Scott 2009). Within-rule discretion is constrained by the rules that outline the possible choices available to the street-level bureaucrat. Discretion that exists due to a lack of regulations is constrained by social and organizational norms and values.

Not all discretion is created equal, and the magnitude of discretionary decisions can vary significantly from one street-level occupation to another. Though critically important, the discretionary decisions made by welfare eligibility workers or teachers can be viewed as more routine than the potentially life-and-death decisions made by street-level public safety workers.

The magnitude of these discretionary decisions is not firm, and they are dependent to a great extent on the situation in which the street-level bureaucrat and client are immersed. Not every discretionary decision made by workers on either end of the spectrum is of

Figure 12.1 **Shaping Street-Level Behavior**

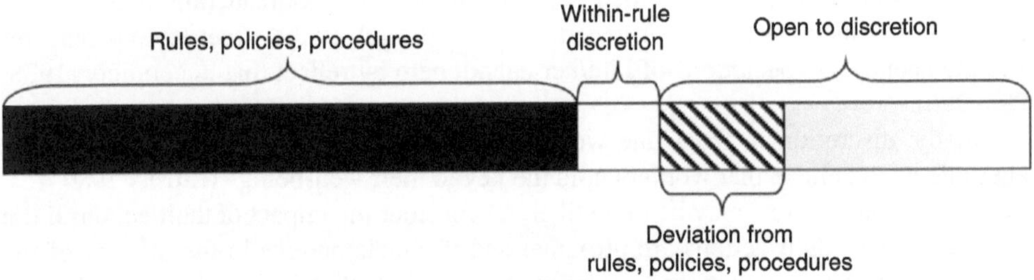


equal magnitude. To the street-level bureaucrat working in his or her respective organization, the magnitude of their decision might appear great; they see at first hand the effect of a decision on clients and have intimate knowledge of the process and needs of clients. To the citizen requesting services, the magnitude of these decisions is, in many cases, immense.

This discussion of discretion to this point has assumed that all individuals are equally able to exercise discretion. However, this glosses over some truly important differences between individuals working on the front lines. Beyond the rules created to shape behavior and varying types of constraints on discretion, individuals differ in their attitude toward discretion. As Thompson notes, "there are obvious differences in the abilities of individuals to exercise discretion; both education and experience emphasize such differences and produce trained incapacity in other areas" (1967, 117). Moreover, individuals are more or less likely to exercise discretion based on their personal feelings about risk, uncertainty, and ambiguity (Thompson 1967).

## Management and Autonomy in Street-Level Work

In most cases, street-level bureaucrats work in an environment free of direct supervision by their superiors, a factor that complicates management of behavior. This is due in part to the location of service interactions and in part to the nature of street-level work. The work of street-level bureaucrats occurs in the public domain (sometimes literally "in the street"), in office buildings, or in the homes of citizens (Prottas 1978). This, combined with the bottom-heavy nature of street-level occupations and the inability to maintain a corps of supervisors equal to the number of street-level workers, creates a problem of what Prottas calls "compliance observability," an inability to observe the actions of street-level bureaucrats and determine the extent to which they are following rules, policies, and procedures (1978, 298). The problems of supervision posed by service location are complicated by the nature of human services generally. Indeed, "the social programs that public service bureaucracies are charged with administering are almost invariably complex and controversial. As a result, the goals of these agencies are multiple, contradictory, and vague" (Hill 1974, 44). Although it is often possible to reduce organizational goals to a manageable list, the ability to pinpoint a specific definition of those goals is often elusive and, if accomplished, might not always garner full and complete support from all stakeholders.

After these issues have been acknowledged, the difficulty in supervising street-level employees becomes apparent. Attempting to compare front-line workers' actual behavior and ideal behavior is a daunting task, one that is complicated if one takes into consideration the position of the street-level bureaucrat at the organizational boundary and in control of information (Prottas 1978). As Prottas notes,

> what the street-level bureaucrat should be doing necessarily depends on what clients are in fact doing. The hierarchy may say that if a client has characteristic X then the street-level bureaucrat will be expected to do Y. The characterization of clients is the street-level bureaucrat's job and so the agency must depend on the street-level bureaucrat for almost all of its information about clients. Therefore the hierarchy can judge street-level bureaucrat behavior only on a contingent basis, i.e. only in relation to the way a client has been slotted, but it is dependent on the street-level bureaucrat to first slot that client. The agency then needs a way of independently judging the appropriateness of the client's slotting. (1978, 297)

In the realm of critical incidents, the autonomy of SLBs can vary by the magnitude of the incident. As mentioned above, smaller, more "routine" incidents are generally free of managerial supervision, thus allowing for greater discretion by front-line workers. Likewise, disasters and crises, which often pose problems of geography, complexity, and scale, are often so large that close management of every aspect of response is impossible. Interestingly, greater supervision might be possible in incidents of moderate complexity and scope, which may bring with them the managerial resources of larger incidents due to the initial incident uncertainty. Managers and leaders, however, are only one source of influence shaping the behavior of SLBs in responding to critical incidents. A discussion of other sources of influence is thus warranted.

### Sources of Influence on the Front Lines

The nature of discretion and resolution of these difficult problems of street-level service can be understood only by examining sources of influence on the front lines (see Figure 12.2). As Riccucci notes, "there is a long tradition of research in the social sciences on the multiple interests that interact during implementation to influence and shape policies as they devolve down hierarchical lines" (2002, 902). Weatherly notes that the actions of SLBs are also certainly responsive to a number of other influences over which the policy-maker and administrator may only have limited or no control. The pyramid-shaped organization chart depicting at the bottom the front-line worker as passively receiving and carrying out policies and procedures dispensed from above is a gross oversimplification. A more realistic model would place the front-line worker in the centre of an irregularly shaped sphere with vectors of differing size and directed inward. (1979, 9)

Figure 12.2 **Sources of Influence on Street-Level Bureaucrats**

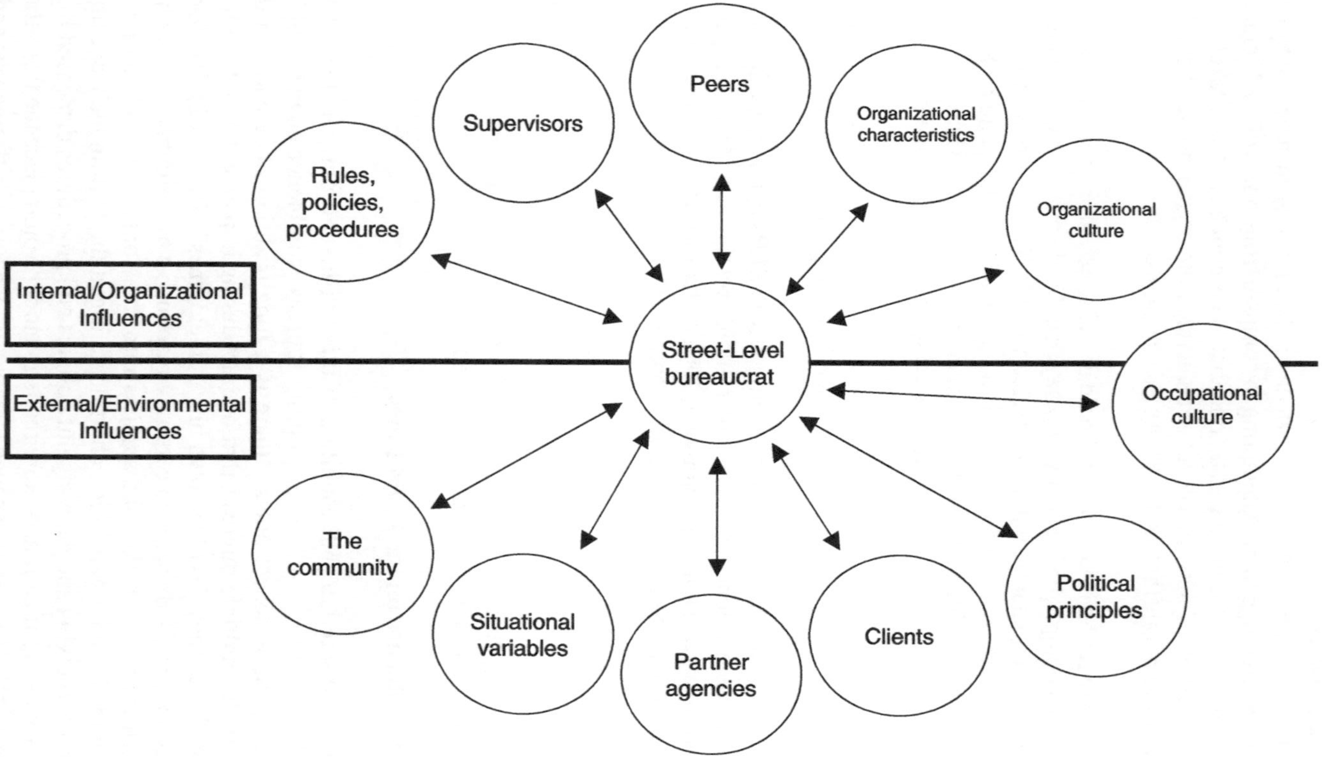


*Source:* Adapted from Vinzant and Crothers, 1998.

Thus, it is important to understand the nature and magnitude of the many sources of influence at work on street-level bureaucrats that are sometimes interwoven and potentially conflicting. This is not to suggest that these various types of pressures on street-level bureaucrats are inseparable and impossible to study empirically.

Although much of the research presented below is based on nonemergency service interactions that do not carry the same urgency, uncertainty, or time constraints as critical incidents, they are nonetheless applicable to the efforts of front-line workers responding to critical incidents. In fact, the two sources of influence highlighted here may be viewed as more salient in complex, uncertain situations due to the ability to respond quickly through the use of routinized behavior in response to complex, uncertain events.

The numerous sources of influence on street-level bureaucrats can be divided into categories by their location relative to the organization, with several sources apparent both inside and outside the organization. Sources of influence within the organization include coworkers, supervisors, formal organizational structure, and organizational culture (Vinzant and Crothers 1998, 12–14). The nature and quality of interactions with peers, direct supervisors, and executive-level supervisors can affect how a front-line worker views problems and the processes used to make decisions. Managerial influence on street-level bureaucrats has been studied empirically and was found to vary from less (Riccucci 2005) to more (May and Winter 2007). Organizational factors, including structure and culture (Schein 2004; Trice and Beyer 1993; Weick 1995), can influence the behavior of street-level bureaucrats by several avenues and have been found to be important among SLBs, depending on the rule-bound nature of the occupation and cultural characteristics (Kelly 1994; Scott 2009).

Sources of influence outside the organization include situational variables, clients, individuals from other organizations, political principles, the media, and the community as a whole (Vinzant and Crothers 1998, 15–16). Situational variables, such as the environment in which the service is being performed, the actions of other street-level bureaucrats, and client-specific factors, can influence the behavior of street-level bureaucrats by altering the both the decision-making process and the desired outcomes of the incident. Political principals have also been found to be particularly influential in street-level work. They include executive agencies at the federal, state, and local levels that promulgate operating rules and standards for training and education for these positions (Vinzant and Crothers 1998), and legislative bodies can influence the behavior of street-level bureaucrats more indirectly (Gilboy 1992).

Gilboy (1992), using observations of service interactions and informal interviews of front-line officials, examines "casework" as a form of indirect influence on front-line officials at the Immigration and Naturalization Service (now called Immigration and Customs Enforcement). Perhaps the most important contribution of this work is the link between anticipated behavior of legislators and the actions (or inaction) of front-line bureaucrats (Gilboy 1992, 276). This form of influence is perhaps even more important in the politically charged and well-documented area of crisis and disaster management. SLBs are keenly aware that their activities are being observed by victims, elected officials and administrators, and the media, thus creating an environment filled with anticipatory behavior.

Two of these sources of influence have been found to be increasingly important in street-level work and are worthy of in-depth examination: client characteristics and occupational culture. Van Ryn and Burke write: "All humans share the generally adaptive strategy of making the world more manageable by using categorizing and generalizing techniques to simplify the massive amounts of complex information and stimuli to which they are exposed" (2000, 814). As mentioned above, the routinization of both tasks and clients in response uncertainty and time constraints becomes a key topic of discussion in an examination of responses to crises and disasters.

**Occupational Culture as Influencing Street-Level Bureaucrats**

Occupational culture, conceptually similar to, and in some cases identical to professional norms and values, permeates the roles that individuals hold within organizations. As Schein states, "As members of different occupations, we are aware that being a doctor, lawyer, engineer, accountant, or other professional involves not only the learning of technical skills but also the adoption of certain values and norms that define our occupation" (2004, 10).

Occupational culture functions at the street level just as it does in the offices of doctors and lawyers (Hupe and Hill 2007), something that should not be surprising if we view these bureaucrats as professionals who are "*expected* to exercise discretionary judgment in their field" (Lipsky 1980, 14, emphasis added). Maynard-Moody and Musheno echo this and note that street-level workers "identify strongly with their occupation and similar occupations" (2003, 58). Indeed, occupational culture can be seen as of great importance on the front lines in that "discretion is filled by rules professionals impose upon themselves" (Hupe and Hill 2007, 282). Merging several concepts of the construct allows for occupational culture to be defined as the norms, values, beliefs, rules, practices, and technical skills of a profession, as well as the "myths, rituals, symbols, and rites" of the occupation (Schein 2004; Trice and Beyer 1993, 180).

Trice and Beyer claim that "the most highly organized, distinctive, and pervasive sources of subcultures in work organizations are people's occupations" (1993, 178). The socialization process for new members of a profession can directly shape the norms, values, beliefs, and assumptions of the individuals. This is especially true for those occupations that involve "an intense period of education and apprenticeship," a characteristic of many public safety agencies (Schein 2004, 20; Trice and Beyer 1993). This intense period of socialization typically "result in members' internalizing rather detailed sets of expectations for their behavior in their work roles" (Trice and Beyer 1993, 179). These expectations can take the form of incentives that shape the nature of peer relationships (Prottas 1978). Street-level bureaucrats might find that they gain in prestige, power, or social standing when they conform to occupational cultural norms.

Occupational culture can bring ideas and perspectives on job functioning from the environment into the organization. Occupations, in this sense, are "imported subcultures. They carry ideas that have their origins outside organizations. Furthermore, the ideas they carry are . . . focused on work-related issues" (Trice and Beyer 1993, 178). In addition, as in many other occupations, "street-level bureaucrats in many agencies are members of professional or quasi-professional associations. These associations generate their own definitions of proper behavior for their members" (Prottas 1978, 296).

The importance of these professional norms goes beyond their direct impact on behavior. Prottas writes:

> Since bureaucracies can rarely provide satisfactory measures of their own outputs . . . , in their own defense they must frequently point to the expertise of their members rather than the success of their endeavors. But once it admits that its workers have special skills and that those skills provide a justification for the agency's activities, two conclusions follow: one, the agency admits that the norms of [the occupation] are a legitimate component in the definition of street-level bureaucrat behavior, and two, it admits that the [front-line worker's] understanding of those norms has some validity based on their expertise. In this way they admit, at least, to a potential limitation of their right (aside from the capacity) to define appropriate street-level behavior. (1978, 296–297)

Thus, beyond the direct impact on behavior, the expertise of street-level bureaucrats serves to reify their own position of power within agencies. Indeed, as Trice and Beyer note, occupational culture can serve as a method of exerting power against or in place of administrative or managerial control (1993, 185). Occupational subcultures are also a source of power in that they can be self-supporting and work to ensure their own continued existence within an organization. Thus, "when members of occupations control recruitment, hiring, and induction into employing organizations, their cultures are likely to influence those of employing organizations" (Trice and Beyer 1993, 220). Moreover, the existence of these professional skills, norms, and values exerts pressure on methods of employee evaluation (Thompson 1967). After it is acknowledged that occupational cultures have specific and specialized skill sets, some power to conduct evaluation and performance measurement is ceded to those individuals within the same "occupational family" (Thompson 1967, 113).

Occupational culture can also be a source of conflict among emergency service organizations responding to crises (Comfort 1990). As noted, the subcultures of public safety institutions can be particularly strong, so the potential for conflict on processes and even outputs and outcomes during collaborative response efforts is palpable.

The salient effects of occupational culture on street-level bureaucrats have been examined empirically. Sandfort (2000) examined the impact of managers and leaders on the behavior of front-line workers in public welfare eligibility determination agencies and in private contracted welfare-to-work agencies. Occupational culture was found to be of great importance in shaping the behavior of street-level bureaucrats. Although acknowledging the role of managers and leaders in shaping the behavior of street-level bureaucrats, Sandfort claims that there is "a gap between management ideals and daily front-line realities" (2000, 740). Rather, she theorizes that other sources of influence may be more important and that "a more complex process motivates front-line workers and sustains front-line practices in these organizations" (2000, 740).

Two important findings result from Sandfort's efforts. First, she states that "in both the public welfare bureaucracy and privatized welfare-to-work systems, front-line practices strain the premises of the public management techniques designed to direct and influence them" (2000, 740). The management practices intended to shape and direct the behaviors of front-line workers did not function as intended in either the public or private agencies. Second, street-level bureaucrats play an important role in shaping the norms, values, and beliefs about the nature of their own work, work context, relationships with coworkers, and work-related issues. In Sandfort's words, "street-level workers in both contexts exert more agency than is conventionally recognized. Through daily experiences, staff generate collective schema that help them to understand their work and efficiently utilize organizational resources" (2000, 731). In summary, "front-line staff within the public welfare bureaucracy depend on their collective knowledge and organizational resources to develop the structure within which they operate" (2000, 746).

Riccucci (2005), in a study of the importance of management on street-level bureaucrats in welfare agencies, finds that occupational culture is of great importance in understanding behavior on the front lines. While noting the importance of management, Riccucci argues that direct managerial influence has a smaller impact on the behavior of front-line workers than other sources of influence. Instead, "workers at the front lines often pursue goals that are consistent with their work norms, familiar routines, professional standards, and socialization experiences" (2005, 2). Thus behavior is often shaped not by direct managerial intervention but by the salient experiences of workers as they struggle simply to do their job. These norms, routines, and standards can be consciously or unconsciously incorporated into the front-line worker's repertoire of action (2005, 5). Thus, "it is the work ethics, norms, professionalism, and occupational culture at the street level of bureaucracy that influence the behaviors of front-line staff, *not* management, leadership, and organizational factors" (2005, 74).

In a study of front-line health-care providers, Isett, Morrissey, and Topping (2006) find that managerial changes to the policies at the macrolevel are less important in shaping the perspectives of health-care providers. Specifically, they state that "the attitudes and perceptions of street-level bureaucrats, such as frontline health care workers in the managed care setting, are not likely affected by large-scale systems or regime changes because their day-to-day activities remain fairly constant" (2006, 223). Rather, their perspectives are shaped mainly by the nature of their daily tasks. What is important to street-level bureaucrats is *how* they complete their daily work. *What* is being done or *why* it is being done is less important (2006). The second important source of influence, client characteristics, is to be subject to simplification and categorization for many of the same reasons.

## Client Characteristics as Influencing Street-Level Bureaucrats

Clients and the street-level bureaucrat's perception of the client's identity have been found to significantly influence front-line worker behavior. Front-line workers provide a service and in so doing must "first reduce that citizen and his or her demands into a simple and patterned package of processable attributes" (Goffman 1973, in Prottas 1978, 79–109). Street-level bureaucrats must assess the client's needs and determine how the agency can or cannot assist the individual. Prottas writes that "in this sense, the clients of a public service bureaucracy play the role of the organization's raw materials; the real work of the employees is remolding clients to define their relationship to the agency as consumers" (1978, 289). Thus perceived client identity is linked to the street-level bureaucrat's evaluation of the client's worthiness to receive certain types of service provided by the agency (Maynard-Moody and Musheno 2003).

The categorization of a client determines the types of assistance that person will receive, as well as the responsibilities of the organization in assisting the individual. As the principal component in the categorization process, the street-level bureaucrat wields substantial power by being able to "dominate the client by controlling the alternatives and information coming from the bureaucracy, s/he can influence the bureaucracy by controlling the alternatives and information coming from the client" (Prottas 1978, 294). These interactions between street-level bureaucrats and clients are meaningful in that they have an effect on both the client and the organization. The interactions essentially embody a "duality. . . . They label a citizen so as to define what treatment the citizen is to receive from the bureaucracy. At the same time, they bind the bureaucracy to do a specific series of tasks and to expend specific resources" (Prottas 1978, 291).

To accomplish this evaluation of the client, street-level workers, according to Maynard-Moody and Musheno, want to get a sense of who citizen-clients are, get a fix on people, to decide how to handle them. . . . In figuring out who a person is, they . . . [assign] them to a social identity or social belonging that carries with it significant meaning and consequence. (2003, 78)

Thus, client identity is the assignment of clients to a social identity or a social group based on an observation of ascribed characteristics, such as gender, race or ethnicity, age, socioeconomic status, or attribution of membership in some other type of social or cultural grouping.

The influence of clients on bureaucratic discretion on the front lines was also found significant in the experimental study of welfare eligibility workers conducted by Scott (2009). The study used multiple cases to depict clients with differing levels of need to evoke varying levels of compassion for the client's situation. The results showed that, on average, those clients who inspired a higher need for compassion received higher levels of assistance.

In a study of street-level workers in a police department, vocational rehabilitation centers, and school counseling offices, Maynard-Moody and Musheno (2003) find that four dimensions are important in shaping perceptions of client identity and worthiness. The first is an evaluation of the client's responsibility for needing the services provided by the agency. "If clients are responsible for their disabilities—driving drunk and causing a car crash, for example—they are deemed less worthy than someone disabled by birth, disease, or accident" (Maynard-Moody and Musheno 2003, 103). However, Maynard-Moody and Musheno find that client motivation to improve their situation, the second dimension of worthiness, is more important than the reason or reasons for their need (2003,

104). Indeed, the "motivated citizen-client is nonetheless deemed morally superior to the unmotivated. Conversely, unmotivated clients, regardless of their need or circumstance, are deemed unworthy" (Maynard-Moody and Musheno 2003, 104).

A third dimension is the presence or lack of virtuous intent by the client. The "morally worthy citizen-clients do not try to con or scam workers in the system" (Maynard-Moody and Musheno 2003, 104). Thus intention to manipulate the system is an important defining trait. The final dimension of worthiness deals with the distribution of time, effort, services, and funds, all of which can be expressed as a material investment in the client. Maynard-Moody and Musheno note that this material dimension of worth is a summative judgment: if citizen-clients have genuine needs, are of good character, and are motivated to respond to treatment, then they are likely to repay society for street-level workers' investment of time, effort, and money (2003, 106).

This points to the importance of street-level bureaucrats in the context of a democratic system of governance. At the most basic level, they "are making normative judgments about who gets what from government: they are simultaneously fusing the performance of the state with the construction of the social order" (Maynard-Moody and Musheno, 2003, 106). These issues highlight the difficulties of front-line service, where "identity and moral judgments are bound up with the quotidian work of the state" (Maynard-Moody and Musheno 2003, 8).

Beyond the walls of counseling centers, classrooms, and welfare offices, client identity and perceptions of worthiness have been found to be important in shaping behavior in disasters and crises. Stivers (2007) claims that bureaucratic action during the response to Hurricane Katrina, while not overtly racist, was in some cases guided by systemic inequalities that severely disadvantaged the African American and poor residents of New Orleans. This is not to say that the first responders consciously enacted the policies and procedures of their employer in a manner that was meant to have a disproportionately negative impact on these citizens. Rather, they may have made decisions supported by or cued by "patterns of discriminatory treatment harden[ed] into structural differentiation" (Stivers 2007, 49). Like occupational culture, client characteristics are only one source of influences that shape street-level bureaucratic behavior.

## Responding to Multiple Sources of Influence

The decision-making process on the front lines often requires considering these types of pressures, each of which can "have greater or lesser influence in different circumstances" (Vinzant and Crothers 1998, 17). Although many of the empirical studies described above considered one or several of these sources of influence, a more complete understanding of the nature of street-level behavior and exercise of discretion must be open to additional sources of influence. In addition, the situational nature of influence must not be taken lightly. Indeed, the relative impact of the multitude of sources of influence can vary based not only on occupation but also on the nature of the incident. Discussing a particularly difficult case presented to a police officer, Vinzant and Crothers determine that "ultimately, in making the choices he made, the officer, in effect, interpreted a mix of laws, rules, preferences, and values and decided how they ought to be applied in this particular case" (1998, 19). In summary, "what workers often *do* is balance multiple and sometimes conflicting standards and expectations in making judgments about what to do in specific cases" (Vinzant and Crothers 1998, 19).

## CONCLUSION

Street-level bureaucrats operating on the front lines of response are an important and often overlooked class of public servant in the scholarly literature on crises and disasters. These individuals are vested with potentially substantial discretion and can be influenced by rules (broadly construed), organizational factors, clients, and situations. Their efforts constitute policy implementation in action and often have a tangible impact on the citizens with whom they interact. This impact is noticeable in the normal course of governance and even more so when crises and disasters interrupt normal social, political, and economic systems.

In the end, variation in both discretion and sources of influence is the norm. Although many sources of influence on the behavior of street-level bureaucrats might not be felt as these individuals act during a potentially chaotic and uncertain critical incident, others might be more salient for that very reason. Uncertainty and complexity can push an individual to adopt behaviors that have been developed, honed, and practiced before the incident. This class of public servant is certainly worthy of empirical research with a focus on the relative impact of influence on decision making in all types of critical incidents (Kapucu and Van Wart 2006). Continued research is also necessary because situations and events rarely line up with plans devised for response. As Waugh and Streib write, "[responders] have to innovate, adapt, and improvise because plans, regardless of how well done, seldom fit circumstances" (2006, 132).

## REFERENCES


Bigley, G. A., and K. H. Roberts. 2001. "The Incident Command System: High-Reliability Organizing for Complex and Volatile Task Environments." *Academy of Management Journal* 44 (6): 1281–1299.

Boin, A., and P. 't Hart. 2003. "Public Leadership in Times of Crisis: Mission Impossible?" *Public Administration Review* 63 (5): 544–553.

Bovens, M., and Zouridis, S. 2002. "From Street-Level to System-Level Bureaucracies: How Information and Communication Technology Is Transforming Administrative Discretion and Constitutional Control." *Public Administration Review,* 62 (2): 174–184.

Comfort, L. K. 1985. "Integrating Organizational Action in Emergency Management: Strategies for Change." *Public Administration Review* 45 (Special Issue): 155–164.

———. 1990. "Turning Conflict into Co-Operation: Organizational Designs for Community Response in Disasters." *International Journal of Mental Health* 19 (1): 89–108.

———. 2007. "Crisis Management in Hindsight: Cognition, Communication, Coordination, and Control." *Public Administration Review* 67 (special issue): 189–197.

Comfort, L. K., and T. W. Haase. 2006. "Communication, Coherence, and Collective Action: The Impact of Hurricane Katrina on Communications Infrastructure." *Public Works Management and Policy* 11 (1): 1–16.

Comfort, L. K.; Y. Sungu; D. Johnson; and M. Dunn. 2001. "Complex Systems in Crisis: Anticipation and Resilience in Dynamic Environments." *Journal of Contingencies and Crisis Management* 9 (3): 144–158.

Davis, K. C. 1975. *Police Discretion.* St. Paul: West.

Farazmand, A. 2007. "Learning from the Katrina Crisis: A Global and International Perspective with Implications for Future Crisis Management." *Public Administration Review* 67 (special issue): 149–159.

Gilboy, J. A. 1992. "Penetrability of Administrative Systems: Political "Casework" and Immigration Inspections." *Law and Society Review* 26 (2): 273–314.

Hill, D. 1974. *The Art of "Muddling Through": Policy-Making and Implementation in the Field of Welfare.* Madison: University of Wisconsin Press.

Howitt, A. M., and H. B. Leonard. 2009. *Managing Crises: Response to Large-Scale Crises.* Washington, DC: CQ Press.

Hupe, P., and M. Hill. 2007. "Street-Level Bureaucracy and Public Accountability." *Public Administration* 85 (2): 279–299.

Isett, K. R.; J. P. Morrissey; and S. Topping. 2006. "Systems Ideologies and Street-Level Bureaucrats: Policy Change and Perceptions of Quality in a Behavioral Health Care System." *Public Administration Review* 66 (2): 217–227.

Kapucu, N. 2006. "Interagency Communication Networks During Emergencies: Boundary Spanners in Multiagency Coordination." *American Review of Public Administration* 36 (2): 207–225.

Kapucu, N., and M. Van Wart. 2006. "The Evolving Role of the Public Sector in Managing Catastrophic Disasters: Lessons Learned." *Administration and Society* 38: 279–308.

———. 2008. "Managing Catastrophic Events Making Matters Worse: An Anatomy of Leadership Failures." *Administration and Society* 40 (7): 711–740.

Kelly, M. 1994. "Theories of Justice and Street-Level Discretion." *Journal of Public Administration Research and Theory* 4 (2): 119–140.

La Porte, T. R. (ed.) 1975. *Organized Social Complexity: Challenge to Politics and Policy.* Princeton. NJ: Princeton University Press.

Lipsky, M. 1980. *Street-Level Bureaucracy: Dilemmas of the Individual in Public Services.* New York: Russell Sage Foundation.

McLoughlin, D. 1985. "A Framework for Integrated Emergency Management." *Public Administration Review* 45 (special issue): 165–172.

May, P. J., and S. C. Winter. 2007. "Politicians, Managers, and Street-Level Bureaucrats: Influences on Policy Implementation." *Journal of Public Administration Research and Theory* 19: 453–476.

Maynard-Moody, S., and M. Musheno. 2003. *Cops, Teachers, Counselors: Stories from the Front-lines of Public Service.* Ann Arbor: University of Michigan Press.

Maynard-Moody, S.; M. Musheno; and D. Palumbo. 1990. "Street-Wise Social Policy: Resolving the Dilemma of Street-Level Influence and Successful Implementation." *Western Political Quarterly* 43 (4): 833–848.

Moynihan, D. P. 2008. "Learning Under Uncertainty: Networks in Crisis Management." *Public Administration Review* 68 (2): 350–365.

Prottas, J. M. 1978. "The Power of Street-Level Bureaucrats in Public Service Bureaucracies." *Urban Affairs Review* 13: 285–312.

Riccucci, N. M. 2002. "Implementing Welfare Reform in Michigan: The Role of Street-Level Bureaucrats." *International Journal of Public Administration* 25 (7): 901–921.

———. 2005. *How Management Matters: Street-Level Bureaucrats and Welfare Reform.* Washington, DC: Georgetown University Press.

Salerno, S. M.; K. D. Wrenn; and C. M. Slovis. 1991. "Monitoring EMS Protocol Deviations: A Useful Quality Assurance Tool." *Annals of Emergency Medicine* 20 (12): 1319–1324.

Sandfort, J. R. 2000. "Moving Beyond Discretion and Outcomes: Examining Public Management from the Front Lines of the Welfare System." *Journal of Public Administration Research and Theory* 10 (4): 729–756.

Schein, E. H. 2004. *Organizational Culture and Leadership* (3d ed.). San Francisco: Jossey Bass.

Schneider, S. K. 1992. "Governmental Response to Disasters: The Conflict Between Bureaucratic Procedures and Emergent Norms." *Public Administration Review* 52 (2): 135–145.

Scott, E. 2009. "Hurricane Katrina." In *Managing Crises: Responses to Large-Scale Emergencies,* ed. A. M. Howitt and H. B. Leonard, 13–74. Washington, DC: CQ Press.

Stivers, C. 2007. "'So Poor and So Black': Hurricane Katrina, Public Administration, and the Issue of Race." *Public Administration Review* 67 (1): 48–56.

Thompson, J. D. 1967. *Organizations in Action.* New Brunswick, NJ: Transaction.

Trice, H. M., and J. M. Beyer. 1993. *The Cultures of Work Organizations.* Upper Saddle River, NJ: Prentice Hall.

Van Ryn, M., and J. Burke. 2000. "The Effect of Patient Race and Socio-Economic Status on Physicians' Perceptions of Patients." *Social Science and Medicine* 50: 813–828.

Vinzant, J. C., and L. Crothers. 1998. *Street-Level Leadership: Discretion and Legitimacy in the Front-Line Public Service.* Washington, DC: Georgetown University Press.

Waugh, W. L. 1994. "Regionalizing Emergency Management: Counties as State and Local Government." *Public Administration Review* 54 (3): 253–258.

Waugh, W. L., and G. Streib. 2006. "Collaboration and Leadership for Effective Emergency Management." *Public Administration Review* 66 (Special Issue): 131–140.

Weatherly, R. 1979. *Reforming Special Education: Policy Implementation from State Level to Street Level.* Cambridge: MIT Press.

Weick, K. 1995. *Sensemaking in Organizations.* Thousand Oaks, CA: Sage.

Wenger, J. B., and V. M. Wilkins. 2008. "At the Discretion of Rogue Agents: How Automation Improves Women's Outcomes in Unemployment Insurance." *Journal of Public Administration Research and Theory* 19 (2): 313–333.

CHAPTER 13

# An Earthquake Mitigation Process

## A Case Study of Qinglong County's Pre-Incident Predictions, Preparation, and Mitigation Actions

JEANNE-MARIE COL

Most critical incidents are negative experiences for people, institutions, and society. After the standard operationalization of critical incidents, one expects social panic and trauma, significant injury, loss of life and property, infrastructure damage, lack of coordination and plenty of confusion, and a diminution of public trust in institutions that are expected to take care of people and society.

The case study of Qinglong county's official actions before, during, and after the Great Tangshan Earthquake in China in 1976 defies this stereotype of a critical incident. Instead, the Qinglong county officials correctly predicted, widely prepared, mitigated against possible damage, responded to help people in surrounding areas, and rebuilt their infrastructure and buildings with cooperation and confidence. The actions of the officials and citizens of Qinglong county are even more impressive in contrast to the overall catastrophic consequences of the earthquake, which measured 7.8 on the Richter scale and whose epicenter was in the city of Tangshan. More than 246,000 people were killed in this earthquake, the highest death toll from a natural disaster in the twentieth century. In Qinglong county, 115 kilometers from the epicenter, no deaths were attributable to the earthquake and its aftershocks despite the collapse of 180,000 buildings.

This chapter describes the Great Tangshan Earthquake and presents an analysis of the public administration and emergency management lessons learned. These lessons include effective coordination among public administrators, scientists, and citizens and feature proactive policies, local government initiative, thorough implementation, delegation, information sharing, and citizen participation. These lessons are evident during all phases of the emergency management process: preparedness, mitigation, response, and recovery.

The methodology of data collection and analysis is as follows. Most significantly, the author returned to Qinglong county five times, and many documents were translated into English. Each research visit permitted the author to dig deeper, "peeling away the layers of the onion" concerning the roles and perceptions of the officials and citizens. The author interviewed retired officials and translated more than a hundred personal accounts, which were written by the citizens on the tenth anniversary of the earthquake in 1986 and assisted in confirming the sequence of official and citizen actions. The chronology of events in Qinglong county begins two years before the earthquake and continues through the short-term aftermath. The events are described by "who, what, when, and impact" on incident management.

Aspects of the case study are used to illustrate the fifteen factors identified as relevant to critical incident analysis: social trauma, fear, emotional effects, societal norms, public trust, citizen participation, length of event, loss of life/injury, property damage, official communications, degree of prediction, scope, intergovernmental coordination, positive aspects, and media coverage. Finally, conclusions are drawn concerning the relevance of a "best practice case study" for advancing analysis of critical incidents.

## CASE STUDY: QINGLONG COUNTY AND THE GREAT TANGSHAN EARTHQUAKE

Northeastern China is composed of four provinces (Hebei, Liaoning, Shanxi, and Shandong), one autonomous region (Inner Mongolia), and two special municipalities (Beijing and Tianjin). Within these jurisdictions, there are 431 counties (*China Daily* 2003). Qinglong county is one of these counties. In the summer of 1976, the Tangshan city region in northeastern China suffered a powerful earthquake; more than 246,000 people in the region died as a result of the 1976 earthquake and its aftershocks. However, no one in Qinglong county died from the earthquake (Col 2007).

Several factors account for this seeming miracle in Qinglong county. One of the most important factors was the scientific understanding and preparatory actions of what was then called China's National Bureau of Earthquakes and is now the China Earthquake Agency (CEA). The CEA generates four types of earthquake predictions: long-term, medium-term, short-term, and imminent predictions that are revised annually at local, regional, provincial, and national levels. Long term refers to more than two years; medium term is one to two years; short term is a few months, and imminent is one to fourteen days. When the CEA gathers and analyzes sufficient data to indicate a coming earthquake, the CEA notifies the State Council (China's national cabinet), which in turn issues an alert or a warning to the localities likely to be affected (Li 1991).

In 1974, two years before the Great Tangshan Earthquake, the State Council issued Document 69 to warn the region of the increased risk of seismic activity in the coming decade (Col and Chu 1999) and to alert its leaders to prepare for the possibility of earthquakes. Specifically, Document 69 recommended that municipalities and provinces in the area, including Hebei Province, where Tangshan and Qinglong county are located, prepare for earthquakes measuring more than 6 on the Richter scale.

Document 69 also called for increased local monitoring, public education, and strengthening of infrastructure, including buildings in the area. It also encouraged monitoring in factories by citizens (i.e., lay monitoring) and in mines and schools by local science volunteers. By 1976, most counties in Hebei had established an earthquake office and launched a variety of preparedness activities. Their proactive reaction to Document 69 reflected their knowledge of an earthquake experienced in 1975 in nearly Haicheng. Based on the 1974 alert, several communities in the Haicheng area were evacuated before the 1975 earthquake.

In response to Document 69, the administrative leaders of Qinglong, who were also political leaders of the county's Communist Party, established an earthquake office and hired a twenty-one-year-old graduate of a technical school as the earthquake disaster mitigation officer. He intensified public education by distributing thousands of copies of booklets and posters, presenting slide shows and movies in villages, towns, and cities, including a short earthquake preparedness film as a reminder to be presented in movie theaters before the main feature. He was also responsible for increasing the county's earthquake monitoring stations (sites) from six to sixteen, nine of them under the supervision of local middle schools and high schools.

At the same time, at the national level of government, CEA officials increased earthquake monitoring in northeast China and organized many regional training sessions for earthquake management professionals at which advice was dispensed as to how to prepare for earthquakes. At one of the sessions, attendees, including the Qinglong earthquake officer, were informed that professional earthquake monitoring teams and detection centers were reporting abnormal earthquake-related signals for the Beijing-Tianjin-Tangshan-Bohai-Zhangjiakou region, indicating the possibility of a significant earthquake between July 22 and August 1976.

Session attendees were encouraged to examine readings from county-level earthquake monitoring equipment managed by the lay public. The officials were also encouraged to intensify earthquake preparation measures, such as examining buildings and other public works in critical condition, conducting public education teach-ins, and promoting general awareness of possible approaching earthquakes.

When the head of the earthquake office returned to Qinglong on July 21, he learned that four sites were registering unusual earthquake-related signals. On July 23, a special evening meeting of the county's Standing Committee of the Chinese Communist Party (SC CCP) was convened to discuss the earthquake officer's findings and their significance. On July 24, 1976, the Qinglong County SC CCP issued an "official alert" to all residents to evacuate buildings and to move to outdoor tents. It was decided that village officials should return to their communities immediately, accompanied by a leading county-level official.

It is important to note that, despite the overwhelming possibility of a catastrophe, the social fabric was strong enough for all officials to continue their work, rather than checking on the families' well-being, so as to accelerate earthquake and flood preparation. Several photographs taken at the time depict schoolchildren sitting on benches and participating in outdoor classes, and merchants selling their goods outdoors on makeshift shelves under plastic sheeting. Most residents of the county were also sleeping in tents outdoors.

On July 25 and 26, two and three days, respectively, before the earthquake, each community in Qinglong held an emergency meeting to prepare and instruct villagers in disaster damage reduction. Community leaders identified vulnerable buildings and reservoirs and assigned guards to report infrastructure failures. Most villages had overnight watch guards on duty circulating among the buildings to enforce the evacuation order. County and village broadcasts instructed people who were still in their houses not to close their doors and windows at night so that they could leave their homes as soon as they felt the ground shaking. Residents were also told to avoid being close to tall buildings and power lines. On July 27, 1976, one day before the earthquake, a leading county official spoke at a countywide agricultural meeting on the earthquake situation and on mitigation measures.

At 3:42 A.M. on July 28, 1976, the Great Tangshan Earthquake struck. More than 180,000 buildings in Qinglong were destroyed, yet only one person died—and he died of a heart attack—while in Tangshan and in all the surrounding counties, more than 246,000 people died.

The seeming miracle of Qinglong county was the result of several factors: the involvement of all levels of government, clear national policies, effective administration delegated to the lowest level, adaptation to changing conditions, open and operational channels of communication, and citizen participation.

## LESSONS LEARNED FROM QINGLONG COUNTY

### Lesson 1: All Levels of Government Must Be Involved in Disaster Management and Mitigation

In the Great Tangshan Earthquake of 1976, each level of government performed a critical role in earthquake management and mitigation. National officials aggregated data from stations throughout the region and issued earthquake predictions. Ongoing monitoring of earthquake precursors by China's professional earthquake staff (the CEA) was shared in periodic regional meetings. In all communities subject to earthquake risk, the CEA, through national, provincial, and municipal officers, conducted emergency preparedness workshops.

As mentioned earlier, Document 69, issued in 1974 for the northeast region, was part of a continuing national program of applied earthquake science maintained by the CEA. It regularly included outreach, training, and communication with subnational governments, produced synthesized reports on risk and vulnerability, and promulgated official directors for earthquake mitigation to regional and local administrators. The leader of the Northeast Analysis and Prediction Group sent teams, several of whose members later died in the earthquake, to northeast China to gather additional information in July 1976. He attended a regional training meeting scheduled in Tangshan, during which he reminded the county-level participants of the range of measureable precursors, illustrated his talk with current data on precursor signals, and encouraged them to check their local stations for unusual signals and anomalies. Although he could not make an official prediction, he encouraged local officials to check their locally generated data. The subtle message was to increase disaster preparations if local data indicated a likely imminent event.

Below the national level, institutions in Qinglong acted on local data, within the context of national policy. Acting on the recommendations of Document 69 and further instructions from the CEA, Qinglong officials added staff to disaster management, checked infrastructure for structural vulnerabilities, and increased the number of precursor monitoring stations—both professionally managed and citizen-staffed.

### Lesson 2: National Policies Must Be Proactive and Widely Understood

By the 1970s, CEA had developed a vast network of precursor monitoring sites, employing a wide variety of scientific methods. These sites, with professional scientists, equipment and communications systems, were supplemented by intensive lay monitoring by local volunteers who were assisted by the CEA but organized by local governments.

Professional and lay monitors met annually under the auspices of the CEA in information-sharing meetings at the local, provincial, and national level. Out of these meetings emerged long-term, medium-term, short-term, and imminent predictions. Continuous monitoring and communications kept the entire network up to date on the probability of earthquake events, occasionally leading to imminent predictions.

### Lesson 3: Operational Decision Making and Organizing Must Be Delegated to the Lowest Level of Government Possible

In 1976, earthquake alerts were required to be issued only at the national level but depended on information gathered locally. There were many recorded precursor signals but no foreshocks. The national-level scientists could not agree on a definite prediction and therefore did not issue an imminent warning for the Tangshan area. In contrast, Qinglong county officials, analyzing their locally generated data, observed that precursor signals were becoming increasingly strong and decided to issue a county-level warning. Strictly speaking, the county's earthquake warning was not permissible under national law at the time. Thus, the local officials circumvented national instructions and made a decision that they believed would be useful to people under their jurisdiction. In interviews in 1996, the officials indicated that they were aware of the risks of disobeying national instructions but were convinced that they were taking a calculated risk for the benefit of people in the county (Col and Chu 1999).

In Qinglong, many officials took extraordinary actions with an urgency that communicated seriousness to colleagues and community members. For example, the head of the Qinglong science committee, when leaving for a meeting in Beijing, instructed his staff not to wait for his return for permission to act but, rather, to take action based on incoming local data. Despite the high regard for top-down hierarchical management, these Qinglong officials delegated decision making to the officials closest to the problem. During a telephone conference with the leading officials of all twenty-three towns, the head of Qinglong county administration emphasized that all officials would be responsible for preparing people in their area and would be held accountable for their action or inaction.

### Lesson 4: Local-Level Administrators Must Adapt Plans and Mitigation Strategies to Meet Changing Conditions as the Disaster Develops

On July 24, without a specific instruction from national government officials (but under the auspices of Document 69 of 1974), the County Committee declared an earthquake alert and a countywide evacuation of buildings, with special attention to auditoriums, movie theaters, theaters, and schools. Aware that this alert would interrupt some economic and social activities, the committee decided to err on the side of citizen safety, risking criticism if no earthquake occurred. In later interviews, these officials told the author that they relied on Document 69 to give them adequate "administrative cover" for their evacuation order, which was "not specifically authorized" (Col and Chu 1999).

The SC CCP alert included specific instructions for public administrators and citizens to work side by side. The SC CCP established a command center that was staffed around the clock. First responders and other officials were required to stay at their posts twenty-four hours a day. Some officials were assigned to work with citizens at monitoring stations and report daily to the command center.

County officials used existing and parallel mechanisms by shifting activities to earthquake preparedness. On the morning of July 25, in order to intensify communications, the county took over a meeting of more than 800 agriculturalist officers, informed them of the earthquake risk, provided a "short course" on earthquake preparation and mitigation, and sent some of them home with materials to distribute to their towns and villages.

Throughout the run-up to the earthquake, there was inevitable tension between bureaucratic norms that encourage, among other things, stovepipe thinking, and emergent norms that evolved from conditions that required coordination and cooperation. In Qinglong, in every instance, emergency norms replaced bureaucratic norms. Official after official made decisions that were based on factual reality rather than theoretical manuals (Schneider 1995).

### Lesson 5: All Channels of Communication—Vertical and Horizontal—Must Be Open and Operational, Thereby Increasing Community Understanding, Sharing of Perceptions of Risk, and Ease of Coordinating Actions

On July 25 and 26, citizen meetings were held at the county and village level. All residents had the opportunity to contribute to the analysis of the situation and to decisions about actions to be taken. The author's interviews in three villages, twenty years after the earthquake, revealed that each village had a forum in which people shared their understanding of the earthquake threat and the need for joint action (Col and Chu 1999).

Choices were made in the forum, and local officials and volunteers enforced those decisions, including evacuating buildings and living in tents. If the decisions had been made in a less open manner and citizens had been told to evacuate without community-wide discussion, some community members would no doubt have questioned the motives of the enforcers. With open discussion, even those who disagreed with the decision perceived the community consensus, and social pressure encouraged follow-through on disaster mitigation practices.

In order to facilitate interlevel coordination, teams dispersed throughout the county were each composed of one county-level official and one local-level official of the target community. Their goal was to inform county resident by the end of July 26 (thirty-six hours). Each community needed to have a command center, twenty-four-hour communications, and patrols to keep people from entering their homes and buildings. The communications system supported 24/7 announcements so that each community would know every detail of trends in surrounding areas.

By activating the countywide information network, county officials were able to broadcast lectures on earthquakes and to alert all precursor monitoring stations to report anomalous signals and to maintain overnight watch duty until August 5 (the end of the prediction window). Furthermore, communication between national and local communities was maintained before and during the earthquake response. CEA professional scientists visited lay monitoring stations to discuss recorded precursor signals with the citizen monitors. Regional and national response units worked alongside Qinglong residents to assist the victims in Tangshan.

## Lesson 6: Citizens Must Participate in All Stages of Preparation and Execution of Emergency Management Measures

To the extent that citizens participate in the emergency management functions, especially preparation, mitigation, and response, they will embed "disaster awareness" into their everyday lives. While it is common for citizens to help neighbors and victims in the aftermath of a major disaster, it is less common for citizens to participate directly in preparatory activities. For instance, in Qinglong in 1976, local citizens enthusiastically engaged in both precursor monitoring and emergency preparations with little sign of panic. More generally, vulnerability reduction is most effective and sustainable when implemented by the vulnerable people themselves.

Beginning in 1974, when the county was first notified about the possibility of earthquake(s) over the next few years, the citizens of Qinglong received pamphlets, watched movies, read posters, participated in drills, and held community discussions about earthquake dynamics. The public learned about lay detection of precursors, such as anomalies in water level, color, temperature, chemistry, and quality; release of gases such as radon; unusual animal behavior; and weather changes. Individuals in each jurisdiction accepted responsibility for participating in monitoring their environment over the coming years (Schneider and Ingram 1993). Many institutions became involved. Employees in factories and mines were trained to monitor for earthquake precursors using instrument readings or observational data. In addition, students in schools became involved in operating simple scientific equipment and in interviewing neighbors on observable earthquake-related phenomena.

When Qinglong declared the earthquake alert on July 24, local residents remained calm as the following unfolded:

- Schools moved furniture outside and held classes in the open air.
- Merchants constructed shelving outdoors under plastic sheets and continued selling their goods.
- Rural populations built makeshift tents and camped outdoors.
- Urban populations practiced exiting their homes and offices quickly and slept near exits at night.
- Citizens assembled in teach-ins to share earthquake preparedness strategies.
- Communities established patrols to monitor preparatory activities.

Even schoolchildren contributed to successful community decision making. Assisted by three physics teachers, a group of students at a secondary school developed a database on earthquake precursors. The students collected data from monitoring equipment and conducted interviews with local people. They noted changes in the water level, temperature, and chemistry. They also noticed that normally nocturnal yellow weasels were running around in broad daylight in large numbers. The weasel activity became more numerous and frequent on July 27.

Although an earthquake teach-in was scheduled for the morning of July 28, the students, based on incoming data, persuaded their teachers to hold the teach-in early in the evening of July 27, after which the school buildings were declared off-limits and patrols were set up to prevent reentry to them. Because the teachers followed the advice of the students, the buildings were completely evacuated before the earthquake hit.

In situations where community members are passive and only tangentially hear about disasters, they are ill prepared to take action against the evolving natural phenomenon and are susceptible to panic. Partial citizen action involves those who read about natural disasters and respond to warnings. Such citizens are aware of the community's interest in being protected and expect the government to warn them of danger from natural disasters. By contrast, Qinglong's citizens went far beyond relying on the government to take care of them. They were measuring changes in their environment and discussing the changes at the dinner table and in the workplace. Their involvement in daily monitoring of earthquake precursors led to a community-wide sense of risk awareness and joint community responsibility (Comfort 1999).

People can learn to be more active through experiencing natural disasters and through focused learning (e.g., public disaster education campaigns during a warning [prepare] or an alert [evacuate or take cover] period.) The occurrence of dramatic natural disasters, though often hugely tragic, can make a lasting impressing on community members and push "disaster awareness and mitigation" to the top of the policy agenda (Birkland 2006).

During the Tangshan Earthquake, officials and citizens had a significant success that went unpublicized in China and unknown in the outside world. Within China, officials chose not to publicize the success because it was not a sanctioned practice for local governments to declare their own earthquake alerts. The success was achieved by local people taking account of locally perceived precursor signals and acting on this information in a decisive and timely manner. Effective synergy between evolving conditions and administrative actions occurred because local organizations believed that they could deviate from national guidelines or that they could tolerate the consequences of sanctions for having deviated from a national guideline.

The sequence of actions associated with information sharing follow. Ten of the approximately sixty jurisdictions that were notified actually checked their precursor measurements. Of the ten jurisdictions, only two followed through with collecting local data, making the leadership aware of the precursor signals, preparing the citizens, and taking actions including timely evacuation. Of these two jurisdictions, only Qinglong county completed its preparation and mitigation activities before the earthquake struck. One might conclude that many jurisdictions will attempt to predict and mitigate earthquakes but that it is difficult for all guidelines and practices to come together to successfully do so.

## METHODOLOGY OF DATA COLLECTION AND ANALYSIS

Research on the Tangshan Earthquake was carried out under the auspices of the United Nations from 1995 to 2000, with multiple visits to Qinglong county. An initial familiarization visit was followed by a more intensive set of interviews, including those with retired public administrators. The State Science and Technology Commission (now the Ministry of Science and Technology) facilitated my visits to Qinglong county. The State Earthquake Bureau (now the China Earthquake Agency—CEA) facilitated introductions to provincial, city, and county officials.

Qinglong officials opened their archives, including a unique collection of essays written by county residents recollecting their experiences on the tenth anniversary of the earthquake. These essays were translated into English by several Chinese colleagues and made possible comparison between interview comments and the statements written ten years earlier. After two years of research and analysis, the chronology of Qinglong county events was translated back into Chinese and circulated among the interviewees in order to ensure accuracy.

Throughout this five-year period, the Chinese Academy of Science (CAS) organized expert panels to explain the fine points of the science of earthquakes and, to provide an understanding of some background on earthquakes, the CEA and CAS arranged for additional interviews at earthquake epicenters throughout China.

## FACTORS (DEFINITION ATTRIBUTES) IN CRITICAL INCIDENT ANALYSIS

Although the Qinglong case study illustrates most factors of critical incident analysis (CIA), there is a positive slant to the analysis because the incident had a positive outcome there, despite incident management failure in surrounding areas. Much can be learned from a process in which useful decisions were made at every decision point. Although "best practices" incident management processes are rare, they provide clues to more effective performance in urgent situations.

### Relatively Brief Occurrence

Most critical incidents and natural disasters have a specific time frame, and earthquakes are no exception. For a truly large earthquake (magnitude 6 or larger), there is the major earthquake and its aftershocks as well as a series of foreshocks. The time span of the event is usually a few weeks, but a truly large earthquake's aftershocks can continue for months. Generally, these aftershocks continue to destabilize buildings and infrastructure and make risky the tasks of response and recovery. The recovery stage can last for ten or more years. Research visits twenty years after the event found the city of Tangshan completely rebuilt, except for the preserved sites, such as the epicenter, a large library building, and a heavy industrial site. The sites chosen for preservation were places where few people were killed; buildings such as university dormitories in which all occupants were killed were rebuilt from scratch.

### Significant Loss of Life/Injury

As mentioned earlier, more than 246,000 people were killed, with more than 7,000 families were wiped out. Hundreds of thousands of others were injured. Even in Beijing, more than 300 kilometers from the epicenter, about 1,000 people died. Even now, many disabled victims continue to have access to rehabilitation and assisted living facilities.

In Qinglong, there were only a few injuries among those who wandered close to buildings despite the evacuation.

### Significant Property Damage

Earthquakes are particularly destructive to buildings and infrastructure. The Great Tangshan Earthquake leveled Tangshan, a city of 1 million, and left industrial plants that had produced railroad cars with twisted steel. To assist in recovery, a public works specialist was brought in from another part of China, and the city was largely rebuilt within ten years. As part of the recovery, Tangshan preserved seven sites that illustrate the earthquake's destruction as well as building a large earthquake museum. This museum and the related sites serve as a reminder of the destructive power of large earthquakes. As schoolchildren and others visit these sites, they are reminded to be prepared for future earthquakes and to be vigilant in mitigation efforts, such as enforcement of building codes.

In Qinglong, aside from the 180,000 buildings that collapsed or suffered damage, a portion of the Great Wall of China there collapsed except for the towers. Those buildings had to be rebuilt, but the county's citizens had all survived and were able to contribute to the reconstruction.

*Scope of Incident*

Earthquakes come in all sizes and happen around the world every day. Small earthquakes, that is, those that measure from 1 to 3 on the Richter scale, are recorded only by seismographs. Those measuring from 4 to 5 can be felt near the epicenter but seldom result in deaths or damage. Those that measure 6 or greater cause damage ranging from localized to widespread. The earthquake near Indonesia in 2004, which measured 9, killed thousands and caused a tsunami that traveled as far as Africa. The earthquake near Chile in 2010, which also measured 9, killed hundreds and caused a minor tsunami felt as far away as Asia. Even the earthquake in Haiti, which measured 6, killed thousands and led to widespread damage, due in part to its shallow epicenter, the urban location of the epicenter, and the poorly constructed buildings throughout the country. The Great Tangshan Earthquake measured 7.8.

Earthquake specialists have created the Mercalli scale of earthquake intensity to measure the level of damage. The epicenter experiences the greatest damage (up to level 9) with semiconcentric circles of lesser damage that emanate from the epicenter (from level 1 to level 11). The greatest damage in the Tangshan Earthquake was at the epicenter, in the center of the city. Qinglong county was within the concentric circles of Mercalli 6- and 7-level damage zones, while Beijing was within the 6-level zone.

*Declaration of a State of Emergency*

The declaration of a state of emergency was accomplished in two stages. First, the State Council issued Document 69 in 1974, which required communities to strengthen leadership, develop lay public monitoring of precursors, and mobilize public awareness. These responsibilities were taken seriously in some locations and neglected in others. Qinglong county took these responsibilities seriously by (1) appointing a young college graduate to be in charge of earthquake preparedness, (2) funding the establishment of additional sites to monitor earthquake precursors, (3) sending professionals for training and information sharing, (4) using posters, films, and public meetings to communicate with the general public about sensing earthquake precursors and preparing their buildings and evacuation plans.

The second important declaration of a state of emergency was issued by Qinglong county on July 24, 1976, at 8:30 P.M. at an emergency night meeting of the administrative committee, four days before the earthquake. This document alerted the officials and citizens of the possibility of an imminent earthquake and an extremely large earthquake in the near future. The alert included specific actions to be taken by various officials, organizations, and citizens.

What is important about the Qinglong county alert is that it was not authorized by the national government. Although the national government had the same information as the county, the State Earthquake Bureau determined that the information was not definitive and did not to issue a warning to the people in the Tangshan area. Despite the silence of the national government, and the Cultural Revolution imperative to maintain high economic production, the Qinglong county officials risked their reputations, and perhaps their lives, to declare an emergency on their own at the local level. This position was vindicated more than twenty years later, when the State Council authorized local governments to declare earthquake alerts.

## Expectation and Prediction

Some natural disasters, such as hurricanes and volcanic eruptions, can be predicted. But the science of earthquake prediction is still rudimentary. For centuries, Chinese public administrators kept records of earthquakes and the events that seem to occur simultaneously with them. Although seismologists in most countries dominate earthquake science, the field of earthquake prediction in China is multidisciplinary and relies on tracking macro and micro precursors that are associated with earthquakes.

One characteristic of Chinese society is the strong alliance between professional scientists and the general public. Citizens in their communities are encouraged to participate in scientific data collection by tending professional monitoring equipment (sensing micro precursors) and watching for earthquake precursors in their environment. Scientists at all levels install equipment in localities and recruit volunteers to record and report the data. Moreover, some precursors, denoted as macro precursors, are perceptible to observant viewers. Ordinary citizens are generally aware of possible precursors and know whom to call to report anomalies. Sometimes referred to as "mass monitoring" in a derogatory manner by Western scientists, this citizen participation in perceiving and acting upon threats in their environment is viewed more favorably in the current trend toward citizen engagement.

## Equitable Distribution of Government Services

Although in 1976 China was dominated at the national level by the Cultural Revolution, characterized as authoritarian and totalitarian, localities, especially outside the major cities, were highly democratic in two ways. First, in 1976 the benefits and services of government were broadly and equitably distributed throughout society, so there was not a wide distinction between the wealthiest and the poorest in communities. Second, citizen participation was practiced extensively, as exemplified by the role of lay monitors in earthquake prediction.

The professional earthquake community holds annual and occasional conferences at all levels, in a nationwide effort to produce long-term, medium-term, short-term, and imminent earthquake predictions. In my travels throughout China (earthquake-prone areas), I have found local offices for earthquake prediction and the integration of volunteers in their work. Local offices hold meetings with volunteers, regional offices hold meetings among local offices, provincial and municipal offices hold meetings with regional offices, and the national CEA holds an annual meeting of earthquake officials for the purpose of deciding the timing and content of earthquake predictions. The community volunteers are an integral part of this national system.

## Intergovernmental Coordination

In response to the Tangshan Earthquake, all levels of government were mobilized, including the national army. It is common in unitary governments for the national army to be responsible for disaster management. National armies have the equipment to move debris, trained officials to organize security, logistical strength to deliver relief supplies, and even expertise in rebuilding communities. China follows this model.

The earthquake caused widespread destruction of local and regional first-responder assets, including police, hospitals, public works, and fire departments. It took several days for army units to reach the epicenter in Tangshan, but water, food, and medical help reached the area quickly from residents in Qinglong after it was determined that no one there had been killed.

## Media Coverage

While it is normal for media to cover large natural disasters extensively, the newspapers in China in 1976 "knew" to keep coverage to a minimum. In Chinese culture, natural events are often associated with sociopolitical events. The media in modern China have been tightly controlled by the government, and the State Council feared that widespread knowledge of this disastrous earthquake would cause political and social instability.

Another significant factor was the state of earthquake prediction in 1976. A year earlier, an earthquake was predicted in Haicheng, a nearby northeastern area, and local officials evacuated buildings and thereby greatly reduced the death toll. But in the summer of 1976, the State Earthquake Bureau found the data for Tangshan inconclusive and decided not to issue an alert. After the earthquake struck, officials in the SEB and the central government were embarrassed about their lack of action compared with the huge scale of the disaster and did not want to give the impression that they could have predicted the earthquake. The officials feared that citizens throughout the country would lose faith in the government or even begin to distrust the government. Formally, the government implemented a massive response to the Tangshan disaster and supported rapid reconstruction of the city. Informally, the government was obliged to hide the fact that local officials and citizens had predicted the earthquake and taken evacuation seriously while its own behavior had been more modest.

The need to suppress the story of Qinglong's decisions (the alert) and actions (preparation and mitigation) lasted for decades. In my research on the Tangshan earthquake, I found only one sentence in one book that mentioned these successful actions. Based on that one sentence, I persuaded my boss, a UN Undersecretary-General to send me to China to investigate whether there was a "hidden" best practice in emergency management. When I presented my research at an international meeting in Beijing in 1997, I was asked to go to the hotel lounge to "visit" with the minister of the CEA, who told me that I should take down my posters and cease discussing the Qinglong experience. We agreed that I would discuss it only in the context of other successful Chinese earthquake predictions, of which there were many. In the next few years, the CEA enabled me to visit earthquake sites in the Central, Northern, Southern, and Western China. My initial visit to Qinglong county was sanctioned not by the CEA but by the State Committee for Science and Technology. This higher body was interested in researching best practices in disaster management.

Eventually, the Qinglong experience became widely known both inside and outside China. In 2008, at the time of the large earthquake in Sichuan, faculty at the Communist Party School contacted me for information about Qinglong county. By this time, several of my articles and key materials were already translated into Chinese. On anniversaries of the Tangshan earthquake, I am still contacted by radio and television stations for interviews.

## Social Trauma and Fear

Throughout the affected area, especially in Tangshan, people were traumatized by the earthquake. In focus groups in 1995–96, I saw people reliving the experience and mourning their lost relatives and friends. As the aftershocks continued for months, people in Tangshan lived in fear of further losses.

By contrast, in Qinglong the level of trauma was relatively lower, as county officials and citizens celebrated the effectiveness of their decisions and actions. Although they had to rebuild so many buildings, they were happy to be alive, experiencing only the inconveniences of continuing to live in tents for several months during the period of the most severe aftershocks.

## Reconstruction and Remembrance

In Tangshan, officials, first responders, and citizens felt helpless as they experienced the devastation of the earthquake. Officials and neighbors dug among the rubble to find those who were injured but still alive. The central government sent the army to restore order, but it took time for the equipment and personnel to be positioned for effective relief and recovery. Furthermore, the government dispatched a nationally known and experienced public works planner to be in charge of the recovery, and by 1986, most of Tangshan had been rebuilt. This expert was still serving as a deputy mayor for public works in 1996, and he was respected widely for his design and management of the reconstruction.

One of the key actions of the reconstruction was to preserve seven sites intact in their collapsed state throughout Tangshan including the epicenter, a school, a university library, and a heavy industry factory that was building railroad cars, not buildings in which people perished, such as their homes. In addition, the city established an earthquake museum that showcases the science of earthquakes as well as pictures of people and places in 1976. This memorial and its associated preserved sites remind succeeding generations of schoolchildren and scientists of the risk of earthquakes in their area and the need for building code enforcement and other preparedness and mitigation strategies.

Qinglong, however, preserved its memories in essays written by officials and citizens. The residents of Qinglong were proud of their officials and their neighbors who took charge of prediction, mitigation, and preparedness. As they were already settled in their tents, and they had established schools and stores outdoors, they did not panic and were able to transition into response and recovery modes right away. The rebuilding, however, took longer than in Tangshan, as Qinglong is a relatively poor area of China.

### Positive Effects on Societal Norms and Public Trust

Tangshan officials sought to create a culture of remembrance and preparedness, in order to maintain a high level of awareness among the residents of the earthquake risk. More commonly, cities clean up the debris and forget the tragedy. It is a public norm to talk about earthquake preparedness in Tangshan and neighboring areas. Annually, on the anniversary of the 1976 earthquake, a public ceremony is held to remember the dead and to remind everyone to be prepared for future earthquakes. This remembrance represents a citizen-government partnership to maintain vigilance concerning earthquake disasters. Such remembrance reinforces a culture of preparedness and social trust in the public institutions involved in disaster management.

Tangshan built an earthquake museum and preserved seven earthquake-damaged sites in order to instill the social norm of earthquake awareness, preparedness and mitigation. Official and lay monitoring sites are found strategically around the city. The city supports the officials and equipment, and encourages volunteers to participate in monitoring macro and micro earthquake precursors. This pattern is followed throughout China, led by local governments and the CEA at the national level.

Unfortunately, earthquake prediction still depends upon awareness and decisiveness (and luck). For example, I met one official in southern China who was in charge of earthquake prediction in a county. While he was analyzing data in 1997, he noticed anomalies in two parts of the county. The officials felt that they could not evacuate two towns. They evacuated the wrong target, leaving a more populated town at the mercy of the earthquake.

By contrast, in Qinglong county there is widespread confidence that they can predict another earthquake. This confidence spills over into other areas, such as economic development. In 1995–2000, the officials shared with me their development strategies and which ones were becoming successful. They feel comfortable "taking a chance" on a strategy. They do not feel immobilized, like many poor areas of developing countries. Local government officials are held in high esteem, and citizens continue to participate in lay monitoring for earthquakes.

## ROLE OF "BEST PRACTICE CASE STUDIES" IN CRITICAL INCIDENT ANALYSIS

The field of critical incident analysis is overwhelmingly focused on negative incidents with negative consequences. Many case studies chronicle "what not to do." Analyses point to areas needing improvement, especially emergency management. It is rare to be able to research a case in which, at each decision point, officials, citizens, and even children made the correct decision in the emergency. Many case studies lead to one critical factor that, if handled correctly, would have avoided the incident or mitigated the incident's negative effects.

The Qinglong county case study is an example of a rare successful case. While conducting the research, I uncovered several minor instances in which officials and citizens tried to avoid the preparedness and evacuation orders, such as trying to return to their evacuated houses or businesses. In every case, county officials and the neighbors patrolled their areas and maintained the evacuation, as well as securing community assets such as reservoirs.

## CONCLUSION

The main lesson of Qinglong county in the Tangshan Earthquake is that it is possible to predict, prepare, and mitigate large disasters. A chronology of events shows no sloppy public administration and no one passing the buck. Every official and every citizen participated in the success. If anyone tried to avoid the evacuation, his or her neighbors intervened to correct that behavior.

Furthermore, the case of Qinglong is a ringing endorsement of citizen participation in civic society. Without the volunteers monitoring earthquake precursors, there would have been no basis for making a decision. Without preparation and study, the officials would have been discouraged from making a prediction that was not handed down by a higher-level authority. Without citizens active in their communities, leaders would not have been able to enforce the evacuation. And without the success of Qinglong county, it would have taken longer for victims at the epicenter in Tangshan to receive their first supplies of water, food, and medical care.

## REFERENCES

Birkland, T. A. 2006. *Lessons of Disasters: Policy Change after Catastrophic Events.* Washington, DC: Georgetown University Press.

Chen, Y.; K. L. Tsoi; F. B. Chen; Z. H. Gao; Q. J. Zou; and Z. L. Chen. 1988. *The Great Tangshan Earthquake of 1976: An Anatomy of Disaster.* New York: Pergamon Press.

*China Daily.* 2003. "Natural Disasters Cost China 100 Billion a Year." October 8.

Col, J-M. 2007. "Successful Earthquake Mitigation in Qinglong County During the Great Tangshan Earthquake." *Chinese Public Administration Review* 4 (1/2): 114–123.

Col, J.-M., and J. J. Chu. 1999. "Integrating Public Administration, Science and Community Action: A Case of Early-Warning Success in Qinglong County for the Magnitude 7.8 Tangshan Earthquake." In *Handbook of Crisis and Emergency Management,* ed. Ali Farazmand, 581–616. New York: Marcel Dekker.

Comfort, L. K. 1999. *Shared Risk: Complex Systems in Seismic Response.* New York: Pergamon Press.

Li, J. F. 1991. "Social Responses to the Tangshan Earthquake." University of Delaware Disaster Research Center, Preliminary Paper No. 165. Http://udel.edu/DRC/preliminary/165.pdf (accessed August 16, 2007).

Ma, Z. J., et al. 1989. *Earthquake Prediction: Nine Major Earthquakes in China (1966–1976).* New York: Springer.

Schneider, A., and H. Ingram. 1993. "Social Construction of Target Populations: Implications for Politics and Policy." *American Political Science Review* 87 (2): 334–347.

Schneider, S. K. 1995. *Flirting with Disaster: Public Management in Crisis Situations.* Armonk, NY: M. E. Sharpe.

CHAPTER 14

# Jokela and Kauhajoki

## Experiencing School Shootings in a Nordic Welfare Society

ATTE OKSANEN, PEKKA RÄSÄNEN, AND JOHANNA NURMI

In November 2007, an eighteen-year-old male student killed eight students, staff, and faculty and finally shot himself at Jokela high school in Tuusula, Finland. This small town, near Helsinki, was immediately the focus of headlines around the world. The shock value of the incident was enormous, in part because the tragedy took place in a Nordic country known for its welfare resources provided by the state. Only ten months after this incident, another school shooting rampage took place in Finland—this time in Kauhajoki, also a small town near a big city. In this case, a twenty-two-year-old male culinary arts student killed nine students and a teacher before turning the gun on himself. The impact of these school shootings has been profound in Finland, a country with a population of only 5.3 million.

The common view is that before the early 2000s school-shooting rampages tended to take place only in the United States. Indeed, only a few incidents took place in Europe before 2000, the most notorious examples being at Aarhus University in Denmark in 1994 and Dunblane, Scotland, in 1996. In Finland, a fourteen-year-old student shot two fellow students in Rauma in 1989. But these and other less-known incidents did not have any clear international roots, while the Jokela and Kauhajoki shootings did. This is evident from the fact that the perpetrators of the latter shootings had idolized the perpetrators of the Columbine and Virginia Tech massacres, among others. Since the Columbine shootings, similar violence at schools has become an international phenomenon (Larkin 2009, 1316). Several shootings took place in Germany, in particular, in the 2000s, the most lethal in Erfurt in 2002 (Scheithauer et al. 2008).

The recent school shootings have caused intensive public debate on mental disorders, lack of therapeutic treatment, school bullying, and the dangers of the Internet, leading to social self-examination (Newman et al. 2004, 14). Since the shootings, the Jokela and

Kauhajoki communities have been under constant surveillance by both the media and officials from the welfare sector. Both communities have received considerable public funding (between one and two million euros per year) for the aftercare of the tragedies. In Jokela, for example, almost two years after the incident, forty additional municipal workers are addressing aftercare. The resources allocated for the youth work in Jokela increased after the shootings, which created expectations among the general public for quick results in terms of well-being of the young in Jokela. After a tragedy, especially one in which many young people in Jokela were seriously traumatized, however, it would be unrealistic to expect any kind of rapid results.

In the aftermath of the shootings, dozens of pupils in Finland tried to attract public attention by making threats about a new shooting incident and leaving messages on Internet bulletin boards and discussion forums. Three months after the shootings at Jokela, at least seventy examples occurred in the Finnish schools. Most of the offenders were boys age fourteen to fifteen, but some were girls and older boys (Puustinen 2008). After the shootings in Kauhajoki, yet another wave of threat messages swept the country. By March 2009, a total of 250 threats were reported to the police (Heikkinen 2009). In spring 2011, officials of the Finnish police said that there had been over 500 threats reported to Finnish police (Tasala 2011).

Despite the wide media coverage on school shootings, with its many explanations, not much has been written academically. Explanations in the media usually center on a shooter's psychological state and broad cultural problems, such as antisocial rebellion, family disintegration, or a decline in culturally shared values (Newman et al. 2004, 20). Most academic studies in psychology and sociology have concentrated on explaining the reasons for school shootings. Very few studies have addressed the question of what happens to local communities after shootings and the proximate and long-term effects of these cases (Muschert 2007, 68–69, 71, 75).

International research on critical incidents often emphasizes the role of local communities. Most of these studies underline a rise in social cohesion and cooperation after disasters (e.g., Drabek 1986; Quarantelli and Dynes 1977; Sweet 1998; Turkel 2002). Some researchers, however, claim that high levels of group solidarity do not automatically follow dramatic incidents (Collins 2004, 55; Ryan and Hawdon 2008). Disasters might themselves produce new social problems and conflicts between different groups, and some economic and psychological factors also reduce the likelihood of social solidarity (Webb 2002, 89).

As social disasters, school shootings are problematic for communities because offenders are usually members of the community and the roots of violence emerge from within the community. In this respect, the shootings in Jokela and Kauhajoki were slightly different from each other. The offender in Jokela had lived in the village for over ten years, whereas the perpetrator in Kauhajoki had moved to the town only a year before the shootings. In both cases, however, the shooters attacked their own school and community. This clearly caused not only grief but also fear, insecurity, and collective guilt among the inhabitants of the communities. Violent acts are particularly difficult to deal with for local residents, who must continue their daily life in the same place where the incidents occurred.

In this chapter, we focus on local residents' experiences of school shootings in Finland. In 2008 and 2009, we collected comparable survey data from the local communities Jokela ($n = 330$) and Kauhajoki ($n = 319$). Both data sets were collected approximately six months after the shootings and represent local populations age eighteen to seventy-five. We concentrate on questions such as whether the shootings were considered single tragedies and whether they could have been prevented. In addition to surveys, we have interview data from nine professional experts who worked in Jokela and Kauhajoki during the crisis. In general, this chapter explores the possible implications that recent critical incidents may have for Finland as a Nordic welfare society.

## UNEXPECTED VIOLENCE IN A WELFARE SOCIETY

Societies in the various European countries differ from one another in terms of economic and sociopolitical environment. Finland, Denmark, Norway, and Sweden, called the Nordic welfare states, are characterized by having a good education system, a low rate of poverty, and low income inequality (e.g., Esping-Andersen 1990; Erikson et al. 1987).

Indeed, the Nordic countries share a commitment to full employment for men and women, equality in terms of free participation in higher education, and a promise of universal social benefits for all citizens. It is often noticed that in comparison with other countries, Nordic countries have a high degree of state involvement, a high degree of equality, a high level of taxation, and a high level of public spending on welfare (Greve 2007; Kautto et al. 2001).

Moreover, in other respects, Nordic countries are often considered examples of societies that hold supportive attitudes toward the state and social policy programs. Nordic countries also score high in international surveys on generalized trust and social capital. This means that in spite of welfare benefits by the state, informal social support between individuals is also high in Nordic countries. Compared to other countries, citizens there have more active social networks related to their jobs and a higher participation rate in different non-governmental organizations, associations, and informal clubs. Only social ties between family members and friends remain weak in Nordic countries, especially in Finland, which has the lowest score on this metric in a recent study comparing twenty-one countries (Kääriäinen and Lehtonen 2006).

Nordic countries do very well when it comes to questions of security. In a recent survey, Finland scored the lowest of all EU member countries in terms of fear of crime and feelings of insecurity on the street after dark (van Dijk and et al. 2005, 64–68). Furthermore, every Nordic country rates high on measures of public trust in police and other state institutions. In Finland the number of police officers is the lowest in the European Union, but the trust in police is the highest (Kääriäinen 2008). In this sense the whole issue of safety is different in Nordic countries compared with the UK or the United States, for example. That is, the visibility and proximity of the police are not related to trust in the police (Kääriäinen 2008, 148; see also Hawdon et al. 2003; Holmberg 2004).

Yet Finland is different from other Nordic countries in having exceptionally high homicide rates. In fact, the rate of homicides per capita in Finland is two to three times that of other West European countries. Although gun ownership is widespread in Finland by European standards, the murder weapon most typically used there is a knife. Another Finnish peculiarity is that both offenders and victims are typically middle-aged and long-term unemployed men, and in 76 percent of the homicides, both the victim and the alleged assailant are under the influence of alcohol. In the United States, lethal violence is often portrayed as stemming from poverty, illegal drugs, gun violence, youth gangs, and ethnic discrimination—none of which are relevant to the Finnish homicides (Savolainen et al. 2008).

Typical Finnish homicides happen on the margins of society and concern individuals who are already socially isolated. This may partly explain why generalized trust is high in spite of high murder rates. At the same time, murders of young people are not more numerous in Finland than in other West European countries (Savolainen et al. 2008). In this light, it is easy to understand the shock value that murders of young people have in Finland. Discussion of youth violence started at the end of 1990s after a series of murders of young people. The most dramatic case was at the Myyrmanni Shopping Centre in Myyrmäki, near Helsinki, in which a nineteen-year-old man set off a homemade bomb, killing seven people and wounding almost two hundred others.

From a political perspective, murders of young people are considered a serious problem. After the Kauhajoki shootings, in particular, arguments flared over whether politicians had done enough in the wake of the incident in Jokela. For example, the gun law remained untouched. Furthermore, actions by police were widely discussed. The day before the Kauhajoki shootings, a police officer in the town questioned the killer because he had uploaded videos onto the Internet in which he was shooting with a gun and saying to the camera, "You will be next." The police subsequently released him. After the tragedy, families of the victims sued the police officer in question for involuntary manslaughter, but the police officer was released by the court in the spring of 2011.

The Finnish school shootings have highlighted some weak spots in Nordic welfare states. Although Nordic countries have succeeded in dealing with many traditional problems, such as social marginalization or poverty, they are not as effective when it comes to prevention of spectacular violence such as school shootings or extreme targeted attacks as seen in Norway in July 2011. In this sense, school shooting rampages have initiated a new phase of violence, which might have an impact on generalized trust and feelings of security among citizens.

## DATA AND METHODS

The empirical part of this chapter examines how local residents experienced the Jokela and Kauhajoki shootings, from a point of view of insecurity and uncertainty. Our starting point is the negative feelings that violent events evoke in the residents. In addition, we aim to describe some key characteristics of the communities when the shootings took place. We believe that this kind of description will help us to understand our observations in the two communities more profoundly. We explore the following research questions:

Research question 1. What kind of communities were Jokela and Kauhajoki before the shootings?
Research question 2. How did the local residents of Jokela and Kauhajoki react to the shootings?
Research question 3. Were the sociodemographic profiles of experiences in Jokela and Kauhajoki similar or different?

Our interpretations are based on two types of data. The first data type was derived from two mail surveys of the adult population living in the vicinity of Jokela ($n = 330$) and Kauhajoki ($n = 319$). These data were collected in May–June 2008 and March–April 2009, approximately six months, respectively, after the shootings. The surveys used simple random sampling and had response rates of 47 (Jokela) and 46 (Kauhajoki) percent. Although response rates were below 50 percent, the sample is sufficiently representative of local residents age eighteen to seventy-five of the two towns.

The survey questions focused on local residents' subjective perceptions of the shootings, their evaluations of the community and their experiences of social solidarity in the neighborhood. In this chapter, we examine the local respondents' experiences by using two dependent measures. These items were elicited with the questions "Do you think that the incident was a single tragedy?" and "Do you think that the incident could have been prevented?" Both of these questions could be answered "Yes" or "No." Respondents' gender, age, residential history, and depressive mood are used as the independent variables.

Residential history measures the number of years the respondent has lived at his/her current residence in the Jokela or Kauhajoki area. Depressive mood was measured using the Finnish version of the Beck's thirteen-item depression inventory (see Kaltiala-Heino et al. 1999; Raitasalo 2007). Reliability of the depression scale was found to be high in both Jokela (Cronbach's $\alpha = 0.87$) and Kauhajoki ($\alpha = 0.87$). However, the original variable was dichotomized in order to distinguish those with even slight symptoms ("At least slight depressiveness") from those with no symptoms ("No depressiveness"). In both data sets, approximately 85 percent of the respondents did not have any symptoms. Age and residential history variables were categorized into usable categories. The data analysis is based on descriptive quantitative techniques (namely, examinations of frequency distributions and cross tabulations).

Regarding the second data type, we use thematic interviews of nine professionals or volunteers in the communities of Jokela and Kauhajoki. All of those interviewed worked or lived in Jokela or Kauhajoki before, during, and after the shootings and were thus considered experts with significant information on local reactions to the crises. The interviews were conducted in Jokela about fourteen months after the shootings and in Kauhajoki a year after the incident.

The group of experts interviewed included officials from social services and youth work, a therapist working at the Church of Jokela, and volunteers from two non-governmental organizations. The main focus of each interview was what the interviewees witnessed regarding the local residents' grieving strategies. Because the interview data enable us to evaluate only the experts' views of the residents' experiences, the interviews provide an incomparable source of detail on the overall atmosphere in Jokela and Kauhajoki.

Before turning to questions related to the experience of the shootings in Jokela and Kauhajoki, we briefly explore the communities as they were before the shootings.

## JOKELA AND KAUHAJOKI AS LOCAL COMMUNITIES

During the interviews, the professionals and volunteers were first asked to describe their community and then encouraged to consider the possible changes in the community from the shootings.

Jokela is a small village of 6,000 inhabitants in the south of Finland, within metropolitan Helsinki. Although Jokela is a part of the municipality of Tuusula (which has a population of 36,000), all the interviewed experts depicted it as an individual village with its own distinct character. Traditionally, it was predominantly populated by industrial workers and was a place where everyone knew his neighbors. In the past ten years, however, population increases have changed Jokela from a little village to a small commuter town, and a lot of its population is employed elsewhere in the Helsinki area.

At the same time, however, all the interviewed experts agreed that residents of Jokela have a strong collective identity as Jokela inhabitants. Of the four population centers of Tuusula, Jokela was also reported by the experts as having the most local activities and the highest level of community cohesion. It was considered a good and peaceful place to raise children. Despite its image as a quiet suburban neighborhood, it was also commonly known before the shootings that there were problems with young people in Jokela, such as persistent bullying at schools and drug use.

Kauhajoki is located in the Ostrobothnia region in western Finland, 300 kilometers from Helsinki. With a population of over 14,000, Kauhajoki is larger than Jokela both geographically and in terms of population. It was described as a lively town with several surrounding villages. All the interviewed experts characterized Kauhajoki as having an outstanding program of civic organizational activities, revolving above all around sports. These organizational activities were described as the main source of community cohesion in the area.

Kauhajoki is a geographically decentralized community, and it has a steady flow of students from all over the county coming to study at its vocational and polytechnic schools. These factors present challenges to the social cohesion of the town. One of those interviewed stated that it should be easy for every resident of Kauhajoki to find a hobby among the different activities but suspected that joining these activity circles or clubs where everyone knows one another might not be easy for an "outsider" who is new to town.

The main difference between Jokela and Kauhajoki is that Jokela was seen essentially as a close-knit community, whereas Kauhajoki was not depicted as having such a strong collective identity. Jokela was portrayed as a socially and geographically distinct area, while it is administratively part of Tuusula. Kauhajoki, with its remote villages and incoming students from neighboring towns, was characterized more by its connectedness and cooperation with other communities of the region. However, both communities were described as safe and peaceful small towns, where no one could ever have predicted or imagined a school shooting would take place.

## EXPERIENCING JOKELA AND KAUHAJOKI SHOOTINGS

Because Jokela and Kauhajoki are relatively small communities, it is likely that many of their inhabitants knew at least one of the victims or the perpetrator or at least their family members. We explored this issue in the surveys. As much as one-third (34 percent) of the respondents in Jokela said they were acquainted with someone who died in the shootings. By contrast, less than one-fifth (18 percent) of the respondents in Kauhajoki reported the same thing. The observed difference can be explained by the fact that none of the victims of the shootings in Kauhajoki actually came from Kauhajoki but were only attending school there.

So, how did the respondents experience the shootings? Were the experiences similar in both communities? Table 14.1 lists the results of our dependent measures for Jokela and Kauhajoki.

As the table shows, the proportion of the respondents who considered the incident a single tragedy is considerably higher in Jokela (60.2 percent) than in Kauhajoki (28.7 percent). This is not necessarily surprising, however. In 2007, the Jokela case in many ways represented the first school-shooting rampage in Finland. But what is more interesting is the fact that nearly one-third of the local respondents found the Kauhajoki shooting a single tragedy.

Naturally, it is not possible to elaborate further why so many respondents answered this way. We can only assume, for example, that the term "single tragedy" was understood as a local tragedy, which cannot happen again in the same community. Another possibility is that "single tragedy" was understood as a synonym for a unique tragedy in terms of global disasters, such as 9/11.

With regard to the second item, the response rates in both towns are rather similar. Nearly 44 percent of the Jokela respondents believed that the incident could have been prevented. In Kauhajoki, the proportion of believing this was almost 41 percent. This is an interesting finding, in particular if we bear in mind that the Jokela incident was characterized by more respondents as a single tragedy compared to Kauhajoki. Although some groups have suggested that the actions of a single police officer could have prevented the incident in Kauhajoki, this is not reflected in our data. On the contrary, in Kauhajoki the respondents believed that nothing could have prevented the shootings in Jokela.

One might argue that respondents' experiences of the shootings were heavily influenced by when this research commenced. This explains in particular the results regarding the first item. Both sets of data were collected within six months after the shootings. However, our second item reveals that the respondents are generally rather pessimistic when it comes to preventing school shooting rampages. This finding is interesting in the context of welfare state research because our results suggest that the majority of the local residents do not believe that the school shootings can be prevented. In other words, trust

Table 14.1

**Respondents' Experiences of the Jokela and Kauhajoki Shootings**
Percentages (N)

| | Jokela | Kauhajoki |
|---|---|---|
| Was the incident just a single tragedy? | | |
| Yes | 60.2 (197) | 28.7 (89) |
| No | 39.8 (130) | 71.3 (221) |
| Could the incident have been prevented? | | |
| Yes | 43.7 (136) | 40.6 (128) |
| No | 56.3 (175) | 59.4 (187) |

with regard to school shooting prevention is low. This interpretation contradicts many of the basic beliefs regarding general trust of Nordic welfare state institutions.

Next, we examine the sociodemographic variation in the responses. The aim of this analysis is to address whether the experiences in Jokela and Kauhajoki exhibit a similar profile.

## EXPERIENCING THE SHOOTINGS IN DIFFERENT POPULATION GROUPS

Although the shootings shocked these communities, individual evaluations of the shootings varied. We compare experiences by respondents' gender, age, residential history, and depressive mood. The effect of each variable is examined separately in the Jokela and Kauhajoki data. Our analysis starts from the item measuring whether the shooting incident was understood as a single tragedy. Table 14.2 displays the results.

Differences by gender were tested first. As seen in Table 14.2, the effects are significant across the data sets (at $p < 0.05$). More men than women reported that the shootings were a single tragedy. The magnitude of the difference is approximately ten percentage points in Kauhajoki and fourteen percentage points in Jokela. This finding is interesting because the overall percentages vary considerably between Jokela and Kauhajoki. In general, this suggests that women are more likely than men to believe that similar tragedies can happen again.

Examination by age failed to reveal significant differences. However, the results indicate that the oldest age group (those sixty and over) are the least likely to consider the shootings a single tragedy. Differences between younger age groups are relatively small and do not form a clear pattern. Therefore we can argue that age has a noticeable impact on respondents' experiences.

Residential history shows more interesting results, although the effect is not significant in the Kauhajoki data. In Jokela, however, respondents who have lived in the area for more than ten years are most likely to understand shootings as a single tragedy—nearly 70 percent in this category (more than ten years), while in all other categories response

Table 14.2

**Experiencing Shootings as a Single Tragedy**
Percentages (N)

| Was the incident just a single tragedy? (YES) | Jokela | Kauhajoki |
|---|---|---|
| Gender | $\chi^2$ 6.15* | $\chi^2$ 4.32* |
| Male | 67.1 (112) | 34.3 (47) |
| Female | 53.5 (83) | 24.5 (39) |
| Age | $\chi^2$ 4.89 (ns) | $\chi^2$ 5.60 (ns) |
| < 30 years | 58.3 (14) | 24.4 (10) |
| 30–44 | 52.2 (48) | 25.8 (17) |
| 45–59 | 60.4 (67) | 24.4 (27) |
| 60+ | 68.1 (62) | 38.6 (32) |
| Residential history | $\chi^2$ 7.92* | $\chi^2$ 1.7 (ns) |
| < 1 year | 57.1 (16) | 22.2 (6) |
| 1–5 years | 52.4 (54) | 28.1 (16) |
| 6–10 years | 54.7 (35) | 23.1 (9) |
| +10 years | 69.1 (90) | 31.1 (57) |
| Depressive mood | $\chi^2$ 6.72** | $\chi^2$ 0.03 (ns) |
| 'No depressiveness' | 62.6 (166) | 28.2 (71) |
| 'At least slight depressiveness' | 42.9 (21) | 28.6 (14) |

*Note:* ** = $p < 0.01$; * = $p < 0.05$; (ns) = $p > 0.05$ (2-sided Pearson $\chi^2$)

rates remain below 60 percent. It is also possible to interpret the second data set similarly, even though there are no significant differences. Is it possible to argue that those who have stayed for a long period of time in the area prefer to see unexpected incidents as unique? Perhaps, but it is more important that many older respondents have long histories of residency. This may partly explain our observations here.

The last variable tested was depressive mood. There is a body of research in psychology literature suggesting that people with depressive feelings are more likely to have traumatic experiences and tend to be sensitive to violent acts in general (see Beck and Alford 2009). Given this, it is interesting to see that depressiveness did not have any effect on experiences in Kauhajoki. In Jokela, however, those with at least slight depressive mood reported less often that the incident was a single tragedy. The effect is clearly significant ($p < 0.01$). We interpret this finding to be in line with earlier research on psychological depression and violence. Furthermore, it is possible that the reason depressive mood failed to be significant in Kauhajoki depends on the general distribution of the variable (only 29 percent reported that the shootings were a single tragedy).

The second item measured was whether the shooting incident could have been prevented. The results are listed in Table 14.3. With regard to gender, there are no significant effects. Percentages are nearly identical for both sexes. Age, by contrast, does have some impact in the Kauhajoki data. It appears that the youngest respondents (those under thirty) are more likely to report that the incident could have been prevented. The oldest

Table 14.3

**Experiencing Shootings as an Incident That Could Have Been Prevented**
Percentages (N)

| Could the incident have been prevented? (YES) | Jokela | Kauhajoki |
|---|---|---|
| Gender | $\chi^2$ 0.00 (ns) | $\chi^2$ 1.65 (ns) |
| Male | 44.1 (71) | 39.1 (54) |
| Female | 44.1 (64) | 41.4 (70) |
| Age | $\chi^2$ 3.65 (ns) | $\chi^2$ 7.00* |
| < 30 years | 29.2 (7) | 56.1 (23) |
| 30–44 years | 48.3 (42) | 43.5 (30) |
| 45–59 years | 41.5 (44) | 39.4 (43) |
| 60+ years | 48.2 (41) | 32.2 (28) |
| Residential history | $\chi^2$ 3.88 (ns) | $\chi^2$ 9.84* |
| < 1 year | 26.9 (7) | 59.3 (16) |
| 1–5 years | 47.5 (48) | 50.9 (29) |
| 6–10 years | 46.8 (21) | 45.0 (18) |
| +10 years | 42.5 (51) | 34.2 (64) |
| Depressive mood | $\chi^2$ 5.71* | $\chi^2$ 1.76 (ns) |
| 'No depressiveness' | 39.8 (101) | 42.1 (106) |
| 'At least slight depressiveness' | 58.7 (27) | 32.0 (16) |

*Note:* * = p < 0.05; (ns) = p > 0.05 (2-sided Pearson $\chi^2$)

age group is the least likely to believe this. The differences are quite weak, however. In addition, the Jokela data do not show a similar pattern. Therefore, age does not explain effectively the response reactions to the shootings.

Period of residency appears to be significant ($p < 0.05$) in Kauhajoki but not in Jokela. Those who have resided for a short time in Kauhajoki are more likely to report that the shootings could have been prevented, but this interpretation cannot be applied to Jokela data. The results thus suggest that history of residency does not play an important role when trying to understand local residents' experiences. In other words, period of residency may have an impact, but it varies from one context to another.

The last variable examined was depressive mood. As Table 14.3 indicates, it is the only significant variable in the Jokela data ($p < 0.05$). When compared to those with depressiveness, the respondents without depressiveness report less often that the shootings could have been prevented. The response rates are nearly 40 and 60 percent, respectively. With regard to the Kauhajoki data, however, depressive mood does not have a significant effect. Furthermore, it is not possible to argue that the reason depressive mood fails to be significant here depends on the general distribution of the variable. This is because the overall percentages in Jokela and Kauhajoki are nearly alike with regard to our second variable. Thus we must conclude that depressiveness cannot predict respondents' experiences in Kauhajoki even though it does so in Jokela.

Overall, the results above indicate that the sociodemographic profiles of the experiences differ between Jokela and Kauhajoki. These communities are not similar in size, community structure, or in terms of the residents' experiences. So, what conclusions do our findings suggest?

## CONCLUSIONS

School shootings had a dramatic impact on the local communities of both Jokela and Kauhajoki. A high proportion of community members in both towns said they were acquainted with someone who died in the shootings. Our analysis showed that shootings were experienced differently based on respondents' gender, age, residential history, and depressive mood. In addition, there were differences in how locals experienced shootings in Jokela and Kauhajoki.

What is noteworthy in the results is the general unresponsiveness in the face of violence in Finland. The problem with school shootings is that the offender is usually a member of the community, as was the case in Jokela—this makes it much more difficult for residents to cope with the situation. Questions such as whether the case was single tragedy and whether it could have been prevented reveal how the local community approached the shootings. The majority of the respondents in Jokela and Kauhajoki thought that the shootings could not have been prevented. It was most often younger people and people with a shorter residential history who thought the opposite. In Jokela, by contrast, people with depressive disorders were more likely to believe that the shooting could have been prevented.

In both communities, women were less likely to experience shootings as a single tragedy. In Jokela, residents with a shorter residential history and depressive disorders were more likely not to regard the cases as single tragedies. In other words, we found unresponsiveness and insensitivity among residents who are more rooted in the community and do not express any psychological problems.

School shootings and rage murders in workplaces have been difficult to understand and manage. As a consequence, general attention is often concentrated on the possible psychological profile of the killer, instead of profiling social contexts that are vulnerable to such incidents (see Ames 2007). Our analysis also reveals that there is a lack of sensitivity in the response to school shootings from the very people who might have the resources to solve local social problems. Some of the interviewed experts were concerned that social problems are not openly discussed in Finland.

One of the consequences of school shootings was that the Jokela and Kauhajoki areas were given additional government funding especially for youth work, because youths were at the center of these crises. In Kauhajoki, before the shootings, responsibility for youth-oriented social work for the entire town was handled by only two people; after the crisis this was increased to five. At the same time, however, youth work and mental health-care professionals argued that preventive measures were not adequate at the national level. Most of the public funds were allocated to psychological crisis counseling at local communities.

We believe that the recent school shootings have created a challenge—at least in terms of the welfare state's safety net. Traditionally, citizens of Nordic countries have high trust in state institutions, which might reduce their communal activity. In other words, safety and security issues are something to be handled by the police and state, not by the citizens themselves. Nonetheless, some criminologists contend that Finland has been unsuccessful in coping with even traditional homicides (Savolainen et al. 2008, 85). Therefore, school shootings and threats pose a new problem of insecurity for Nordic welfare societies. In the future, it is possible that insecurity will affect generalized trust between individuals and increase the need for policing more targeted at preventing these kinds of violence.

## REFERENCES

Ames, M. 2007. *Going Postal: Rage, Murder and Rebellion in America.* London: Snowbooks.

Beck, A. T., and B. A. Alford. 2009. *Depression: Causes and Treatment.* Philadelphia: University of Pennsylvania Press.

Collins, R. 2004. "Rituals of Solidarity and Security in the Wake of Terrorist Attack." *Sociological Theory* 22 (1): 53–87.

Drabek, T. E. 1986. *Human System Responses to Disaster.* New York: Springer.

Erikson, R.; E. J. Hansen; S. Ringen; and H. Uusitalo. 1987. *The Scandinavian Model. Welfare States and Welfare Research.* Armonk, NY: M. E. Sharpe.

Esping-Andersen, G. 1990. *The Three Worlds of Welfare Capitalism.* Cambridge, UK: Polity.

Greve, B. 2007. "What Characterizes the Nordic Welfare State Model?" *Journal of Social Sciences* 3 (2): 43–51.

Hawdon, J.; J. Ryan; and S. Griffin. 2003. "Policing Tactics and Perceptions of Police Legitimacy." *Police Quarterly: The Journal of the Police Section of the Academy of Criminal Justice* 6 (4): 469–491.

Heikkinen, M. 2009. "Kauhajoen jälkeen poliisin tietoon on tullut 250 koulu-uhkausta" [A Total of 250 Threats Have Been Reported to the Police After Kauhajoki]. *Helsingin Sanomat*, March 8, 2009.

Holmberg, L. 2004. "Policing and the Feeling of Safety: The Rise (and Fall?) of Community Policing in the Nordic Countries." *Journal of Scandinavian Studies in Criminology and Crime Prevention* 5 (2): 205–219.

Kääriäinen, J. 2008. "Why Do the Finns Trust the Police?" *Journal of Scandinavian Studies in Criminology and Crime Prevention* 9: 141–159.

Kääriäinen, J., and H. Lehtonen. 2006. "The Variety of Social Capital in Welfare State Regimes—A Comparative Study of 21 Countries." *European Societies* 8 (1): 27–57.

Kaltiala-Heino, R.; M. Rimpelä; P. Rantanen; and P. Laippala. 1999. "Finnish Modification of the 13-Item Beck Depression Inventory in Screening an Adolescent Population for Depressiveness and Positive Mood." *Nordic Journal of Psychiatry* 53 (6): 451–457.

Kautto, M.; J. Fritzell; B. Hvinden; J. Kvist; and H. Uusitalo. 2001. *Nordic Welfare States in the European Context.* New York: Routledge.

Larkin, R. W. 2009. "The Columbine Legacy: Rampage Shootings as Political Acts." *American Behavioral Scientist* 52 (9): 1309–1326.

Muschert, G. 2007. "Research in School Shootings." *Sociology Compass* 1 (1): 60–80.

Newman, K. S.; C. Fox; D. J. Harding; J. Mehta; and W. Roth. 2004. *Rampage: The Social Roots of School Shootings.* New York: Basic Books.

Puustinen, M. 2008. "Jokelan jäljet" [Signs of Jokela]. *Opettaja* 7/2008: 8–12.

Quarantelli, E. L., and R. R. Dynes. 1977. "Response to Social Crisis and Disaster." *Annual Review of Sociology* 3: 23–49.

Raitasalo, R. 2007. Mielialakysely: Suomen oloihin Beckin lyhyen depressiokyselyn pohjalta kehitetty masennusoireilun ja itsetunnon kysely [Mood Questionnaire. Finnish Modification of the Short Form of the Beck Depression Inventory Measuring Depression Symptoms and Self-Esteem]. Helsinki: Social Insurance Institution.

Ryan, J., and J. Hawdon. 2008. "From Individual to Community: The 'Framing' of 4–16 and the Display of Social Solidarity." *Traumatology* 14 (1): 43–51.

Savolainen, J.; M. Lehti; and J. Kivivuori. 2008. "Historical Origins of a Cross-National Puzzle: Homicide in Finland, 1750 to 2000." *Homicide Studies* 12 (1): 67–89.

Scheithauer, H.; R. Bondü; S. Meixner; H. D. Bull; and C. Dölitzsch. 2008. "Sechs Jahre nach Erfurt–Das Berliner Leaking-Projekt. Ein Ansatz für die Prävention von School Shootings und Amokläufen an Schulen." [Six Years after Erfurt—the Berlin Leaking Project. An Approach to the Prevention of Shootings and Mass Killings in Schools] *Trauma & Gewalt* 2 (1): 2–13.

Sweet, S. 1998. "The Effect of a Natural Disaster on Social Cohesion: A Longitudinal Study." *International Journal of Mass Emergencies and Disasters* 16 (3): 321–331.

Tasala, M. 2011. "Uhkaukset eivät laannu" [Threats Continue]. *Talentia* 2/2011, 19.

Turkel, G. 2002. "Sudden Solidarity and the Rush to Normalization: Toward an Alternative Approach." *Sociological Focus* 35 (1): 73–79.

Van Dijk, J.; R. Manchin; J. van Kesteren; S. Nevala; and G. Hideg. 2005. "The Burden of Crime in the EU. *Research Report: A Comparative Analysis of the European Crime and Safety Survey* (EU ICS) 2005." Gallup Europe, Brussels.

Webb, G. R. 2002. "Sociology, Disasters, and Terrorism: Understanding Threats of the New Millennium." *Sociological Focus* 35 (1): 87–95.

CHAPTER 15

# Critical Incident Trauma Intervention

## A Sense of Coherence Approach to Psychological First Aid with Student Survivors

PATRICIA LEVY

Violence has increasingly become a part of the reality of American campus life. In response, university counseling centers have focused primarily on offering a basket of supportive counseling services to students coping with interpersonal critical incidences of victimization, including stalking, hate crimes, sexual assault and harassment, attempted suicide and suicide, rape, and partner violence (Carr 2005; Griffin 2007). However, since the sniper shootings at the University of Texas at Austin in 1966 to the more recent shooting attacks at Northern Illinois University in 2008, the rise in mass violence on-campus attacks has resulted in the murder of faculty, staff, and students alike (Associated Press 2007; *The Guardian* 2007; Schlueter 2008). Thus universities have been put in the spotlight for institutional accountability and responsibility for the provision of appropriate protection, effective control of emergency situations, and the rendering of early-phase psychological first aid to survivors.

Without adequate formal orientation, training, and experience, university administrative interdisciplinary teams across the country have rushed to meet this challenge through the establishment of what have been largely untested crisis management policies and protocols that deal with responses to specific situations. In particular, knowledge of the dynamics and range of immediate, acute, traumatic effects of mass violence dealing with defined "survivors" and the intercepting role of psychological first aid are not clearly defined. Application of an effective proactive cross-cultural crisis intervention management plan containing the unique aspects of individual campus community contexts will require extensive creative preparation. How a college goes about defining crisis and rendering its response needs to include symbols of identity, communal strength, and meaningfulness. In addition, the personal style of administrators can play an integral role affecting the manner and timing of on-campus mass emergency response (Zdziarski and Rollo 2007).

Much like fire drills, staged emergency response run-throughs may be carried out by universities. These practice situations can include designated areas of student and staff/faculty evacuation and can be accompanied by containment measures activated by campus and community police interventions.

The entire student body, faculty, and staff should participate in the formulation of an effective preparatory intervention model to cope with survivors of a mass violence attack. However, to do so effectively, these actions necessitate an understanding of (1) underlying concepts of motivated terror, (2) institutionally defined crisis, (3) theories of acute stress and trauma linked to survivor resiliency, and (4) the administering of psychological first aid (PFA).

## THE NATURE OF PERPETRATED VIOLENCE

During their time at college, young people may be living on their own for the first time, entering and building an academic "home away from home." Parents whose children attend smaller institutions might expect more personalized interactions that provide "caring" environments supplemented by individualized resources. In contrast, larger universities, although possibly more impersonal in their relationships with students, might also be a greater source of services, including sophisticated protective measures to ensure on-campus physical security of their sons and daughters (Zdziarski and Rollo 2007). In this way, in both settings, academic institutions can find themselves expected to fulfill a symbolic parental commitment to their students.

Large sweeping lawns with wide open spaces can impart a psychological, social, and academic sense of freedom and movement denoting the very nature of a learning campus community. These peaceful landscaped scenes portray open boundaries rather than soft targets composed of vulnerable campus populations. Typically, college crisis committees have focused on the aftermath of attacks rather than on prevention efforts (National Center for Crisis Management 2004). Establishing a safe perimeter might require a number of adjustments, including the location and activation of protective devices such as key campus entry checkpoints, automated locks in the classroom, camera surveillance, or increased plainclothes law enforcement and electronic security surveillance.

Thus, the seriousness with which college administration takes threat into account as an issue can be reflected in terms of how a crisis event is defined, followed by a suitably designed managed crisis response plan. Because of a lack of exposure, an attitude of underlying passivity can predominate at institutions that have never experienced this particular type of crisis. Virginia Tech, for example, was reported to have been unprepared to cope with the magnitude of the mass violence attack that it experienced (Blythe 2007). Likewise, many higher education institutions across the country are presumably unprepared as well.

Albert Bandura has defined terror as "a strategy of violence designed to promote desired outcomes by instilling fear in the public at large" (1998, 162). Self-sanction by terrorists to commit acts of mass violence justifies terrorism as an acceptable course of vengeful action in the name of a higher moral purpose (Bandura 1998). In addition, viewing the killing of human beings as objects allows for the psychological "disengagement" of the perpetrator from his or her act of terror. This approach allows for disconnection of responsibility for the consequences. In the terrorist's mind, the victims deserve to die (Reich 1998). In the event of hostage taking, understanding the terrorist's mindset may help a hostage negotiator in addition to emergency responders to design an appropriate intervention.

The World Health Organization has defined violence as "the intentional use of physical force or power, threatened or actual, against oneself, another person, or against a group or community, that either results in or has a high likelihood of resulting in injury, death, psychological harm, maldevelopment or deprivation" (as cited in Krug, Dahlberg, Mercy, Zwi, and Lozano 2002, 2). Targeted, premeditated murder stands in stark contrast with the lack of a sense of immediacy and unsure preparation evidenced by many universities' reactions to the possibility of a mass violence attack. Unless a university acknowledges the reality of a continuing threat, its function to ensure its community members' safety cannot be fully realized.

Definitions of what constitutes a crisis can reflect an institution's self-perceived broad responsibilities and its lack of experience in dealing with these events on a mass scale. As Zdziarski and Rollo (2007) have pointed out, each university's cultural and organizational framework can influence its understanding of response. In addition, an understanding of emergency management not only presents an opportunity to formulate a response to mass violence and disaster but also precipitates an organization's readiness to learn and grow from such situations. The following examples illustrate this point.

> A crisis is defined as any unplanned event that significantly threatens the health and welfare of the multiple members of the Ball State community; causes operation disruption, or physical or environmental damage, and harm to the university's public standing. (Ball State University 2008)

> However, it is impossible to define with any preciseness all emergencies or crisis situations that may develop or the most appropriate responses to such situations. Therefore, the plan strikes a balance between being comprehensive on the one hand, yet flexible enough to allow for desired responses to unforeseen events on the other. (Fort Hays State University 2008)[1]

> The goals of the *CRISIS MANAGEMENT TEAM* (CMT) are as follows:
>
> Coordinate the University's response to critical incidents involving students while paying special attention to the safety and security needs of all members of the University community (Texas A&M International University 2007).

The above-stated policies address issues of safety, security, and institutional responses to crises after they have occurred. However, an emphasis on prevention along with a stated commitment to the preparation and provision of services to traumatized victims of campus violence do not seem to have been included as a major focus of the stated goals of the above-referenced universities. In the face of violence, survival during an attack may depend not only on the immediate reactions of the victims but also on the ability of the institution to respond immediately and effectively. Efficient application of the crisis management plan by the university's crisis committee can not only limit casualties but reduce the acute traumatic stress experienced.

## SENSE OF COHERENCE: MAKING SENSE OF TRAUMA

Literature written from psychosocial, emotional, and spiritual points of view concerning survivor adaptation to threat during and after a traumatic event, respectively, falls into two categories: acute stress reaction (ASR) and long-term post-traumatic stress syndrome disorder (PTSD). ASR disorder emphasizes the transitory nature of an initial reaction of serious mental and physical suffering, which lasts for an immediate three-day period after a critical incident of violence (Tanelian and Jaycox 2008). In contrast, PTSD has been defined by the Diagnostic and Statistical Manual of Mental Disorders DSM-IV as a person's exposure to a severe threat situation that leads to overwhelming and intense feelings of shock, horror, and helplessness (APA 2000) over a long period. Not included in this explanation is the trauma that friends, family, and former survivors who have not actually experienced an incident vicariously feel both immediately and in the long term.

There has been a paucity of research conducted on the short- and long-term effects of mass violence attacks on campuses. Little attention has been paid to individual and group short- and long-term adjustments to the acute and post-traumatic stressors that can be experienced. Moreover, intervention offered through university counseling resources has usually followed psychiatric mental-health modes of applying assessment, diagnosis, and consequent treatment in dealing with these conditions. However, a unique and proven model through which to interpret and intervene with university survivors coping with the immediate and long-lasting effects of a mass violence attack still does not exist.

The DSM-IV pathogenic model is based on the assumption that most victims can successfully recover after the initial ASR stage and that they will not experience PTSD. Thus, this model assumes that resulting stress, continuing anxiety attacks, and long-term difficulties in dealing with a terrorist assault in psychological terms, form an exceptional reaction to an extraordinary event that signals pathology. For example, in a recent RAND study on PTSD in service members (Tanielian and Jaycox 2008), one conclusion drawn indicates that the majority of U.S. soldiers returning from combat situations in Iraq and Afghanistan did not appear to suffer from long-term PTSD.

However, it is important to remember that the nature of a mass violence attack requires survivors to make immediate "sense out of chaos" (Antonovsky 1987, xvii). Human beings seem to have an innate instinct to make efforts at self-repair while trying to cognitively integrate life crises. In contrast to the pathogenic view, a more realistic view might be not whether a trauma survivor who experienced an acute stress reaction subsequently suffers from PTSD but, rather, the degree to which this person, even one who continues to have some symptoms of PTSD, can successfully function on a daily basis. For almost everyone affected by an act of terrorism, severe traumatic distress generally does produce in its aftermath responses such as "intrusive recall and sadness" (Shalev and Ursano 2003, 1). Literature on resiliency has provided evidence that coping with situations of anxiety and stress does not "*lead inevitably to vulnerability, failure to adapt, and psychopathology*" (Wolin and Wolin 1993).

Few students, staff, or faculty have had prior exposure or training on methods of appropriate coping in the face of an act of terror. In fact, initial responses can depend upon general instinctive fight-or-flight mechanisms that may or may not be appropriate to the facts of the event. As Shalev and Ursano have pointed out,

> A mismatch between situational demands, personal resources and survival mechanism is one reason for failure (e.g., when withdrawal is used as survival mechanisms in situations from which one can escape or vice versa when scarce resources are wasted in fighting against uncontrollable dimensions of a stress). (2003, 1)

Adaptation to stress, using Lazarus's transactional model (as cited in Ell and Northen 1990, 8), of one's "cognitive appraisal" and one's interactions with the environment, may indeed influence the assessment of one's inner resources to respond to the projected immediate consequences of the traumatic event. Imaginary visualizations during an attack produced in part by fear of horror and possible death may play a role. In addition, prior personal subjective experiences can influence which coping style is chosen by survivors of an attack and by emergency responders alike when exhibiting outward signs of individual hardiness and "courage" (Druss 1995, 88). The capacity to facilitate a strong and positive internal sense of self-confidence through the use of independent and intelligent thinking, self-assertive and individualistic behaviors, effective decision-making, and self-discipline are more likely to result in positive coping and the adoption of healing strategies (Friedman and DiMatteo 1996). How, then, can people try to make sense of stressful events that happen to them? Another proposed model relevant to people's responses to mass violence is Antonovsky's (1984) theory of Sense of Coherence (SOC). Its main conceptual feature is presenting a health, rather than disease, orientation in understanding how people cope with and successfully overcome negative crisis events to go on to lead productive lives. This model advances the theory that interrelated life factors of daily living and well-being can lead to a mediation and reduction of stress. The assumption is that stress is an expected part of life based upon a "health-ease, health disease continuum," composed of "generalized resistance resources" (Antonovsky 1979, 5) that apply throughout one's life.

This interactive global orientation model describes how people face and then integrate critical life stressors comprising three interrelated, dispositional constructs: "comprehensibility," "manageability," and "meaningfulness." Antonovsky defined SOC more specifically as:

> The sense of coherence is a global orientation that expresses the extent to which one has a pervasive, enduring though dynamic feeling of confidence that (1) the stimuli deriving from one's internal and external environments in the course of living are structured, predictable, and explicable; (2) the resources are available to one to meet the demands posed by these stimuli; and (3) these demands are challenges that are worthy of investment and engagement. (1987, 19)

Hypothetically, respondents who score at the higher end of the Orientation to Life (SOC) Questionnaire demonstrate an internal locus of control characterized by strong and persistent feelings of confidence and endurance accompanied by the ability to translate life events into manageable and meaningful challenges. Differentially, low-scoring respondents show evidence of high levels of anxiety together with reduced levels of resistance to stress. This questionnaire has a uniquely appropriate cross-cultural and international application to a wide range of study populations (Hetzel 2005). Thus, psychological well-being can serve as a mediator to stressful situations allowing for coping choices culminating in reduced anxiety, depression, and physical side effects. In addition, subjective perceptions influence the internal locus of control that can lead to the attitudinal formation of values, beliefs, and coping behaviors involved in transactional relationships with external environmental stimuli.

Antonovsky noted that the single most influential element of the model is that of the meaningfulness component, which provides feelings of motivation and energizes the desire to integrate understanding while putting into operation one's abilities to cope with and manage a critical traumatic incident. In social constructivist terms, attempts to encourage people to experience the self and the world within a framework of spiritual and cultural perspectives (Canda and Furman 1999) is an integral component of human modes of adaptation. As Antonovsky has implied, humans have an innate instinct to try to make explanatory sense of their life crises through the use of a "generalized resistance resource" (Antonovsky 1993, 725). Thus, the answer to Job's question "Why me?" spiritually echoes in this model's blended measurement of comprehension, management, and meaningfulness attached to how people process what happens to them.

One study measuring the relationship between Sense of Coherence, anxiety sensitivity due to individual personality traits, and accessibility to social support for American undergraduates (Hart, Hittner, and Paras 1991) revealed that students who scored lower on the Sense of Coherence scale scored higher on Spielberger's trait anxiety scale. This resulting inverse relationship was further illustrated by findings in a study of adolescents living in Israel (Antonovsky and Sagy 1986). The higher the sense of coherence, the more successful research subjects were at mediating anxiety. Both studies imply that people can bring themselves into the "story" of what they do and the manner in which they choose to cope with what happens to them. Confronting a severe range of frightening stressors and, at the same time, simultaneously experiencing high levels of emotional turmoil may seem to be almost impossible for humans to withstand. However, resiliency seems to be a developing and fluid concept of protective factors in situations of high risk. In a Swedish study of adults who grew up under extremely stressful conditions, respondents who demonstrated high-functioning lives evidenced positive self-perceived levels of confidence and an overall global sense of coherence (Dahlin, Cederblad, Antonovsky, and Hagnell, 1990). The study's findings examined adults who had been defined as vulnerable as children who later exhibited traits of "invincibility" (228) as adults.

Wolin and Wolin (1993) in their examination of how people surmount hardship have also commented:

> At best or worst, depending on one's perspective, only about "a third [of children who face dramatic stress] generally succumb; approximately two thirds do not. The purpose of resilience research is to learn how and why [this two thirds] beat the odds" (as cited in Saleeby 2006, 13).

Thus the ability to clarify and structure the resulting efficacy of coping mechanisms may not affect only one's ability to appraise a situation but also one's ability cognitively to cope with it (Shalev 2001). Some people demonstrate a global coping orientation that helps them respond more successfully to a terror attack. At the opposite extreme, others become completely immobilized, unable to react, their cognitive appraisal system having totally shut down. Prior exposure to severe anxiety-producing incidents can also contribute to a smaller coping repertoire from which to draw. For example, studies show that soldiers returning from the wars in Iraq and Afghanistan who had experienced childhood abuse, fear, and other forms of ill-treatment were more likely to have long-term effects of PTSD (Tanielian and Jaycox 2008). This may be a sign of the importance of one's perceptions of self-efficacy, which operate as a buffer to death-related anxiety. As Taubman-Ben-Ari and Findler (2006) have posited in their study of Israeli soldiers' motivation for service, people who portray higher levels of self-esteem may experience reduced levels of anxiety when confronted with threats to their mortality. Although the SOC model differs in emphasis, it similarly concentrates on the powerful impact that a positive global sense of self can have on overcoming life stressors. Because of a lack of research, whether this is the case for survivors of a campus attack remains to be seen.

The SOC measurement of comprehension, management, and meaningfulness can also explain how people process experiences in terror. Those with a strong SOC are more likely to be able to perceive the onslaught of external and internalized threats of violence as a manageable event, becoming proactive in their coping methods. They are flexible problem solvers who can redirect their energies and formulate revised coping strategies as new input is absorbed. Thus, the successful deflection of adverse environmental threats, while aiming to reduce stress, can exemplify effective coping with elements under their control (Jonsson and Mattsson 2003). However, there is some question about whether these findings could be applied to survivors of cumulative traumas, as in a longitudinal SOC study by Snekkevik, Anke, Stranghelle, and Fugl-Meyer (2003). In subjects who had experienced multiple severe traumas, SOC findings were unstable over a year. However, initial SOC measurement can provide a focal point as to whom and on what basis to offer psychological first aid to victims of a terror attack.

## PSYCHOLOGICAL FIRST AID

Responding to a crisis event can, by definition, require putting into operation a repertoire of responses that go far beyond people's normal coping abilities (Griffin 2007; James and Gilliland 2005). Compared to a natural disaster, a mass violence attack evokes higher levels of psychological distress (Friedman, Ritchie, and Watson, 2006). Perhaps the unpredictability of terrorist attacks creates and reinforces stronger feelings of vulnerability and helplessness than do natural disasters.

From a salutogenic point of view, people generally encounter major life stressors (e.g., a death or divorce) as something on a continuum of discrete happenings, understand the traumatic effects and tensions of the moment, and successfully overcome them (Antonovsky 1984). However, during a life-threatening attack, anything can occur, and people are liable to react in unexpected ways.

Albert Bandura (1998) describes several of these side effects. One major aim of the attacker is to create among the targeted victims feelings of loss of control, lack of self-efficacy, shock, and fear of being killed or maimed for life. Initially, the human mind cannot grasp being the central target of an attack. The stress of the trauma takes over and can cause confusion, dissociation, and fight-or-flight behaviors (Griffin 2007; Shalev and Ursano 2003). In addition, the victims might experience overwhelming anxiety over the welfare of family, friends, and associates (Levy 2002). Psychological First Aid (PFA) may be critically needed to address these effects.

Griffin has explained the primary goal of PFA as:

> aiding the adaptive coping and problem-solving processes of survivors who appear at risk for being unable to regain sufficient functioning equilibrium on their own. (2007, 138)

Therefore, it is essential for an organized group of first responders to offer PFA immediately. Consequently, campus communities need to have an overall response plan to deal with both primary and secondary victims, for example, family members with feelings of "confusion, fear, anger, guilt, lack of insight, and feelings of powerlessness" (Rackley 1982, 94). A primary goal then is to provide an environment of physical and emotional safety while, at the same time, successfully minimizing the sense of threat (Griffin 2007).

A secondary goal of emergency intervention is first, through the process of "defusing" (Ritchie, Watson, and Friedman 2006, 4), to make available a facilitative supportive presence through which to increase the person's ability to endure, express, and survive his or her feelings of stress. For example, informal conversation might take place with an individual who is preoccupied with thoughts and relived visualizations of the traumatic event to ventilate, to receive appropriate tactile comforting (e.g., holding his or her hand or placing a comforting hand on his or her shoulder), in addition to the provision of opportunities to make small decisions (e.g., whom to contact). These strategies can allow victims in small ways to gradually regain a sense of control as a form of terror management (Feldman and Snyder 2005).

Crisis workers must be prepared to attend to clients who are experiencing a "disrupted state" and are unable to activate effectively their own coping methods to reduce stress (Rackley 1982). Instead, an emphasis needs to be placed on giving survivors a sense that the acute stress reactions that they are experiencing are expected responses to the events that they have endured (Ruzek 2007). During this "rescue phase" (Shalev and Ursano 2003, 5), keeping in mind that survivors through physical relocation may might also be adjusting to a new sense of reality, first responders should concentrate on de-escalating these combined traumatic effects of the acute stress reaction (Griffin 2007).

Using the SOC model as a theoretical basis, first responders should monitor and assess continuing client reactions. In light of the multitude of people's individual and personal contexts, practitioners should concentrate on individual styles of making sense of (Comprehensibility) and handling (Manageability) events from within culturally diverse contexts to better understand the personal meanings that have been assigned by victims (Meaningfulness) to having been caught in a mass violence attack. Given a period of time after the event, it is essential that a strengths perspective regarding successful coping resources be employed during the assessment and intervention. Then, as individuals experience a reawakening of their cognitive, affective, and behavioral processes in self-affirmative ways, assessment and validation of individuals' coping sources of empowerment can help them heal and find meaning in their experiences (Saleeby 2006; Watson, Ritchie, Demer, Bartone, and Pfefferbaum 2006).

From an SOC perspective, to help find some sense of stability and keep functioning, people bring their inner perceptions of self-efficacy and manner of analysis of events into the story of what happens to them. Thus, emergency responders listening to survivors relating the events of a mass campus attack should pay attention to (1) elements of "time" (the chronological periods of events), (2) "space" (the shaping of experience framed by a configuration of relationships), (3) "story" (the narrative styles and language that have transpired in both time and space as evidenced through the process of attached meanings), (4) and reasons for survival related to spiritual beliefs (Anderson and Worthen 1997).

One primary responsibility of responders is their assessment of survivors who might be most at risk. Almost all victims can recall in detail modes of coping reactions to crisis events that they experienced in the past, even though these events might not have been at the same level of threat or trauma as an attack of mass violence. Part of this evaluation process is for emergency counselors to be able to distinguish between behaviors that are solely related to the terror event itself or are, instead, a continuation and intensification of prior mental-health problems. People with prior histories of trauma, unstable coping skills, and episodes of dissociation and hyperarousal (Griffin 2007) might be especially vulnerable to the traumatic effects of an attack, needing possible medical and psychological intervention that requires constant monitoring, outpatient referral, or hospitalization. In addition, students who continue to experience depression, loss of concentration and interest in their studies, and withdrawal from social interaction might need ongoing services (Ruzek 2007).

Thus, offers of long-term intervention and support to victims and their families might be necessary for those undergoing "fear, anxiety, re-experiencing, urges to avoid, and hyperarousal" (U.S. Department of Veterans Affairs 2010). In the long term, such incidents can have lasting psychological consequences of hyperactivity, depression, anxiety, sleeplessness, and other symptoms of PTSD for both primary and secondary victims such as faculty, roommates, university staff, and friends (Griffin 2007). From a gestalt point of view, when people's lives have tipped out of balance and they have become stuck within their boundaries, feelings of hopelessness, stagnation, and deterioration may lead to depression, culminating in feelings of meaninglessness, and purpose lessens (Perls 1969).

## OPEN-DOOR CRISIS CENTER MODEL

Currently, American university crisis management committees are structured hierarchically, and their membership is administratively top-heavy. Although this group orchestrates the university's overall emergency response to an incident of mass violence, there is a paucity of research literature on the design and operation of a crisis center for students during and after a mass violence attack. Campus crisis management plans have not included a centralized physical "port of call" for students to go to during a terror attack. The following proposed campus emergency service model stems from two sources of this author's experiences: one from the United States in the1960s at an anti-mainstream 24-hour clinic for drug users and runaways, and the other from professional experiences and knowledge gained in Israel during the early 1970s and in the 1980s and 1990s as a medical social worker in a hospital through two wars and a terrorist attack.

Social service agencies in Israel and in the United States have, for the most part, historically developed through specified settings to deliver particular services to designated populations in need. This has also held true for the delivery of crises services targeted in response to specific events. Often funding has determined the organizational structure and the shape of the services offered.

The proposed model is called Open Door. During the initial days and the first few weeks after a mass violence attack, this center should be accessible on a 24-hour basis. Using a grassroots model, staffing should consist of a multidisciplinary team made up of rotational security personnel, nurses, psychologists, community social workers, police officers, student affairs representatives, volunteer multicultural student companions, including translators, crisis telephone workers, and an outreach "flying squad" comprising professionals and paraprofessionals to provide crisis services or needed transport. All personnel should have received training and practice specifically relevant to culturally diverse populations of survivors of a mass violence disaster.

The designated director of the Open Door Center should set up an information center immediately. The center should consist of a public reception area including "comfort zones" for survivors, where physical and psychological first aid can be provided, creating an informal atmosphere of a caring and safe community. Volunteers and counselors' main functions are to meet the concrete immediate needs of disabled victims, such as contacting family members. In alignment with PFA goals, one major service goal is to give survivors support and a sense of control. Through defusion, those who wish voluntarily to ventilate about the events that they have experienced should do so, relating where they were when it happened, what they felt, and what else was going on. Social workers skilled in group work can provide any needed facilitation and safe psychological boundaries for interactions among victim populations in the waiting areas (Levy 2002).

Open Door staff should have been trained to keep in mind the SOC model and its components. Consequently, in aiding victims to make sense of their surviving an attack, staff provide PFA according to the following guidelines:

1. To avoid further traumatization, do not pressure survivors to relive their experiences.
2. Immediately after the trauma, listen and give support—not counseling.
3. Avoid being judgmental, and without a proper background of personal history, don't leap to labeling what may be an appropriate response to this extremely traumatic event.
4. Expect that survivors might tell their stories over and over and, with each retelling, change the emphasis, shape, and form of the story (Shalev and Ursano, 2003). It takes a while to recover from the shock and trauma of the event, which may have caused a fragmented sense of self and of the world as the survivor once knew it. These stories can reveal elements of the SOC model in terms of understanding, psychosocial and physical coping mechanisms, and interpretations of meanings attached to survivor experiences.
5. Concentrate on empowerment. One such example that is complementary to the global orientation in Antonovsky's SOC model is Saleeby's CPR model (Saleeby, 2006, 10). Look for areas of:
   C stands for competencies, capacities, and courage
   P symbolizes promise, possibility, and positive expectation
   R signifies resilience, reserves, and resources

All services should be provided at no cost. According to a preestablished protocol of notification and confidentiality, a crisis line established to answer public inquiries should respond to and refer concerned friends and relatives regarding the whereabouts of loved ones. Communication devices such as a message board, fax, phones, and computers should link strategically to internal campus and external community resources. These might include emergency medical services at the scene of the attack, the police and military, and social welfare professionals skilled in reaching out to the community to find family members of survivors of the attack who have been brought in for purposes of identification, and to aid family members whose loved ones may have been killed. Inside the center, a centralized electronic software connection between all emergency responder resources should code incoming survivors, names if known, and any other critical details.

One of the main responsibilities of security personnel and initial screening team staffers at the information center should also be to keep unnecessary people away from entering certain areas of the Open Door Center, so that the staff can then concentrate freely on treating physical and psychological victims while, at the same time providing needed resource and referral services. In response to a mass violence attack, depending on the size of the incident, people can appear from other than the site of the attack. After a bomb blast or a shooting, identifying features of individual survivors may have changed, with hair color distorted by debris, dust, burns, or other wounds. Victims of prior violent attacks might show up, even though they are not current victims, experiencing retraumatization, called a "double-effect" (Levy 2002). Sometimes, two or three days later or three weeks later, a victim at the scene of mass violence might appear at the center requesting help, often with PTSD symptoms.

Open Door's emergency responders might also require PFA themselves, especially in light of repeated trauma exposure, because of the impact on them of the mass violence event, and the numerous and difficult circumstances presented by each of the survivors that they help. Much will depend upon the character, skills, initiative, and confidence of the center's director in guiding, supporting, and responding to the cumulative effects of such a chaotic situation. The literature on various emergency response models has detailed the necessity of allowing defusing of emergency team members for their own emotional and psychological processing. In addition, workers will be much more likely to share with their peers with whom they have shared experiences (Friedman, Ritchie, and Watson 2006) while empowering each other individually and as a team.

## CONCLUSION

In light of the rise in mass violence attacks at American universities, holistic critical incident management plans should emphasize prevention along with a stated commitment to preparing and providing direct services to traumatized victims of campus violence. Antonovsky's Sense of Coherence model should be used as an underlying framework that comprises not only institutional perceptions of response but also psychological first aid interventions for survivors. Because of the lack of empirical research on student survivor coping behaviors, there is an immediate need for further study in this critical area on the interactive effects of overcoming hardship and gaining meaning from trauma for both students and their academic community.

## NOTE

1. The interpretation of Fort Hays State University policy is solely that of this author.

## REFERENCES

American Psychiatric Association (APA). 2000. *Diagnostic and Statistical Manual of Mental Disorders* (4th ed.) (DSM-IV). Washington, DC: American Psychiatric Press.

Anderson, D. A., and D. Worthen. 1997. "Exploring a Fourth Dimension: Spirituality as a Resource for the Couple Therapist." *Journal of Marital and Family Therapy* 23 (January): 3–13.

Antonovsky, A. 1979. *Health, Stress, and Coping.* San Francisco: Jossey-Bass.

———. 1984. "The Sense of Coherence as a Determinant of Health." In *Behavioral Health: A Handbook of Health Enhancement and Disease Prevention,* ed. J. D. Matarazzo, S. M. Weiss, J. A. Herd, and N. E. Miller, 114–129. New York: John Wiley & Sons.

———. 1987. *Unraveling the Mystery of Health: How People Manage Stress and Stay Well.* San Francisco: Jossey-Bass.

———. 1993. "The Structure and Properties of the Sense of Coherence Scale." *Social Science Medicine* 36 (6): 725–733.

Antonovsky, H., and S. Sagy. 1986. "The Development of a Sense of Coherence and Its Impact on Responses to Stress Situations." *Journal of Social Psychology* 126: 213–225.

Associated Press. 2007. "List of Deadliest Campus Shootings in United States." Fox News, April 16. www.foxnews.com/story/0,2933,266368,00.html (accessed October 11, 2008).

Bandura, A. 1998. "Mechanisms of Moral Disengagement." In *Origins of Terrorism: Psychologies, Ideologies, Theologies, States of Mind,* ed. W. Reich, 161–191. Washington, DC: Woodrow Wilson Center Press.

Ball State University. 2008. "Crisis Management Protocol." http://cms.bsu.edu/About/AdministrativeOffices/EmergencyPrepared/CMT/Protocol.aspx (accessed October 2, 2008).

Blythe, B. 2007. "Virginia Tech Shootings: Crisis Magnifies the Significance of Small Weaknesses." Crisis Management International, April 18. www.cmiatl.com/news_article73.html (accessed October 11, 2008).

Canda, E. R., and L. Furman. 1999. *Spiritual Diversity in Social Work Practice: The Heart of Helping.* New York: Free Press.

Carr, J. L. 2005. *American College Health Association Campus Violence White Paper.* Baltimore: American College Health Association.

Dahlin, L.; M. Cederblad; A. Antonovsky; and O. Hagnell. 1990. "Childhood Vulnerability and Adult Invincibility." *Acta Psychiatrica Scandinovica* 82: 228–232.

Druss, R. G. 1995. *The Psychology of Illness: In Sickness and in Health.* Washington, DC: American Psychiatric Press.

Ell, K. O., and H. Northen. 1990. *Families and Health Care: Psychosocial Practice.* New York: De Gruyter.

Feldman, D. B., and C. R. Snyder. 2005. "Hope and the Meaningful Life: Theoretical and Empirical Associations Between Goal-Directed Thinking and Life Meaning." *Journal of Social and Clinical Psychology* 24 (3): 401–421.

Fort Hays State University. 2008. "FHSU Crisis Management Plan." https://bigcat.fhsu.edu/crisis/intro.php (accessed October 11, 2008).

Friedman, H. F., and DiMatteo, R. (1996). "Social Psychology and Medicine." *Journal of Social Issues, 35,* 82–99.

Friedman, M. J.; E. C. Ritchie; and P. J. Watson. 2006. "Overview." In *Interventions Following Mass Violence and Disasters,* ed. M. J. Friedman, E. C. Ritchie, and P. J. Watson, 3–15. New York: Guilford Press.

Griffin, W. 2007. "Psychological First Aid in the Aftermath." In *Campus Crisis Management,* ed. E. L. Zdziarski, N. W. Dunkel, and J. M. Rollo, 145–182. San Francisco: Jossey-Bass.

*The Guardian.* 2007. "A History of Violence." Guardian.co.uk, April 17. www.guardian.co.uk/world/2007/apr/17/usa.internationaleducationnews (accessed October 11, 2008).

Hart, K. E.; J. B. Hittner; and A. Paras. 1991. "Sense of Coherence, Trait Anxiety, and the Perceived Availability of Social Support." *Journal of Research in Personality* 25: 137–145.

Hetzel, A. M. 2005. "Sense of Coherence and Depressive Symptomotology in Parents of Children Diagnosed with Asperger Syndrome." PhD diss., Capella University.

James, R. K., and B. E. Gilliland. 2005. *Crisis Intervention Strategies* (5th ed.). Belmont, CA: Thomson Brooks/Cole.

Jonsson, A. S., and B. Mattsson. 2003. "Post-Traumatic Stress Among Swedish Ambulance Personnel." *Emergency Medical Journal* 20: 79–84.

Krug, E. G., L. L. Dahlberg, J. A. Mercy, A. B. Zwi, and R. Lozano (eds.). 2002. *World Report on Violence and Health.* Geneva, Switzerland.

Levy, P. 2002. "Terrorism and Social Work Practice: Memories of Terrorism in Israel." *BPD Update* 24 (2): 5–9.

National Center for Crisis Management. 2004. "University Crisis Management." www.nc-cm.org/universities.htm (accessed September 25, 2008).

Perls, F. 1969. *Gestalt Therapy Verbatim.* Lafayette, CA: Real People Press.

Rackley, F. 1982. Religion and Terrorism—A Romantic Model of Secular Gnosticism. In *The Rationalization of Terrorism,* ed. D.C. Rapoport and Y. Alexander, 91–100. Frederick, MD: University Publications of America.

Ritchie, E. C., P. J. Watson, M. J. Friedman (eds.) 2006. *Interventions Following Mass Violence and Disasters: Strategies for Mental Health Practice.* New York: Guilford.

Ruzek, J. I. 2007. "Models of Early Intervention Following Mass Violence." In *Interventions Following Mass Violence and Disasters,* ed. M. J. Friedman, E. C. Ritchie, and P. J. Watson, 16–36. New York: Guilford Press.

Saleeby, D. 2006. *The Strengths Perspective in Social Work Practice*, 4th ed. Boston: Allyn and Bacon.

Schlueter, K. 2008. "Campus Violence Increases." *Chicago Flame*, February 25. www.chicagoflame.com/2.9144/campus-violence-increases-1.1295986/ (accessed December 12, 2008).

Shalev, A. Y. 2001. "What Is Posttraumatic Stress Disorder?" *Journal of Clinical Psychiatry* 62: 4–10.

Shalev, A. Y., and R. J. Ursano. 2003. "Mapping the Multidimensional Picture of Acute Responses to Traumatic Stress." In *Reconstructing Early Intervention After Trauma: Innovations in the Care of Survivors,* ed. Roderick Ørner and Ulrich Schnyder, 118–129. New York: Oxford University Press.

Snekkevik, H.; A. G. Anke; J. K. Stanghelle; and A. R. Fugl-Meyer. 2003. "Is Sense of Coherence Stable After Multiple Trauma?" *Clinical Rehabilitation* 17: 443–453.

Tanielian, T., and L. H. Jaycox (eds.). 2008. *Invisible Wounds of War: Psychological and Cognitive Injuries, Their Consequences, and Services to Assist Recovery.* Santa Monica, CA: RAND. www.rand.org/pubs/monographs/2008/RAND_MG720.pdf.

Taubman-Ben-Ari, O., and L. Findler. 2006. "Motivation for Military Service: A Terror Management Perspective." *Military Psychology* 18 (2): 149–159.

Texas A&M International University. 2007. "Annual Institutional Effectiveness Review." www.tamiu.edu/adminis/ie/data/admin/documents/univ-police-2006.pdf (accessed October 11, 2008).

U.S. Department of Veterans Affairs. 2010. "What Is Post-Traumatic Stress Disorder?" National Center for Post-Traumatic Stress Disorder. www.ptsd.va.gov/public/pages/what-is-ptsd.asp.

Watson, P. J.; E. C. Ritchie; J. Demer; P. Bartone; and B. J. Pfefferbaum. 2006. "Improving Resilience Trajectories Following Mass Violence and Disaster." In *Interventions Following Mass Violence and Disasters,* ed. E. C. Ritchie, P. J. Watson, and M. J. Friedman, 37–53. New York: Guilford Press.

Wolin, S., and S. J. Wolin. 1993. *The Resilient Self: How Survivors of Troubled Families Rise Above Adversity.* New York: Villard.

Zdziarski, E. L.; N. W. Dunkel; and J. M. Rollo (eds.). 2007. *Campus Crisis Management: A Comprehensive Guide to Prevention, Response, and Recovery.* San Francisco: Jossey-Bass.

Zdziarski, E. L., and J. M. Rollo. 2007. "The Impact of Crisis." In *Campus Crisis Management: A Comprehensive Guide to Prevention, Response, and Recovery,* ed. E. L. Zdziarski, N. W. Dunkel, and J. M. Rollo, 3–34. San Francisco: Jossey-Bass.

CHAPTER 16

# The Use of a Multi-Attribute Method to Analyze Preparedness for an Incident of Terrorism

RUTH A. MAURER AND MARIAN E. MOSSER

Since the September 11, 2001, terrorist attacks, most local city and county government agencies across the United States have ended up bearing the burden of financing and managing critical incidents. The impact on local governments of federal and state mandated homeland security programs has varied considerably across various levels of government, but what remains constant is that agencies and organizations involved in critical incident management must understand how prepared they are before a disaster strikes. No first-responder organization can afford not to have a preparedness strategy. Local law enforcement officials are charged with overseeing an array of critical duties associated with emergency preparedness, and, therefore, their preparedness is vital to preventing acts of terrorism and minimizing the impact of other critical incidents.

## DESCRIPTION OF THE STUDY

The purpose of this chapter is to complete a quantitative, descriptive study of law enforcement preparedness against terrorism in Florida. First, we define preparedness and identify factors of preparedness. Second, we present data collected from Florida law enforcement terrorism experts about their views on the relative importance of each factor of preparedness to overall preparedness and how each factor influences a level of preparedness. Third, we present data collected from county and municipal law enforcement agencies in Florida to determine the current level of preparedness against terrorism in the state. Finally, we analyze the data using the method presented in this chapter.

Three questions guided the research:

1. What is preparedness and what factors define it?
2. How important is each factor to preparedness?

3. Given the study's definition of preparedness and its identification of preparedness factors, what is Florida law enforcement's assessment of the state's level of preparedness against terrorism?

Preparedness is a local issue, and there is no better state for measuring preparedness than Florida, which has 67 counties and more than 400 municipalities. Also, given Florida's known status as a high-risk state for critical incidents, it offered a vital testing ground for research on the preparedness model to evaluate law enforcement preparedness. Because of Florida's many military installations, visitor attractions, airports and seaports, and extensive coastline, not to mention the traversing of the terrorists pre-9/11 and the anthrax incident in South Florida, terrorism is an important issue there. As homeland security mandates increase security requirements and local budgets are cut, Florida law enforcement agencies are scrambling to find ways to remain prepared for terrorism and other critical incidents.

The study examined law enforcement preparedness for terrorism in Florida, using a more in-depth approach than the critical incident technique introduced by Flanagan (1954). Flanagan used a five-step technique that is often modified for specific settings and research questions: (1) identification of aim, (2) identification of types of events to collected, (3) data collection, (4) data analysis, and (5) project dissemination. However, the critical incident analysis technique is aimed at isolating critical facts and features of an incident, rather then evaluating the interrelated pieces to identify key influencing factors in order to gain a greater understanding of the whole and reframe an assumption and yield transformation (Brookfield, 1987). This study method used experts to determine the importance of each factor of preparedness to overall preparedness.

## THE CATALYST FOR THE RESEARCH STUDY

The study builds on the Federal Emergency Management Agency (FEMA) preparedness cycle, which defines preparedness as "a continuous cycle of planning, organizing, training, equipping, exercising, evaluating, and taking corrective action in an effort to ensure effective coordination during incident response" (2009, 1). The following FEMA definition of preparedness was adopted in this study:

> The leadership, training, readiness, and exercise support, and technical and financial assistance to strengthen citizens, communities, state, local, tribal governments, and professional workers as they prepare for disasters, mitigate the effects of disasters, respond to community needs after a disaster, and launch effective recovery efforts. (2007)

The study's theoretical framework consisted of the integration of disaster management theory and emergency management approaches to preparedness, preparedness factors, and parameters for modeling preparedness. Figure 16.1 depicts the systematic approach of the FEMA preparedness planning cycle.

Figure 16.1 **FEMA Preparedness Cycle National Incident Management System**

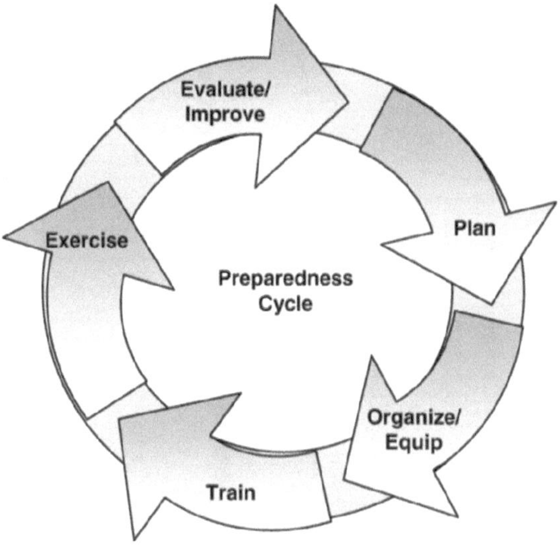


*Source:* FEMA 2009, 1.

At the practical level, preparedness is defined as an ongoing effort to mitigate the impact of disasters of people and property, yet research to quantify preparedness is almost nonexistent. The problem with the FEMA definition is that it does not specify specific factors of preparedness or any measures of effectiveness for preparedness. An extensive review of the disaster and emergency management literature on preparedness allowed us to identify the six main factors of preparedness—that is, criteria intended to maximize the overall effectiveness and productivity of an organization in becoming prepared for disaster or terrorism. The six factors in this model are identified and defined as:

- equipment (e.g., purchase or use of equipment, including personnel protective equipment).
- communications (e.g., coordination, information and intelligence, and investigation sharing with other agencies).
- personnel (e.g., hiring of people for terrorism-related duties, specialized terrorism units, or the reorganization of personnel duties).
- planning (e.g., actions to improve one's response capabilities, planning, and interoperability planning with other agencies).
- procedures (e.g., agency meetings, written policies or procedures, and written procedures with other agencies).
- training activities (e.g., field or tabletop exercises, counterterrorism training, and courses related to weapons of mass destruction, drills, and rehearsals).

A hierarchy decision model using the factors of preparedness is shown in Figure 16.2.

Figure 16.2 **Framework for Determining Factors of Preparedness against Terrorism**

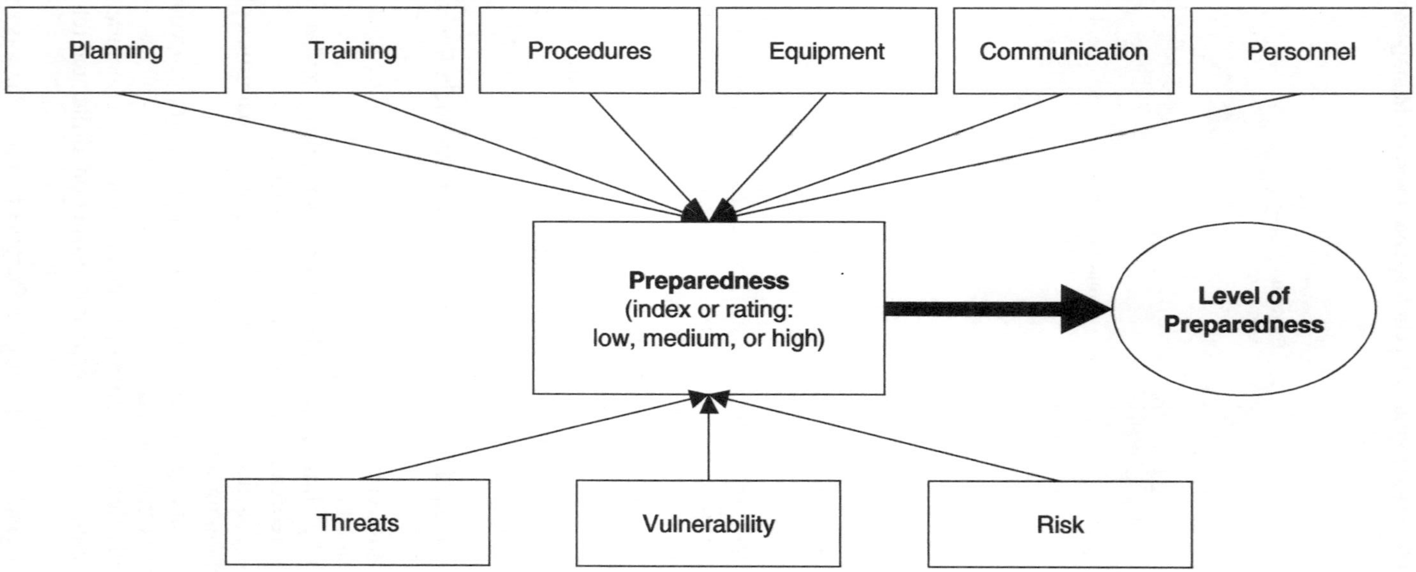

The model was developed to better understand what is meant by preparedness, to identify factors of preparedness, and to measure the factors to determine the level of law enforcement preparedness in Florida.

## THE DECISION-MAKING PROCESS IN GENERAL

The decision making process involves using a clearly defined set of terms and understanding the decision-making process. Asking the right questions and analyzing enough answers can ensure that the final model clearly addresses the research questions. In this research, we considered using the Public Protection Classification (PPC) Program, which evaluates community features of fire-protection systems according to a manual called the Fire Suppression Rating Schedule (FSRS). The PPC method allows the evaluation of fire alarm and communications systems, the fire department, and the water supply in communities. Using the FSRS, data on municipal fire-protection efforts are collected and analyzed, allowing communities to receive a public protection classification rating that can lead to lower fire insurance rates and better public protection. We also considered Kanji's (1976, 1994, 1999) business excellence model. However, after fully exploring the PPC method and Kanji's method, we decided against using them. The PPC method was too detailed and complex to analyze, and Kanji's method would not be fully applicable to law enforcement preparedness for terrorism, though the idea could be customized for measuring preparedness. Instead, we decided to use the Multi-Attribute Decision Making (MADM) approach, which is suitable for simple decisions with a small number of alternatives or factors (Saaty 2000, 2001, 2005). This method allows for the comparison of the six preparedness factors, which are hard to quantify.

## ANALYTICAL HIERARCHY PROCESS IN THE STUDY

Because most decision makers have trouble with more than seven decision criteria at a time, the Analytic Hierarchy Process (AHP) was selected for this application as it is useful when there are multiple attributes (Saaty 1980, 1986, 2000, 2005). The importance of AHP and the use of pair-wise comparisons in decision making are referred to throughout the literature (Saaty 2000). Also, MADM involves structuring and solving decisions and planning problems involving multiple criteria. The quantification and aggregation of experts' beliefs provided important information to the researcher while considering preparedness against a critical incident such as terrorism. Because most decision makers have trouble with more than seven decision criteria at a time, the Analytical Hierarchy Process (AHP) was selected for this use in this study. AHP is one of the more widely applied multi-attribute decision-making methods used by researchers because subjective assessments of relative importance can be converted to a set of overall scores or weights. The methodology of AHP is based on a pair of wise comparisons of how important one criterion C*i* (*communications*) is relative to criterion C*j* (*equipment*). Questions concerning the six preparedness factors such as, Is communications more important than training? were used in this study to establish the weights for criteria. AHP allowed questions to be answered to assess the scores for alternatives on the subjective (judgmental) criteria.

The AHP method was appropriate for this study because participants were faced with a complex decision of rank ordering factors of preparedness against terrorism. It allowed law enforcement officials to weigh each of the criteria (preparedness factors) in a nine-point scale. For each pair of criteria, the study participants were required to respond to a pair-wise comparison question asking the relative importance of the two. The responses were indicated using a nine-point scale expressing the intensity of the preference for one criterion versus another:

1 = Equal importance or preference
3 = Moderate importance or preference of one over another
5 = Strong or essential importance or preference
7 = Very strong or demonstrated importance or preference
9 = Extreme importance or preference (Satty, 2000, 2001, 2005).

The law enforcement terrorism experts were asked to rank order each factor and then complete a structured pair-wise comparison. A pair-wise comparison was assigned a numerical value to the pair corresponding to the predictive importance between the two factors.

Saaty argued that "one of the greatest values of soft measurement with a relative scale is to order elements" (2000, 6). By setting ordered priorities, decision makers can better plan and allocate resources to resolve conflicts and achieve buy-in to implement a decision. Further, weighting each factor within a cluster would allow for the development of a ratio that could be used for checking the consistency of the data (2005, 2000). Saaty, during his calibration research in 1977, developed a scale for the pair-wise comparison using a linear 1–9 scale, which was used here in Table 16.1.

## EXPERT ELICITATION TO POPULATE THE MODEL

The next step in the process was to obtain various expert responses about the importance of each factor for the identification and weighting of the model factors. Expert judgment could be elicited three ways: (1) interactive groups, in which experts are face to face with one another and a moderator; (2) individual interviews, in which one expert is interviewed by the researcher in private and face to face; or (3) Delphi, in which experts are isolated from one another and give their judgments to a facilitator. For this study, the individual interview methodology was used. Seven terrorism experts, who each had more than ten years' experience and were currently active in anti-terrorism or counterterrorism work, were asked to participate. Clemen and Winkler (1999) suggested that eliciting and aggregating individual assessments after group interaction, rather than forcing the group to reach a consensus, allows the sharing of knowledge while avoiding group interaction problems. Further, Clemen and Winkler held that it is better to use experts who have different viewpoints and knowledge to minimize redundant information and maximize the effectiveness of aggregation.

Table 16.1

**Converting Verbal Judgments to Numbers**

| Verbal expressions | Associated numbers |
|---|---|
| Equally preferred | 1 |
| Equal to moderate | 2 |
| Moderately preferred | 3 |
| Moderately to strongly | 4 |
| Strong preferred | 5 |
| Strongly to very strongly | 6 |
| Very strongly preferred | 7 |
| Very strong to extremely strong | 8 |
| Extremely preferred | 9 |

*Source:* Saaty 2000, p. 73.

## THE QUESTIONNAIRE PACKAGE

### The Expert Factors of Preparedness Questionnaire

Each expert was contacted by telephone to request and schedule a one-on-one meeting. At the interview each expert was given a questionnaire packet containing (a) a cover letter describing the elicitation process and outlined the willingness to participate in the study; (b) a list of the six evaluation factors and their definitions (listed on pages 281); (c) a six-page factors of preparedness questionnaire that allowed for comparison of all six factors; (d) a page for participants to list their degree of confidence in their ability to answer the questions; and (e) a feedback form and a note thanking participants for their participation. Participant experts ranked the six evaluation factors listed below in order of importance concerning preparedness for terrorism. They marked the number indicating its importance for terrorism preparedness (from 1 = most important to 6 = least important) in the brackets on the left for each factor and, if two or more factors are equally important, participants were told to assign them the same number.

[ ] Equipment (EQ)
[ ] Communications (CO)
[ ] Personnel (PE)
[ ] Planning (PL)
[ ] Procedures (PR)
[ ] Training (TR)

Experts filled out the six-page questionnaire, which compared one factor to the other five factors. For example, equipment was compared to communications, personnel, planning, procedures, and training. All participants were asked to place an "x" in the box next to the factor that was more important, then determine how much more important the factor was. Figure 16.3 shows an example of the factors of preparedness questionnaire.

Figure 16.3 **An Example of the Factors of Preparedness Questionnaire**

**Equipment vs. Evaluation Factors**

| Which factor is more important? **Equipment** ☐ or **Communications** ☐ | | | | | | | | |
|---|---|---|---|---|---|---|---|---|
| Equally | | Moderately | | Strongly | | Very strongly | | Extremely |
| ☐ | ☐ | ☐ | ☐ | ☐ | ☐ | ☐ | ☐ | ☐ |

| Which factor is more important? **Equipment** ☐ or **Personnel** ☐ | | | | | | | | |
|---|---|---|---|---|---|---|---|---|
| Equally | | Moderately | | Strongly | | Very strongly | | Extremely |
| ☐ | ☐ | ☐ | ☐ | ☐ | ☐ | ☐ | ☐ | ☐ |

| Which factor is more important? **Equipment** ☐ or **Planning** ☐ | | | | | | | | |
|---|---|---|---|---|---|---|---|---|
| Equally | | Moderately | | Strongly | | Very strongly | | Extremely |
| ☐ | ☐ | ☐ | ☐ | ☐ | ☐ | ☐ | ☐ | ☐ |

| Which factor is more important? **Equipment** ☐ or **Procedures** ☐ | | | | | | | | |
|---|---|---|---|---|---|---|---|---|
| Equally | | Moderately | | Strongly | | Very strongly | | Extremely |
| ☐ | ☐ | ☐ | ☐ | ☐ | ☐ | ☐ | ☐ | ☐ |

| Which factor is more important? **Equipment** ☐ or **Training** ☐ | | | | | | | | |
|---|---|---|---|---|---|---|---|---|
| Equally | | Moderately | | Strongly | | Very strongly | | Extremely |
| ☐ | ☐ | ☐ | ☐ | ☐ | ☐ | ☐ | ☐ | ☐ |

## Overview of Experts' Evaluation Process

*Step 1*

The overall evaluation preparedness was broken down into evaluation factors and arranged into a hierarchy. The six essential factors (EFs) were assessed (ranked) against one another.

*Step 2*

Experts compared two EFs as a pair, considering the contribution of the overall topic evaluation, for all combinations of EFS. A pair-wise comparison was performed with a judgment scale, which was not fixed. Each pair-wise comparison was assigned a numerical value corresponding to the relative importance of the two EFs. For this study, the selection of a judgment scale (linear) was key to building the model because it was understood by the respondents.

*Step 3*

Pair-wise comparisons of each EF group were made, and a pair-wise comparison matrix was developed. In an ideal case of exact measurement, the relationship between the weights $w_i$ and the judgments $a_{ij}$ is mathematically expressed by:

$$\begin{bmatrix} w_1/w_1 & w_1/w_2 & \cdots & w_1/w_n \\ w_2/w_1 & w_2/w_2 & \cdots & w_2/w_n \\ \cdot & \cdot & & \cdot \\ \cdot & \cdot & & \cdot \\ \cdot & \cdot & & \cdot \\ w_n/w_1 & w_n/w_2 & \cdots & w_n/w_n \end{bmatrix}$$

in which $n$ = number of factors, $a_{ij}$ = relative weight determined by the pair-wise comparison of the relative importance of the $i$th factor ($i, j = 1, 2, 3, \ldots, n$) $= \frac{w_i}{w_j}$, $w_i$ = weight of the $i$th factor, and $a_{ij} = \frac{1}{a_{ji}}$.

*Step 4*

The consistency of the pair-wise comparison matrix was evaluated. Deviation from consistency, or the consistency index (C.I.), was represented by $\frac{(\lambda_{\max} - n)}{n-1}$. Saaty (1980) suggested that a consistency rating (C.R.) of 0.10 or less is acceptable.

*Step 5*

The overall weights of EFs in the hierarchy were computed, and an overall evaluation was performed.

**The Purposive Sample and Combating Terrorism Survey**

Florida county and municipal law enforcement chiefs were invited to participate in the sixty-two-question Combating Terrorism Survey. The population for this study was ($N = 300$) and included 67 Florida county law enforcement agencies and 233 municipal law enforcement agencies. The names and locations of police agencies were obtained from the Florida Division of Law Enforcement (FDLE). Any discrepancies in police organizations listed but no longer in existence were resolved by contacting the FDLE. The study had an overall response rate of 44 percent. The questionnaire response rate was 46 percent from county law enforcement agencies and 43 percent from municipal law enforcement agencies.

## Analytical Procedures for the Combating Terrorism Survey

The researcher used the general model of surveying all municipal and county law enforcement agencies to obtain their feedback on factors of preparedness using the Combating Terrorism Survey. Participants' answers to the Combating Terrorism Survey were calculated to obtain the *F* factor for each factor of preparedness. After the *F* factors were obtained from the Combating Terrorism Survey, they were entered into the formula. To generate the *F* values, counts from the surveys were used to determine a rating for that entity. The following formula was used to measure preparedness, given values of the factors from the measure preparedness factors in the survey.

$$\Psi = \omega_1 F_1 + \omega_2 F_2 + \omega_3 F_3 + \omega_4 F_4 + \omega_5 F_5 + \omega_6 F_6,$$

where $\omega_i$ is the AHP-determined weight for factor *i*, and $F_i$ is the survey-determined factor value for factor *i*. $F_i$ is the ratio of counted items for the factor to total possible items for the factor. The survey was broken down into the same factors of preparedness categories. To obtain factors of preparedness (the $F_i$ values in the formula) from the survey, the following process was used. First, each question in the Combating Terrorism Survey was reviewed to determine the question's appropriateness to the six factors of preparedness. The second step involved the review of the selected questions for possible points. Possible points for each question were determined by the number of possible positive responses to the question, multiplied by the number of respondents who answered it. Consider, for example, question 38: "Does an interagency disaster preparedness committee, task force, or working group exist in your jurisdiction or region?" This question required a "yes" or a "no" answer. There were 130 respondents who answered this question, generating 130 possible points. Next, the actual points for each question were computed. In our example, of the 130 respondents, 118 answered "yes" to the question, generating an actual (count) of 118.

The third step involved adding the total number of possible points and the total number of actual points (counts) for each factor of preparedness. To accomplish this, the number of possible points was added and the number of actual points was added for each factor of preparedness. For example, to obtain a factor of preparedness score for communications, the researcher analyzed eleven questions to determine the number of possible points and the number of actual points (count).

The fourth step was to take the actual points (count), divided by the number of possible points, to obtain a numerical score for that particular factor of preparedness. This normalization process was done to balance the scores; some factors had more possible points than others. This process was repeated for each of the six factors of preparedness to obtain a score for each.

To obtain the final outcome, the researcher entered the six $F$ factors into the formula, and the resulting value of $\Psi$ was calculated.

## DATA ANALYSIS

To further explain the analytical process, the following example of a decision is presented. Florida faces possible terrorist attacks, so what actions should county and municipal law enforcement agency leaders take to protect the state's people and property from terrorism? Should law enforcement leaders increase the number of personnel assigned to combat terrorism? Should law enforcement agencies increase or improve communications, equipment, training, personnel, plans and procedures, or training? Should law enforcement leaders limit some of the previously mentioned factors of preparedness?

### Determination of Priority of Requirements

The instructions for creating and analyzing the factor of preparedness included spreadsheets and formulas in Excel. The first step in establishing priorities for the factors of preparedness was to make pair-wise comparisons of them, using a pair-wise comparison matrix that was created to calculate their relative value. The matrix format was used because it is a simple, well-established tool that offers a framework for testing consistency, making comparisons, and analyzing priorities in judgment. Several other matrices were created to analyze each participant's responses concerning factors of preparedness.

A response matrix (Table 16.3), a calculation matrix (Table 16.4), and a priority matrix (Table 16.6) were developed for each law enforcement terrorism expert. In the response matrix, the factors, in the column on the left, were compared with the same factors in the row at the top of the matrix. For example, communications was compared to equipment, personnel, planning, and so on. As mentioned earlier, numbers from 1 to 9 were used to represent the relative importance of factors with respect to preparedness for terrorism. Also, when comparing one factor to itself—for example, communications with communications—the comparison was given the score of 1, so the main diagonal of the matrix contains 1s. The first factor was compared with the factors in the top row, during the first comparison, and the reciprocal value was then used for the comparison of the second factor with the first. In Table 16.3, the entries in the shaded cells are reciprocals of their "opposites" in the white cells. The normalized geometric means are the priority values that will be used in the consistency calculation matrix.

One of the strengths of the AHP method is its ability to test for consistency. One example of the calculation of priority values for each factor of preparedness for a respondent is shown in Table 16.4.

To determine the combined weight factor (Table 16.5) for all respondents, the researcher calculated the geometric mean of all the respondents' values for each factor.

Table 16.2

**Step-by-Step Process Using Analytical Hierarchy Process**

| Step no. | Step |
|---|---|
| **For the priority value calculation** | |
| 1 | Fill in the original response matrix, based on the interview data. |
| 2 | Compute the "combined factor" for each row by taking the geometric mean of the entries in the respective row. |
| 3 | Add the column of combined factors. |
| 4 | Compute the "normalized combined factors" by dividing each combined factor from step 2 by the sum obtained in step 3. The result here is the priority value for each factor for this respondent. |
| **For the consistency calculation** | |
| 1 | Copy the original response matrix for each interviewee. |
| 2 | Multiply each entry of each column by the respective value from the column of priority values. That is, multiply each entry in the communications column by the priority value for communications. Do this for all six columns. The result is the matrix of comparison activities. |
| 3 | Sum the entries in each row of the comparison matrix to determine row totals. |
| 4 | Divide each entry in the column that results from step 3 by its corresponding priority value. |
| 5 | Average the results of step 4. This is an arithmetic mean. |
| 6 | Calculate the consistency index by (a) subtracting the number of factors (in this case, 6) from the result step 5 and (b) dividing the result by 2. |
| 7 | Divide the result of step 6 by the comparison value provided by Saaty (1986, p. 84). In this case, the value is 1.24. The result is the consistency ratio. |

*Source:* Table developed by the authors originally for Mosser (2007).

For example, if a respondent preferred personnel over communications and communications over planning, then did that person do so in a consistent manner? In this way, if a respondent at some point judged personnel to be less important than planning, a judgmental error existed, and the response matrix was inconsistent. The consistency matrix was calculated by multiplying the consistency index of each participant's matrix by the priority of the factor used for the comparison, and then adding the quantities. The advantage of using AHP was that it allows individual response scores to be aggregated to obtain a composite group score. So after a consistency ratio was calculated for each of the seven law enforcement terrorism expert respondents, the combined weights for the respondents was determined.

Of course consistency is important to setting priorities for the factors of preparedness. AHP measures the overall consistency of judgments by means of a consistency

Table 16.3

**Response Matrix**

| Comparison activities | EQ | CO | PE | PL | PR | TR | Geometric mean of the row | Normalized geometric means |
|---|---|---|---|---|---|---|---|---|
| Equipment (EQ) | 1 | 0.11 | 0.15 | 0.25 | 2 | 0.22 | 0.35 | 0.04 |
| Communications (CO) | 9 | 1 | 3 | 5.5 | 9 | 2 | 3.73 | 0.44 |
| Personnel (PE) | 6.5 | 0.33 | 1 | 4 | 7 | 2 | 2.23 | 0.26 |
| Planning (PL) | 4 | 0.18 | 0.25 | 1 | 5 | 0.33 | 0.82 | 0.1 |
| Procedures (PR) | 0.5 | 0.11 | 0.14 | 0.2 | 1 | 5.5 | 0.45 | 0.05 |
| Training (TR) | 4.5 | 0.5 | 0.5 | 3 | 0.18 | 1 | 0.92 | 0.11 |
| Column Sum | 25.5 | 2.24 | 5.05 | 13.95 | 24.18 | 11.06 | 8.5 | 1 |

Table 16.4

**An Example of an Individual Respondent Consistency Calculation Matrix**

| Comparison activities | EQ | CO | PE | PL | PR | TR | Priority value |
|---|---|---|---|---|---|---|---|
| Equipment (EQ) | 0.04141 | 0.049 | 0.04 | 0.02 | 0.11 | 0.024108 | 6.892159 |
| Communications (CO) | 0.37273 | 0.438 | 0.79 | 0.53 | 0.48 | 0.216968 | 6.4430915 |
| Personnel (PE) | 0.26919 | 0.146 | 0.26 | 0.39 | 0.37 | 0.216968 | 6.3157359 |
| Planning (PL) | 0.16566 | 0.080 | 0.07 | 0.1 | 0.27 | 0.036161 | 7.3660534 |
| Procedures (PR) | 0.02071 | 0.049 | 0.04 | 0.02 | 0.05 | 0.596661 | 14.534856 |
| Training (TR) | 0.18636 | 0.219 | 0.13 | 0.29 | 0.01 | 0.108484 | 8.7020097 |

ratio, so to determine the group consistency ratio, first, the individual consistency ratio was calculated; then, the group consistency ratio was calculated. It should be noted that Florida Terrorism Expert Interview 2 (FL2) was very inconsistent and thus was omitted from the analysis, causing the group consistency ratio to fall from .565 to .399.

To obtain factors of preparedness from the survey, the following process was used: First, each question was reviewed to determine its appropriateness to each of the six factors of preparedness. Then, each question was assigned points relative to possibility of positive responses and multiplied by the number of respondents who answered the question. The possible points and the total number of actual points (counts) for each factor of preparedness were determined. The actual points were divided by the number of possible points to obtain a numerical factor score for the particular preparedness factor. At the end of the Factors of Preparedness Survey, law enforcement terrorism experts were asked to rate their level of confidence in their ability to rank the factors of preparedness. The results are based on the geometric mean of the individual confidence level. The group confidence level is the geometric mean of the individual Florida interviews. In this study Florida terrorism experts responded with an overall confidence level of 86.1 percent.

Table 16.5

**Determined Combined Weight Values for All Respondents**

| Factors | FL1 | FL2 | FL3 | FL4 | FL5 | FL6 | FL7 | Combined factors | Normalized combined factors |
|---|---|---|---|---|---|---|---|---|---|
| Equipment (EQ) | 0.041 | 0.259 | 0.083 | 0.017 | 0.247 | 0.062 | 0.255 | 0.093 | 0.13 |
| Communications (CO) | 0.438 | 0.471 | 0.044 | 0.034 | 0.161 | 0.138 | 0.142 | 0.138 | 0.194 |
| Personnel (PE) | 0.262 | 0.076 | 0.238 | 0.525 | 0.449 | 0.054 | 0.462 | 0.223 | 0.313 |
| Planning (PL) | 0.096 | 0.039 | 0.142 | 0.213 | 0.047 | 0.357 | 0.039 | 0.096 | 0.135 |
| Procedures (PR) | 0.053 | 0.021 | 0.025 | 0.139 | 0.072 | 0.202 | 0.026 | 0.055 | 0.077 |
| Training (TR) | 0.108 | 0.134 | 0.469 | 0.072 | 0.024 | 0.187 | 0.076 | 0.108 | 0.151 |

*Note:* FL = Florida terrorism expert interview.

## Outcome of Data Analysis of Experts

A scale of relative priorities was developed for the Florida terrorism experts, and then the data priorities were analyzed to provide an overall assessment of the importance of each factor in preparedness (Table 16.6).

After the combined factors were determined, the following formula was updated to calculate law enforcement preparedness for terrorism in Florida:

$$\psi = \omega_1 F_1 + \omega_2 F_2 + \omega_3 F_3 + \omega_4 F_4 + \omega_5 F_5 + \omega_6 F_6,$$

where $\omega_i$ is the AHP-determined weight for factor $i$ and $F_i$ is the survey-determined factor value for factor $i$. The formula was updated to the following:

$$\Psi = .106F_1 + .153F_2 + .362F_3 = .151F_4 + .087F_5 + .141F_6\ .$$

## Outcome of Data for the Combating Terrorism Survey

The measures presented in this study can be used by public safety decision makers who engage in critical incident analysis to determine an organization's level of preparedness (see Table 16.7).

This normalization process was performed to balance the scores; some factors had more possible points than others. This process was repeated for each of the six factors of preparedness.

## Outcome of Data for Overall Preparedness

To obtain the final outcome, the researcher inserted the six $F$ factors in the formula, and the resulting value of $\Psi$ was calculated.

$$\begin{aligned}\Psi &= .106(.290) + .153(.268) + .362(.420) + .151(.269) + .087(.504) + .141(.170)\\ &= .031 + .041 + .152 + .041 + .044 + .024\\ &= .333\end{aligned}$$

A preparedness score of 1, a perfect score, means that all possible points were earned. Florida's score, .333, indicates that the state is about one-third as prepared as it should be.

# LIMITATIONS AND SIGNIFICANCE

Decision makers and law enforcement need to take further actions to better prepare for a terrorist attack, and Florida law enforcement leaders must begin to prioritize preparedness factors. The combating terrorism survey used in this study should also be evaluated for its inadequacies. It was too long and lacked specific emphasis on the

Table 16.6

**Normalized Combined Factors and Overall Priority Ranking**

| Factors (1–6) | FL normalized combined factors | FL Rank |
|---|---|---|
| Equipment (EQ) | 0.106 | 5 |
| Communications (CO) | 0.153 | 2 |
| Personnel (PE) | 0.362 | 1 |
| Planning (PL) | 0.151 | 3 |
| Procedures (PR) | 0.087 | 6 |
| Training (TR) | 0.141 | 4 |

*Note:* FL = Florida terrorism expert interview.

study's established factors of preparedness. This chapter therefore proposes the use of an improved twelve-question survey, not tested for validity or reliability, for use in any further research. This improved questionnaire was reviewed for accuracy and feasibility to analyze preparedness by a current Department of Homeland Security (DHS) expert who did not wish to be identified, who stated, "Based upon my 35 years in law enforcement, working for the DHS/FEMA, and research in the area of terrorism, the six areas listed are valid topical areas that can be used collectively to evaluate an agency or organizations preparedness against terrorism." The questionnaire, titled Assessment of Preparedness Questionnaire, could be used to evaluate preparedness for multiple threats, include natural disasters, manmade threats, or technological threats. The survey contains sixteen questions focused on the six factors mentioned earlier. The questions are structured and closed ended.

## PRACTICAL APPLICATION

Empirical studies on the use of identified models of reflection and critical incident analysis are urgently needed to assist practitioners and scholars alike in both preparedness and response to incidents. Most research involving critical incident analysis continues to be focused on the complexity of emergency management response with no universal systematic approach to preparedness. Defining preparedness and identifying the factors that can measure it for a variety of threats is important for decision makers. The measures presented here can be used to determine a level of preparedness so that decision makers who engage in regular critical incident analysis can link theoretical frameworks on many levels from leadership, best practices of preparedness, and conduct a briefing for the senior leadership and any additional evaluators about the goals and procedures of the organization's assessment. Research on preparedness requires procedures that allow an organization first to define preparedness and then to measure and evaluate factors of preparedness to determine its preparedness level. The research described in this chapter is the first study to develop a model of preparedness for terrorism by defining what is meant by preparedness, identifying factors of preparedness, and then applying weights

Table 16.7

**Factor Score for Six Factors of Preparedness Research Question 3**

| Factor and question no. | Possible points | Number of respondents | Actual points (count) | Factor value |
|---|---|---|---|---|
| Communications (CO) | | | | |
| Q11 | 896 | 128 | 275 | |
| Q13 | 2,210 | 130 | 715 | |
| Q21 | 408 | 102 | 95 | |
| Q22 | 124 | 124 | 70 | |
| Q23 | 1,064 | 133 | 210 | |
| Q24 | 1,064 | 133 | 243 | |
| Q28 | 1,040 | 130 | 106 | |
| Q38 | 130 | 130 | 118 | |
| Q39 | 130 | 130 | 119 | |
| Q41 | 780 | 122 | 122 | |
| Q47 | 390 | 130 | 112 | |
| Total | 8,242 | | 2,215 | 0.268 |
| Equipment (EQ) | | | | |
| Q60 | 1430 | 129 | 345 | |
| Q61 | 130 | 130 | 106 | |
| Total | 1,560 | | 451 | 0.290 |
| Personnel (PE) | | | | |
| Q2 | 122 | 122 | 55 | |
| Q54 | 130 | 130 | 81 | |
| Total | 252 | | 136 | 0.420 |
| Plans (PL) | | | | |
| Q44 | 130 | 130 | 77 | |
| Q45 | 122 | 122 | 50 | |
| Q46 | 854 | 122 | 170 | |
| Total | 1,106 | | 297 | 0.269 |
| Procedures (PR) | | | | |
| Q42 | 266 | 130 | 164 | |
| Q43 | 266 | 133 | 104 | |
| Total | 532 | | 268 | 0.504 |
| Training (TR) | | | | |
| Q51 | 133 | 133 | 44 | |
| Q52 | 133 | 133 | 56 | |
| Q56 | 825 | 75 | 142 | |
| Total | 1,091 | 341 | 242 | 0.170 |

*Note:* Q = question.

to measure the factors to determine a level of law enforcement preparedness in the state of Florida against terrorism. This model can easily be replicated for critical analysis of incidents other than terrorism. The method is practical and flexible and allows decision makers or researchers to understand the complexities of preparedness functions and the interaction and importance of analyzing preparedness. The process can be easily modified to suit specific organizations and allows for a rich contextual exchange concerning preparedness information. The importance of this research is that it provides a model that decision makers can use to rank order factors of preparedness to better prepare for critical incidents.

## SUMMARY OF RESULTS

Rich contextual information about preparedness can be obtained using the AHP method. The method allowed experts to rank order factors of preparedness to better prepare for critical incidents. The analytical results of the study appear to be (a) useful to Florida decision-makers; (b) judgmentally sound with small changes in the judgment of experts; (c) accurate from an expert elicitation perspective; and (d) easily understood and replicated. Preparedness was defined, and six factors of preparedness were identified and rank ordered by law enforcement terrorism experts as to their level of importance. The results rank ordered the factors (most important to least important) as follows: personnel, communications, planning, training, equipment, and procedures. The study results indicate that overall Florida is only one-third as prepared as it should be.

## APPENDIX

### How to Begin Your Own Assessment

An assessment of an organization's preparedness should begin with the full support of the leadership and members of that organization. The leader of the organization should appoint an evaluator to conduct a briefing for the senior leadership and any additional evaluators about the goals and procedures of the organization's assessment. The assessment team should represent leaders of the organization and a cross section of employees, with representation from all departments of the organization taking the survey. The assessment team should consist of five to seven members, depending on the size of the organization.

### Determination of Priority of Requirements

Begin by finding at least five, but not more than seven, experts to determine the relative importance of the six factors of preparedness to the critical incident(s) to be examined, unless your evaluation is on terrorism, in which case you may use the study rankings. Have the experts prioritize the six factors of preparedness. Next, develop a response matrix, a normalized matrix, and a comparison calculation matrix for each expert. Following are the steps for conducting an evaluation.

### Evaluation Steps

Step 1: Fill in the original response matrix, based on the interview data. An example is shown in Table 16.3.

Step 2: Compute the "combined factor" for each row by taking the geometric mean of the entries in that row.

Step 3: Add the column of combined factors.
Step 4: Compute the "normalized combined factors" by dividing each combined factor from step 2 by the sum obtained in step 3. The result here is the priority value for each factor for this respondent.

## Consistency Calculation

Step 1: Copy the original response matrix for each interviewee.
Step 2: Multiply each entry of each column by the respective value from the column of priority values. That is, multiply each entry in the communications column by the priority value for communications. Do this for all six columns. The result is the matrix of comparison activities.
Step 3: Sum the entries in each row of the comparison matrix.
Step 4: Divide each entry in the column that results from step 3 by its corresponding priority value.
Step 5: Average the results of step 4. This is an arithmetic mean.
Step 6: Calculate the *consistency index* by (a) subtracting the number of factors (in this case 6) from the result of step 5 and (b) dividing the result by 2.
Step 7: Divide the result of step 6 by the comparison value in Saaty (1986, 84). The result is the consistency ratio.

After the combined factors are determined, use the following formula to calculate preparedness:

$$\Psi = \omega_1 F_1 + \omega_2 F_2 + \omega_3 F_3 + \omega_4 F_4 + \omega_5 F_5 + \omega_6 F_6,$$

where $\omega_i$ is the AHP-determined weight for factor *i*, and $F_i$ is the survey-determined factor value for factor *i*.

Next, use the Assessment of Preparedness Questionnaire in this chapter and administer the questionnaire to individuals in the organization that you want to evaluate. Using the answers to the Assessment of Preparedness Questionnaire, the *F* factor value for each factor of preparedness can be calculated. The number of possible points for a question will depend on how many respondents answered it. The first step involves adding the number of possible points and the number of actual points (counts) for each factor of preparedness. For example, to obtain a factor of preparedness score for communications, analyze two questions on communications to determine the number of possible points and the number of actual points (count).

The second step is to take the actual points (count), divided by the number of possible points, to obtain a numerical factor score for that particular factor of preparedness. This normalization process is undertaken to balance the scores; some factors had more possible points than others. This process is repeated for each of the six factors of preparedness. To obtain the final outcome, insert the six *F* factors in the formula, which will yield the value of $\Psi$, that is, the preparedness level. A preparedness score of 1 is a perfect score, that is, all possible points were earned.

**Assessment of Preparedness Questionnaire**

1. Please write a number between 1 and 6 next to each item below. Put a 1 next to the item that is "least important" to you in your agency's preparedness. Put a 6 next to the item that is "most important." Please use each number only once.
   ___ a. Communications
   ___ b. Equipment
   ___ c. Personnel
   ___ d. Planning
   ___ e. Procedures
   ___ f. Training

*Communications*

2. Your agency uses formal protocols for command and control, such as the Incident Command System (ICS), during response to an incident, and all of the agency's first responders have been trained in a Federal Emergency Management Agency (FEMA) or National Incident Management System (NIMS) approved course.

| Disagree strongly | Disagree somewhat | Neutral | Agree somewhat | Agree strongly |
|---|---|---|---|---|
| ___ | ___ | ___ | ___ | ___ |

3. In the event of a large-scale emergency, your agency is able to communicate with multiple agencies from multiple jurisdictions including local, county, state, and federal organizations.

| Disagree strongly | Disagree somewhat | Neutral | Agree somewhat | Agree strongly |
|---|---|---|---|---|
| ___ | ___ | ___ | ___ | ___ |

*Equipment*

4. Your agency has the newest equipment (communications, personal protection equipment, hazardous materials [HAZMAT], decontamination, weapons, mobile command post) to fully respond to an emergency, such as an active shooter in a high school or college or a chemical, biological, radiological, nuclear, and explosive (CBRNE) incident.

| Disagree strongly | Disagree somewhat | Neutral | Agree somewhat | Agree strongly |
|---|---|---|---|---|
| ___ | ___ | ___ | ___ | ___ |

5. Your agency has equipment to respond to a CBRNE attack.

| Disagree strongly | Disagree somewhat | Neutral | Agree somewhat | Agree strongly |
|---|---|---|---|---|
| ___ | ___ | ___ | ___ | ___ |

*Personnel*

6. Your agency has 90 percent or more of the terrorism preparedness personnel available and assigned.

| Disagree strongly | Disagree somewhat | Neutral | Agree somewhat | Agree strongly |
|---|---|---|---|---|
| ___ | ___ | ___ | ___ | ___ |

7. Your agency has enough authorized positions for personnel (full-time and part-time) to respond to an emergency situation such as an active shooter at a high school or college or a CBRNE incident.

| Disagree strongly | Disagree somewhat | Neutral | Agree somewhat | Agree strongly |
|---|---|---|---|---|
| ___ | ___ | ___ | ___ | ___ |

*Planning*

8. Your agency conducts meetings and planning sessions (at least quarterly) with other emergency response agencies, including other disciplines.

| Disagree strongly | Disagree somewhat | Neutral | Agree somewhat | Agree strongly |
|---|---|---|---|---|
| ___ | ___ | ___ | ___ | ___ |

9. Agencies involved with your agency in planning. (Mark all that apply.)

___ Local law enforcement agencies
___ Local office of emergency management
___ Local medical facilities
___ Public health department
___ Public works department
___ Local fire department
___ Private security firms

___ Private sector owners of identified local critical infrastructure
___ Port Authority
___ Local/regional airport police
___ State law enforcement agencies
___ State office of public safety or emergency management
___ Federal agencies
___ Other states
___ Other (please specify)

10. Your agency has an updated emergency response plan and contingency plan.

| Disagree strongly | Disagree somewhat | Neutral | Agree somewhat | Agree strongly |
|---|---|---|---|---|
| ___ | ___ | ___ | ___ | ___ |

*Procedures*

11. Your agency's emergency response and contingency plan is updated annually.

| Disagree strongly | Disagree somewhat | Neutral | Agree somewhat | Agree strongly |
|---|---|---|---|---|
| ___ | ___ | ___ | ___ | ___ |

12. Your agency requires employees to review the emergency plan and contingency plan annually.

| Disagree strongly | Disagree somewhat | Neutral | Agree somewhat | Agree strongly |
|---|---|---|---|---|
| ___ | ___ | ___ | ___ | ___ |

13. Your agency has written procedures and has signed mutual aid agreements with other agencies.

| Disagree strongly | Disagree somewhat | Neutral | Agree somewhat | Agree strongly |
|---|---|---|---|---|
| ___ | ___ | ___ | ___ | ___ |

*Training*

14. Your agency has participated in state or federally sponsored training programs within the past twelve months.

| Disagree strongly | Disagree somewhat | Neutral | Agree somewhat | Agree strongly |
|---|---|---|---|---|
| ___ | ___ | ___ | ___ | ___ |

15. Your agency has an ongoing training program (certification training, exercises, table-top exercises) for all employees and follows the Homeland Security Exercise and Evaluation Program protocols.

| Disagree strongly | Disagree somewhat | Neutral | Agree somewhat | Agree strongly |
|---|---|---|---|---|
| ___ | ___ | ___ | ___ | ___ |

16. Your agency has all first responder trained in the NIMS and has formally adopted NIMS protocols.

| Disagree strongly | Disagree somewhat | Neutral | Agree somewhat | Agree strongly |
|---|---|---|---|---|
| ___ | ___ | ___ | ___ | ___ |

17. Please write a number between 1 and 6 next to each item below. Put a 1 next to the item that is not trained and a 6 next to the item that is most trained.

___ a. Incident command systems (communications)
___ b. CBRNE
___ c. Personnel protection gear (chemical suits, boots, gloves, breathing apparatus equipment)
___ d. Agency emergency policies
___ e. Individual health monitoring (vital signs, blood pressure, temperature)
___ f. Interagency response techniques

## REFERENCES


Brookfield, S. 1987. *Developing Critical Thinkers: Challenging Adults to Explore Alternative Ways of Thinking and Acting.* San Francisco: Jossey-Bass.

Chell, E., and L. Pittaway. 1998. "A Study of Entrepreneurship in the Restaurant and Café Industry: Exploratory Work Using the Critical Incident Technique as a Methodology." *International Journal of Hospitality Management* 17: 23–32.

Clemen, R. T., and R. L. Winkler. 1999. "Combining Probability Distributions from Experts in Risk Analysis." *Risk Analysis* 19: 187–204.

Cooke, R. M., and L. H. J. Goossens. 2004. "Expert Judgment Elicitation for Risk Assessments of Critical Infrastructures." *Journal of Risk Research* 7 (6): 643–656.

Corbin, J., and A. Strauss. 1990. "Grounded Theory Research: Procedures, Canons, and Evaluative Criteria." *Qualitative Sociology* 13 (1): 3–21.

Creswell, J. 2003. *Research Design: Qualitative, Quantitative, and Mixed Methods Approaches* (2d ed.). Thousand Oaks, CA: Sage.

Davis, Lois M.; K. Jack Riley; Greg Ridgeway; Jennifer E. Pace; Sarah K. Cotton; Paul Steinberg; Kelly Damphousse; and Brent L. Smith. 2004. *When Terrorism Hits Home: How Prepared Are State and Local Law Enforcement?* Santa Monica, CA: RAND.

Falkenrath, R. 2001a. "Problems of Preparedness: U.S. Readiness for a Domestic Terrorist Attack." *International Security* 25 (4): 147–186.

———. 2001b. "Analytic Models and Policy Prescription: Understanding Recent Information in U.S. Counterterrorism." *Studies in Conflict and Terrorism* 24: 159–181.

Federal Emergency Management Agency (FEMA). 2009. "Preparedness." www.fema.gov/prepared/index.shtm (accessed July 12, 2009).

Flanagan, J. 1954. "The Critical Incident Technique." *Psychological Bulletin* 51: 327–358.

Florida's Domestic Security Oversight Board. 2004. "Domestic Security in Florida: 2004 Annual Report. Making Florida Safer Through Strategy, Structure, Funding and Partnerships." Tallahassee, FL.

Kanji, G. K. 1976. "Permutation Theory in the Study of Robustness of Power in the Analysis of Variance." *International Journal of Mathematical Education in Science and Technology* 7 (4): 401–405.

———. 1994. "Total Quality Management and Statistical Understanding." *Total Quality Management* 5 (3): 105–114.

———. 1999. "Business Excellence Model for Supply Chain Management." *Total Quality Management* 10 (8): 1147–1168.

Mosser, M. E. 2007. "Law Enforcement Preparedness Against Terrorism in the State of Florida." PhD dissertation, Walden University.

Saaty, T. L. 1980. *The Analytical Hierarchy Process*. New York: McGraw-Hill.

———. 1986. "Axiomatic Foundation of the Analytic Hierarchy Process." *Management Science* 32 (7): 841–855.

———. 2000. *The Fundamentals of Decision Making and Priority Theory with the Analytic Hierarchy Process*, rev. ed. Pittsburgh: RWS.

———. 2001. *Decision Making for Leaders*, 3d ed. Pittsburgh: RWS.

———. 2005. *Theory and Applications of the Analytic Network Process*, 3d ed. Pittsburgh: RWS.

Von Winterfeldt, D., and W. Edwards. 1986. *Decision Analysis and Behavioral Research*. New York: Cambridge University Press.

CHAPTER 17

# Improved Policing in an Invisible Community

## A Case Study of the Gay and Lesbian Liaison Unit of Washington, DC

RODDRICK COLVIN

This case study examines the efforts of the Metropolitan Police Department (MPD) in Washington, DC, to address the needs of an invisible community, namely, the lesbian, gay, bisexual, and transgender (LGBT) community. Like many other police departments in the United States, the MPD has had a tumultuous relationship with LGBT members of the community under its jurisdiction. The high level of mistrust in the community and ambiguous directives from the MPD has hindered effective community policing. Over time, the LGBT community not only felt invisible but also received a lower level of law enforcement and many other public services. This insular situation helped to breed and foster criminal acts and activities within this community.

After a number of recent critical incidents and thanks to visionary leadership, the MPD has developed a more practical policing model that has built trust in the LGBT community, improved policing efforts, and sensitized the police force about the unique public service needs of the LGBT community. The primary component of MPD policing reform and strategy was the creation of the Gay and Lesbian Liaison Unit (GLLU).

This case study explains the evolution of the MPD from gay community foe to gay community ally. Based on focus group interviews and archival data, I explain the challenges faced by the LGBT community and the police department in creating a safer community. I discuss the role of critical incidents as catalysts for change in the MPD's approach to policing and the role of leadership in guiding police force changes. I then discuss the operations of the GLLU, its approach to policing, and any measurable outcomes that have resulted from its operations. Finally, the chapter concludes with a discussion on best practices and what can be learned from this case.

## CHALLENGES FOR POLICING INVISIBLE COMMUNITIES

Whether providing social, health-care, or police services, the organizations delivering those services must know something about the community being served. They must be able to identify and communicate with individuals and groups in the community. No doubt, service delivery is much easier to some communities than others. Often, the ability and level of service delivery are based on the visibility of that community and its members. The most difficult communities are those that are invisible to service providers.

Several factors can contribute to a communities' visibility or lack thereof. These factors fall into three general categories: social taboo, community norms, and institutional barriers. In some cases, communities engage in active efforts to remain invisible—for example, individuals and communities who engage in activities considered socially taboo or illegal, including closeted homosexuals, illegal immigrants, drug addicts, sex workers, and gang members. Members of the community engaging in such activities are isolated from the larger society. Because their activities are considered taboo by that larger society and in some cases are illegal, they feel motivated to stay away from the police and other public service providers.

Individuals who engage in activities outside community norms can also become invisible: homeless people, alcohol abusers, and teen mothers. Their activities are not necessarily negative or illegal but are often considered outside the range of acceptable behaviors according to community norms. As Schneider and Ingram (1993) note, the social construction of populations can affect the level of service delivery to them.

Finally, institutional barriers can make some individuals and communities invisible. Some policies or procedures create institutional barriers that restrict information and data about a particular population. For example, if the census does not ask about the sexual orientation of participants, we will have little data about gays and lesbians. In most cases, all three of these factors are interconnected, and all contribute to community invisibility.

Layered on top of invisibility are other factors that can add to the difficulty of effective delivery of public goods and services, such as the level of community permeability. Some communities—for example, immigrant communities, separatist communities, or religious sects—might be considered closed due to language barriers, divisions in culture or belief, or geographic isolation and thus present service providers with an additional challenge. In contrast, the LGBT community, while largely invisible, is fairly easy for nonmembers and service providers to enter and exit.

A second factor that can add to the difficulty of effective service provision is the level of vulnerability. Many communities lack political influence, financial resources, or critical mass, which are often needed to attract public services at an appropriate level. These communities might have a history of being denied public support, which has contributed to their vulnerability. For example, people with mental or physical disabilities were neglected for much of American history. Only after interventions by the courts and passage of the Americans with Disabilities Act (1990) did tangible and measurable service provisions flow to these communities.

From a public perspective, delivering services to invisible, closed, and vulnerable communities is the most challenging. Immigrants who do not speak English and lack proper documentation are an example of such a community. Their lack of English skills makes the community closed. Their undocumented immigration status makes them invisible, both as a social taboo and institutional barrier. Because they might also lack political influence, resources, or a measurable population (undocumented workers might be numerous, but there is no reliable way to measure), they are also vulnerable (see Table 17.1).

Policing the largely invisible LGBT community also presents challenges. Unlike members of a racial or ethnic community, LGBT people are not necessarily identifiable on sight. Unlike some religious groups or immigrant communities, LGBT people do not necessarily live in one community or geographic area. LGBT people are integrated into every part of the broader society. They can be found among the racial, ethnic, and religious in our society, among the elderly and the young, and among the rich and the poor. Some LGBT people are out, known, and visible as sexual minorities. Others are closeted, unknown, and invisible within the LGBT and larger community. Overall, we can classify the LGBT community as largely invisible, open, and somewhat vulnerable.

As mentioned earlier, police departments and LGBT communities in the United States have had a long and uneasy history. From beat cops harassing bar owners and patrons, to sting operations in public sex environments, to inappropriate interventions in same-sex domestic disturbances (often both accuser and accused are arrested), distrust of the police in the LGBT community has been well founded. The situation in Washington, DC, is no different. One of the most famous incidents that exemplifies the poor relations between public service providers and the community took place in 1995. In August of that year, Tyra Hunter, a male-to-female transsexual, was severely injured in an auto accident. First responders on the scene begin treating Hunter. In the process of treating her injuries, one paramedic cut away her pants, revealing her as a biological male. The first responders refused to continue treatment, and she was also initially denied treatment in the emergency room. Several hours later, Hunter died. None of the first responders were held responsible for her death, even though medical experts estimated that had her injuries been properly treated she would have had an 86 percent likelihood of survival (Moser 2003). Although this incident included firefighters and emergency medical technicians (EMTs), given that none of the first responders were arrested or found criminal negligent, the police were viewed as complicit in the incident, since the MPD did not view the first responders' actions as criminal and refused to categorize Hunter's death as a hate crime. Ironically, this incident would eventually help to set in motion efforts to improve police service delivery to the LGBT community, as discussed next.

Traditional policing—responding to requests for help and investigating crime—has not proved effective in the LGBT community. Police officers are less effective when

Table 17.1

**Factors Affecting Service Delivery**

| Community factor | Definition | Example |
|---|---|---|
| Invisible | Not immediately visible to larger community or public service providers | Sex workers |
| Visible | General visible to larger community or public service providers | Asian American communities |
| Closed | Community organizations are not accessible to noncommunity members or service providers | Non-English-speaking immigrant communities |
| Open | Community organizations are accessible to non-community members or service providers | Gay, Lesbian, Bisexual, and Transgender (LGBT) |
| Vulnerable | Lacking proper resources or political power to influence public policy | Mentally or physically disabled |
| Invulnerable | Having proper resources or political power to influence public policy | Wealthy |

conducting investigative work if they do not know who to talk to in the community or where community members live and/or congregate. Geographic policing—patrol or beat-based policing within a specific location—is not effective, because the LGBT community members do not necessarily live only in discrete neighborhoods, such as the Dupont Circle area. In fact, 2000 Census data identified same-sex identified households in all seven police jurisdictions of Washington, DC. The 3,678 known same-sex couples represent 5.14 percent of the district's population. Of course, this accounts only for couples who self-identified on the census as same-sex partners. Many other LGBT people, including those not in couples, are not included in this count, and some scholars believe that the 2000 Census undercounted LGBT people by as much as 62 percent (Human Rights Campaign 2001).

Given the invisible nature of the lesbian and gay people and their dispersion throughout the city, both traditional and geographic policing are not optimal approaches to law enforcement. A more optimal approach will include community members as part of the law enforcement mechanism and provide data and information that can be used to help prevent crime.

## THE ROLE OF CRITICAL INCIDENTS AND LEADERSHIP

Critical incidents and leadership both contributed to change in policing by the MPD. Three critical incidents and two acts of leadership led to the creation of the GLLU.

The first incident, mentioned above, was the death of Tyra Hunter in 1995. Although her death did not immediately trigger changes in policing procedures, it did resonate in a number of ways. First, when no first responders were held responsible for her death, the LGBT community actively protested and kept the issue at the top of the public agenda. Community organizations including Gay and Lesbian Activists Alliance of Washington (GLAA) and Gay Men and Lesbians Opposing Violence (GLOV) testified before investigative committees and targeted the fire chief and police chief as culpable, since both defended their agencies and the other agency of his colleague as acting legally and ethically in the Hunter case. Other gay and gay-allied organizations were also active in seeking justice for Hunter and her family, including the DC Coalition of Black Lesbians, Gay Men and Bisexuals, the Gay and Lesbian Alliance Against Defamation (GLAAD), Transgender Nation, the American Civil Liberties Union, and the National Gay and Lesbian Task Force. Community vigils were held, and media attention (gay and nongay) was continuous and persistent. In this sense, Hunter's death was a critical incident for the LGBT community. As one community activist said, "these days, most people understand that hate crimes are message crimes. Most people know that when a transgendered [sic] person is victimized, it does not just affect her friends and family—it terrifies a whole community of people who can't help feeling they might be next" (Moser 2003, 3).

The second critical incident occurred in 1997. That year, MPD Lieutenant Jeffrey Stowe was investigated and charged by the FBI in U.S. District Court with extortion and embezzlement. Stowe was accused of blackmailing men whom he believed to be living closeted gay lives but were married to women. He identified vehicles that appeared to belong to men with families—vehicles with child safety seats in them—parked outside gay clubs. Starting with license plate numbers, he traced the owners, using law enforcement computer databases and the Internet. In one case, he extorted as much as $10,000 from a blackmailed officer. The entire Washington community was shocked when the news broke about Stowe's extortion activities. The investigations against Stowe reached the highest levels of the police force and ultimately resulted in the resignation of the serving police chief, Larry D. Soulsby, a close friend and former roommate of Stowe's. Having been found guilty, in 1998 Stowe was sentenced to a prison term of forty-eight months. More than the Hunter incident, the Stowe incident intensified tensions between the MPD and the Washington LGBT community. To many LGBT community members, these actions were no different than the bar raids and intimidation tactics of police in the pre-Stonewall era.[1] Sharen Shaw Johnson of the Washington-based Gay Men & Lesbians Opposing Violence captures the sentiment of the LGBT community, stating, "This hardly is surprising . . . members of our community are being victimized or revictimized by the very people entrusted with ensuring their safety: the police" (GLAA 1998).

The third and perhaps best-documented critical incident was the 1998 release of regional data about hate crimes and police department responses. In this year, the two jurisdictions surrounding the District—Virginia and Maryland—reported 169 and 282 hate crimes, respectively, for the previous year (Fillichio, 2006). By contrast, Washington, DC, with a then-population of over half a million people, officially recorded two hate/bias related crimes for the entire year (see Table 17.2). (Meanwhile, the community-based organization

GLOV documented eighty-six victims of antigay violence in the previous year [GLAA 1998]). The comparative data highlighted a disparity that could not be easily explained away by politicians or police administration. Both the LGBT community and the MPD knew that the numbers reported were too low to be true. The LGBT community, the police, and the broader community also suspected that a large percentage of the unreported crimes were committed against members of the LGBT community. The unreported hate crimes caused an outcry within the LGBT community. Given the close proximity of the three jurisdictions, the disparities suggested poor performance by the MPD and poor oversight by the political entities, embarrassing both groups.

Taken together, these three critical circumstances created an opportunity for changes in MPD policing. Two acts of leadership helped to facilitate these changes. First, in 1999, two lesbian MPD officers (Officers Kelly McMurray and Berdette Williams) put forward a proposal to improve relations with the LGBT community, designed to address two main issues: increased training for police officers about hate/bias related crimes and building trust in the LGBT community to encourage the reporting of such crimes (Fillichio 2006). The officers believed that if the police were provided additional training about hate/bias crimes and could build trust by vigorously investigating such claims, both the community and police force could benefit.

The then-chief of police, Charles Ramsey, was sensitive to the current condition of the LGBT community (the Hunter case was finally settled with the district in 1999) and the hate/bias crime reporting disparity. In this context, in 2000, the chief appointed the two lesbian officers as the liaisons to the LGBT community. The new liaison post coordinated internal training for police officers about hate and bias crimes and conducted outreach to the LGBT community about the importance of reporting such crimes. The measurable benefits of this unit were immediately evident. During the first year, it was instrumental in solving the murder of a gay student at Gallaudet University via investigative work within the Gallaudet LGBT community. In addition, as a result of the unit's training and outreach, reporting of hate/bias crimes doubled in the first year.

Despite the early success, the liaison post and its staff faced a number of challenges, including structural, institutional, and cultural challenges within the MPD. The liaison officers were of relatively low rank (institutional), which they felt, in addition to their being women (cultural), prevented the unit from realizing its full potential. After a year of operation as an informal liaison post (structural), the founding officers made additional recommendations to the MPD and Chief Ramsey. The recommendations included formalizing the post into an actual unit with resources, adding male officers (who could speak to fellow male officers), and having leadership for the unit with appropriate rank in the MPD. Or, as Brett Parson (the officer assigned to command the unit), told the MPD, to make the unit successful, "you need a gay boy, you need a gay boy that has some rank, and you need a gay boy that has rank and doesn't give a damn what people think about him" (personal communication with the author, 2009). By fall 2001, the unit had its first openly gay, full-time, male police officer to head the liaison unit. In additional to human and financial resources, the group received a physical location in Dupont Circle and a lead male officer who held the rank of lieutenant.

Table 17.2

**Washington, DC, Hate Crimes Reported 1998–2009**

| Type of Hate Crime | 1998 | 1999 | 2000 | 2001 | 2002 | 2003 |
|---|---|---|---|---|---|---|
| Ethnic | 0 | 0 | 1 | 0 | 3 | 1 |
| Racial | 1 | 0 | 2 | 3 | 2 | 9 |
| Religious | 1 | 1 | 0 | 2 | 0 | 2 |
| Sexual Orientation | 0 | 3 | 2 | 5 | 9 | 17 |
| Disability | 0 | 0 | 0 | 0 | 0 | 0 |
| Total | 2 | 4 | 5 | 10 | 14 | 29 |

| Type of Hate Crime | 2004 | 2005 | 2006 | 2007 | 2008 | 2009* |
|---|---|---|---|---|---|---|
| Ethnic | 4 | 1 | 3 | 2 | 2 | 1 |
| Racial | 14 | 6 | 8 | 3 | 5 | 3 |
| Religious | 2 | 7 | 5 | 6 | 0 | 0 |
| Sexual Orientation | 29 | 28 | 36 | 26 | 30 | 24 |
| Disability | 0 | 0 | 1 | 0 | 0 | 0 |
| Total | 49 | 42 | 53 | 37 | 37 | 28 |

*Please note that the 2009 totals are preliminary and have not yet been reported to the FBI. Preliminary data is subject to change.

Taken together, the Hunter incident, the Stowe incident, and reporting disparities in hate crimes, in addition to two acts of leadership—one at the police bottom organization and one at the top—helped to create an environment that was ready for an experimental change in community policing.[2]

This situation highlights the role of critical incidents and its potential impact on individuals, institutions, organizations, and communities (even invisible communities). For example, we see the effects on individuals, organizations, and institutions when the MPD's decided to create the unit to service the lesbian and gay community. Additionally, we can see its effects on the invisible LGBT communities from the statewide hate/bias crime data. In fact, the data release had influence on multiple communities and institutions (regional police departments, LGBT communities, GLOV, and the states of Maryland and Virginia).

## MPD'S GAY AND LESBIAN LIAISON UNIT

The Metropolitan Police Department's Gay and Lesbian Liaison Unit (GLLU) is staffed by openly gay, lesbian, and transgender members of the agency and their allies. The GLLU has three main mission objectives: to provide educational outreach to the LGBT community, to educate fellow police officers about the LBGT community, and to actively participate in day-to-day crime-fighting responsibilities (Harvard University Kennedy School of Government 2006) (Table 17.3).

Table 17.3

**Critical Incidents that Lead to the Creation of GLLU**

| Year | Critical Incident | Effect |
|---|---|---|
| 1995 | Death of trans-person Tyra Hunter | LGBT community is energized and mobilized |
| 1997 | Lt. Stowe accused of gay-related blackmailing | Washington D.C. community shocked, Chief of Police resigns |
| 1998 | Regional hate crimes data released | Washington community and police are embarrassed; Police officers step forward with a corrective plan |

To meet these objectives, the GLLU engages in several community and policing activities, including advising the chief of police on issues involving the LGBT community; serving as MPD representatives at community gatherings, civic meetings, government functions, and other events; participating in the District's Hate Crimes Task Force; visiting gay-owned/operated businesses, corporations, associations, and other groups that seek the support or assistance of the GLLU; providing support and assistance to law enforcement investigations that involve members of the LGBT community; providing support to other city agencies with regard to law enforcement within the LGBT community; engaging in recruitment efforts for both sworn and reserve police officers by operating an MPD recruitment booth at local gay-pride festivals; training and educating the LGBT community and members of the MPD; offering safety training tailored to members of the LGBT community; and serving as a resource for other police agencies nationwide (GLLU 2008a).

Unlike other units or liaison officers, the GLLU was specifically designed to be community focused and highly visible. The model emphasizes collaboration between community and law enforcement to improve community policing and public safety.

Due to the efforts of Sergeant Brett Parson (lead officer for the unit), the GLLU currently has six full-time officers, plus numerous part-time officers and volunteers from the community. Although the unit is headquartered in Dupont Circle, and its outreach and police work is limited to one part of the district, it responds to calls from all over the city and all seven police districts. GLLU's support number is staffed twenty-four hours a day, seven days a week. No other police department in the United States currently has an LGBT unit designed to operate like the GLLU.[3]

## MEASURING THE EFFECTIVENESS OF THE GLLU

One of the challenges of providing public goods and services to an invisible community is the inability to collect and maintain information about the community and establish measures of efficiency and effectiveness. That is, it is often difficult to know how well a particular public service is meeting the needs of the target population. In the case of Washington, DC's LGBT community, accurate measures remain elusive, but the MPD has tried to adapt and compensate. The data are elusive because of the relative invisibility of the community. No data exist that can provide us with a sense of the size of the LGBT community, nor do we have information about the likelihood that LGBT people will report crime or want to interact with the MPD. We do have a few indirect measures of success that can shed some light on the efforts of the police department and the GLLU.

There are some good measures that we can apply to this population. First, the creation of the GLLU raised the profile of internal community-related crime, like same-sex domestic violence. In 2000, no same-sex domestic violence cases were reported to MPD. To date, more than 460 cases of same-sex domestic violence have been investigated by the GLLU (Parson 2009). The change can be attributed to the GLLU's efforts at training officers to recognize the signs of such violence; increasing the levels of trust by the unit and individuals and community organizations; and guiding victims of domestic violence through the criminal justice system (Parson 2009).

In addition to increasing attention to same-sex violence issues, the GLLU has been instrumental in increasing the number of hate/bias crimes reported to authorities. Officers were taught to recognize the signs of such crimes and how to complete the related paperwork to ensure proper investigation. Since 1998, the reporting of gay-related hate/bias crimes has increased from 0 in 1998 to between 30 and 50 a year. The increase in reporting is a positive sign that suggests growing and sustained trust of the MPD and the GLLU. As former chief Ramsey stated, "for the law enforcement community, that [increased reporting] is real progress—not because of the occurrence, but because they were reported, a clear indication, that gays and lesbians who've been victimized once know they will not be victimized again when they report the crime to us" (Fillichio 2006). In 1998, according to GLOV, thirteen of twenty-three LGBT reports of LGBT-related crimes were met with indifference or verbal abuse from MPD officers. Today, not only has the relationship improved, but GLOV now refers victims directly to the GLLU. As Ramsey suggested, community members now believe that GLLU will take their reports more seriously.

Finally, we can measure the success of the GLLU by reviewing its homicide case closure rate. Nationally, murders are solved at a rate of about 72 percent (DC MPD 2009). In Washington DC, the success rate drops to about 50 to 54 percent (DC MPD 2009). Since the Gallaudet murder in 1999, the GLLU has been involved in a number of murder investigations in which the victims, suspects, or witnesses are LGBT. The case closure rate for homicides that GLLU has assisted with is 95 percent (Parson 2009).

While no solid measures exist to fully account for the GLLU's impact on policing, training, and building trust in the LGBT community, we can use several indirect measures. Based on the increased awareness of same-sex domestic violence, the increasing number of reported hate/bias crimes, and the closure rate for LGBT-related murders, the GLLU appears to have had a positive effect on both the community and the police force.

## LEARNING FROM THE GLLU EXPERIENCE

The case of the GLLU offers an exciting and interesting example of providing public goods and services to an invisible community. Although all communities are different, we can draw some generalizations from this case. At a minimum, we can use this case to better understand how policing in the LGBT community can be enhanced. Based on this case, I propose five important observations.

### 1. Reach Out to Community Groups and Other Service Providers

Communities, visible and invisible, usually have some institutions or service providers. This might be a church that provides meals to the homeless community, a support organization for people with addictions, or an advocacy organization that pushes for social change. These groups might not appear to be natural allies, but they are critical for identifying invisible communities and helping to assess the needs of the communities. Their work within the community is invaluable in organizational attempts to establish trust. With regard to the MPD, it selected officers who were lesbian or gay to serve in the unit. More importantly, the GLLU worked closely with community groups, especially during its inception. The GLLU now has a long-standing working relationship with both the GLAA and GLOV.

### 2. Internal Training

In some cases, working with invisible communities requires additional knowledge, skills, or abilities. Invisible communities are often invisible for a reason. Members might engage in "antisocial" or illegal behavior. This means that service providers might have stereotypes and assumptions about the community in need of goods and services. This training can be a valuable tool for service providers and an outreach tool for building community trust.

### 3. Keep Track of Critical Incidents, Both Large and Small

Critical incidents can occur at any time anywhere. Their full impact on various people, communities, and institutions is rarely evident until it is too late to prevent the incident from rising to the level of "critical." However, by establishing an approach to keep track of emerging issues and their effects, a service provider can more quickly respond to potential problems or issues. Patterns in critical incidents are a valuable indicator for better prediction of problems, especially within law enforcement.

### 4. Consider the Leadership

As seen with respect to the MPD, leadership can emerge from any part of an organization. Fostering a climate that encourages service providers to collaborate and be inventive can help to produce leadership and initiative in the organization. Such leadership and expertise will be important if a larger critical incident occurs or the organization wants to be proactive in providing services.

### 5. Develop Good Measures

After determining the problems in the community to be addressed and collaborative partners in the community, measures of success and failure should be developed. Most often, it is valuable to consider outcomes as the mission or objectives are also under consideration. At a minimum, pre- and post-data can be collected to provide information about the effectiveness and efficiency of the service. This means developing protocols that collect data about various community issues and concerns, and having well-developed protocols that collect data after issues and concerns have been addressed. With this approach both agency and community are working together to prevent incidents rising to the level of critical.

## CONCLUSION

This chapter highlights how critical incidents and leadership created an opportunity for the MPD to better serve one of its invisible communities. In order to improve policing services, the MPD had to reduce the level of mistrust within the LGBT community. It did this by being visible and helpful in the community, increasing the number of properly trained police officers, and meeting with community leaders and organizations. As the lead police officer, Sergeant Parson at the GLLU noted that biggest challenge was "walking the tightrope between advocates for the community and enforcers." This challenge is also one of the major successes of the GLLU. The unit has developed an approach to law enforcement that encourages the community to view it both as a community advocacy body and as a public service unit responsible for enforcing laws. If it were to move too far in one direction, it would become a captured agency, bending to community will on policing matters, and if it moves too far in the other direction, it becomes a traditional policing unit without the trust or support of members in the community.

By taking advantage of critical incidents, the MPD has created a response that has had a positive effect on the LGBT community and the broader Washington community. The efforts of the MPD and successes of the GLLU show that some invisible communities can be made more visible. This increased visibility made it possible to improve goods and services delivered. In this case, better community policing.

## NOTES

1. The Stonewall riots were a series of violent conflicts between lesbians, gay, and transgender people, and police officers in New York City. The first night of rioting began on Friday, June 27, 1969, when police raided the Stonewall Inn, a gay bar in Greenwich Village. Over the course of

six days, lesbians, gay men, transgender people, and allies gathered to protest the unfair treatment of sexual minorities. At the time, it was not uncommon for the police to raid such places, and to collect bribes from the owner in some cases (D'Emilio 1991). Police would enter a bar and use any number of reasons to justify arrests on indecency charges including: kissing, holding hands, wearing clothing of the opposite gender, or even just being in the bar during the raid (Duberman 1992). The riots are commonly used as the benchmark, marking the beginnings of the modern gay rights movement.

2. Officers Kelly McMurray and Berdette Williams' efforts to conduct outreach and training within MPD represented efforts at the lower-level of the organization, and Chief Ramsey's willingness to embrace a formal community policing unit represents leadership at the top of the organization.

3. Many other jurisdictions have gay and lesbian liaisons, but none have a unit dedicated to outreach and positive change. More often the role of liaison is focused on public or departmental relations, not community outreach and training.

## REFERENCES

Amnesty International USA. 2005. *Stonewalled: Police Abuse and Misconduct Against Lesbian, Gay, Bisexual and Transgender People in the U.S.* New York. www.amnestyusa.org/outfront/stonewalled/report.pdf (accessed on September 25, 2009).

Atlanta Police Department. 2005. "GLBT Liaison." www.atlantapd.org/glbtliaison.aspx (accessed on October 1, 2009).

Belkin, A., and Jason McNichol. 2002. "Pink and Blue: Outcomes Associated with the Integration of Open Gay and Lesbian Personnel in the San Diego Police Department." *Police Quarterly* 5: 63–95.

Bernstein, M., and Constance Kostelac. 2002. "Lavender and Blue: Attitudes About Homosexuality and Behavior Toward Lesbians and Gay Men Among Police Officers." *Journal of Contemporary Criminal Justice* 18: 302–327.

Buhrke, R. 1996. *A Matter of Justice: Lesbians and Gay Men in Law Enforcement.* New York: Routledge.

Bune, K. L. 2004. "D.C. MPD's Gay and Lesbian Unit Is Model to Emulate." *NY Cop On Line Magazine* 2 (1). www.nycop.com/Winter2004/CIMS/CJ_Connection2005/GLLU_DCPD/body_gllu_dcpd.html (accessed October 1, 2009).

Burke, M. 1993. *Coming Out of the Blue: British Police Officers Talk About Their Lives in "The Job" as Lesbians, Gays and Bisexuals.* London: Cassell.

District of Columbia Metropolitan Police Department (DC MPD). 2009. "MPD Boosts Homicide Closure Rate." News release, June 25. http://newsroom.dc.gov/show.aspx/agency/mpdc/section/2/release/17477/year/2009/month/6/.

D'Emilio, J. 1991. *Sexual Politics, Sexual Communities: The Making of a . . . Lovers: A History of Lesbian Life in 20th Century America.* New York: Penguin.

Duberman, M. 1992. *Cures: A Gay Man's Odyssey.* New York: Plume.

Fillichio, C. 2006. "The new beat." *Public Manager* (Fall). www.thepublicmanager.org/docs_articles/archive/V0135,2006/V0135,Issue03/V35N3_TheNewBeat_Fillichio.pdf (accessed September 25, 2009).

Gay and Lesbian Activists Alliance (GLAA). 1998. "1998 Annual Statistical Report: Gay Men and Lesbians Opposing Violence (GLOV)." Washington, DC.

Gay and Lesbian Liaison Unit (GLLU). 2008a. "About the GLLU." GLLU factsheet, May 27. www.gllu.org/PDFs/facts_GLLU.pdf (accessed on September 25, 2009).

———. 2008b. "Hate Crimes." GLLU factsheet, May 27. www.gllu.org/PDFs/facts_hate.pdf (accessed September 25, 2009).

———. 2009. Focus Group Interview—Gay and Lesbian Liaison Unit. September 23.

Innovations 2006. Government Innovators Network. 2006. "Gay and Lesbian Liaison Unit: 2006 Winner; 2005 Finalist. Innovations in American Government Awards." www.innovations.harvard.edu/awards.html?id=48931/ (accessed September 25, 2009).

Moser, Bob. 2003. "Disposable People." *Intelligence Report* 112 (Winter). www.splcenter.org/get-informed/intelligence-report/browse-all-issues/2003/winter/disposable-people/ (accessed September 25, 2009).

Schneider, A., and H. Ingram. 1993. Social construction of target populations. *American Political Science Review,* 87 (2): 334–347.

CHAPTER 18

# The Dynamics of Critical Incidents in the Age of New Media

## Examining the Role of the Self-Appointed or Media-Anointed Expert

STEVEN M. GORELICK

> A Critical Incident is a relatively brief occurrence involving injury, loss, conflict, discovery or change of significant proportion, usually unscripted and unanticipated, with the potential to alter existing societal norms. Critical incidents are usually traumatic, threatening the bonds of trust that bind communities, but may be positive, initiating historic consequents. (Ochberg et al. 2007)

For much of the twentieth century, as radio and television matured as sources of news and entertainment, critical incidents began to take place against the backdrop of an increasingly crowded, mass-mediated, and "noisy" cultural context. News of incidents reached an interested public more quickly, more people were able to witness norm-altering events as they took place, and a growing public chorus of voices could be heard explaining those events with theories, evidence, and claimed expertise.

The century itself—marked by calamity, catastrophe, and social upheaval—could hardly have provided these maturing media with more raw material for public consumption. Some of these events—the Triangle Shirtwaist Factory fire, the sinking of the *Titanic,* or the fiery crash of the Hindenburg—were accidents and catastrophes that are the inevitable result of rapid industrialization and technological change.

Others took place in the context of the century's world wars, revolutions, and genocides that led to the death of an estimated 100 million combatants and civilians. These conflicts would annihilate entire ethnic or religious groups, destroy social bonds that had endured for centuries, and cause widespread trauma that would persist for generations.

Clearly, what British novelist Margaret Drabble (2001) famously called the "beastly century" might also—following the definition above—be characterized as the century of the critical incident.

Both the maturing of the mass media and the increasing availability of small cameras and recording devices meant that, as the century progressed, films and photographs of critical incidents that previously might never have come to light became pervasive in public culture. Catastrophic incidents intended to be secret, while not broadcast on official channels, were often surreptitiously recorded by perpetrators, victims, and bystanders. Indeed, while German soldiers were strictly forbidden to carry cameras during the earliest stage of the Holocaust—the individual, close-range, shooting deaths of close to 1 million Polish and Soviet Jews by so-called Einsatzgruppen—explicit photos taken secretly by those soldiers have continued to surface since the end of the war (Rossino 1999). The century of the critical incident would more and more become the century of the *visible* critical incident.

Like other rapid technological changes, the growth of mass media, and later the Internet, inspired ambivalent reactions. Some envisioned a global village in which the wide diffusion of information would bring nations and cultures closer together and, while not necessarily reducing conflict, at least allow diverse communities to see each other for the first time (McLuhan 1964). Others, fearing that radio and television networks might become behemoths spreading homogenized information to passive audiences, offered the so-called mass-society critique pointing to an increasingly anonymous and alienated citizenry (Rosenberg and White 1957).

Of course, both extreme scenarios masked the subtle and nuanced change that followed. The global village did indeed seem smaller, and, from the earliest days of the television era, journalists like Edward R. Murrow used their first moments on the air to show audiences remote and distant locations. But understanding and empathy between those who could now see each other did not always follow. Increasing global awareness of a variety of humanitarian crises—from starvation in Biafra to genocide in Darfur—did not guarantee an effective response. Compelling visual evidence of a critical incident might raise awareness, but it could not stop other powerful institutions, including governments willing to use military force, from dominating and appropriating various parts of the global village.

At the same time, networks did become behemoths of a sort, and they did monopolize both message and medium. But the same networks that celebrated mass consumption in the era of dishonest quiz shows also used their legitimacy and enormity to nationally broadcast the first televised presidential debates and, later, news of the first presidential assassination of the television age. Clearly, the new medium would have both diverse consequences and opportunities. There may have been no more eloquent statement of these contradictory possibilities than the speech delivered by Edward R. Murrow to the Radio and Television News Directors Association on October 15, 1958:

> This instrument can teach, it can illuminate; yes, and it can even inspire. But it can do so only to the extent that humans are determined to use it to those ends. Oth-

> erwise it is merely wires and lights in a box. There is a great and perhaps decisive battle to be fought against ignorance, intolerance and indifference. This weapon of television could be useful. (1958)

Obviously new technologies would neither preclude nor guarantee any type of social change. Some uses, like the broadcast of the hearings that led to the censure of Senator Joseph McCarthy, would reveal the potential of the mass media to expose threats to democracy. Others, like the quiz show cheating scandals, would show what was possible when commercial pressures to build a large audience led to corruption and criminality.

For present purposes, which is to understand how changing media technologies have affected, and will continue to affect, the nature and social impact of major critical incidents, an earlier and more obscure incident might be the best place to start. In April 1949, three-year-old Kathy Fiscus was walking across a deserted field in the Los Angeles suburb of San Marino when she fell down a water pipe that was 120 feet deep and 14 inches wide. In what would become a familiar tableau for decades to come, a crowd of reporters, public safety officials, and onlookers focused on the hole, while rescuers, aided by heavy equipment and witnessed by a crowd of more than ten thousand onlookers, desperately attempted to rescue her. It took two days to reach Kathy, and the entire saga received both local and national live television and radio coverage. She did not survive the fall and likely suffocated quickly from a lack of oxygen (*New York Times* 1949a, 1949b; United Press International 1949).

The Kathy Fiscus incident almost instantly assumed iconic status and later became the basis of several books, a film, and even an extended segment in Woody Allen's film *Radio Days*. It may not have been clear at the time, but critical incidents of all sorts would now increasingly play out in front of television cameras, and later on the Internet, and be experienced collectively by a public able to experience events unfolding in real time.

This is not to suggest that television originated the role of mass media in collectively experienced catastrophes and incidents. In an earlier, pre-radio era, vivid newspaper reports described how anxious groups of family members, reporters, and onlookers formed crowds in front of New York telegraph offices, waiting for news about the fate of those traveling on the *Titanic* (Herring 1989). Later, radio's power to amplify the impact of a critical incident would become clear during the "invasion from Mars" described on the Mercury Theatre's broadcast of H.G. Wells's *War of the Worlds* on October 30, 1938 (Cantril 1940).

But the sound and images of broadcast television added an unprecedented emotional immediacy and thoroughly changed the nature of how audiences would collectively experience critical incidents. In fact, scholars of mass media almost immediately noticed that the television lens had the capacity to completely reshape how the public saw and interpreted high-profile, national events. One landmark study revealed that the supposed wave of public support for Gen. Douglas MacArthur following his firing by President Harry Truman—including tumultuous and crowded parades—was in fact significantly less enthusiastic than the television coverage implied. One parade in Chicago, depicted in a broadcast as an uncontrollable crowd, was shown to have actually attracted a much smaller audience (Lang and Lang 1953).

Perhaps no critical incident better illustrates the extent to which the incident and the accompanying coverage coalesced into a larger traumatic episode than the 1963 assassination of President John F. Kennedy. A landmark study of the diffusion of news of the assassination revealed that, within an hour, 9 of 10 Americans knew about the incident (Greenberg 1964). The public would now learn about critical incidents with dazzling speed.

The extraordinary trauma of the assassination itself, amplified by the size of the audience and the speed with which the public learned of the incident, combined to create a critical incident seared in the memory of those who watched the continuous television coverage. Even those too young to understand the full magnitude of the events were part of a national audience that two days later watched—most likely for the first and only time in their lives—the murder of one person by another on live television, when his alleged killer, Lee Harvey Oswald, was shot to death by nightclub owner Jack Ruby in the basement of the Dallas Police Department. The second killing was yet another shock that would persist for decades. The desperate search for facts and explanations of the assassination would now be forever stymied by the killing of the one person perhaps most able to provide them.

This paper is an attempt to make sense of how, in the present age of rapid technological change, the nature of these disorienting critical incidents—and how the public perceives and experiences them—has been radically changed because of the availability of new and widely available tools of digital media. While this change has affected many of the individuals and institutions who play a role in critical incidents, the example explored here is the pervasive "experts" who seem ubiquitous in the age of digital communications and who offer opinions on virtually any aspect of a public calamity whenever a quick explanation for traumatic events is needed.

The category of expert includes a wide range of actors, including experts who are chosen by media outlets and those who are self-appointed, those who bear respected credentials and those with no formal credentials speaking solely from personal experience, those with some symbolic or financial interest in the issue being contested and those who are seemingly disinterested. All of these voices can be heard, promoting points of view in a dramatically changed technological environment.

To simply say that technology has changed hardly describes the diverse technologies, interested parties, cacophony of voices asserting expert knowledge, and global visibility that now seem to characterize these incidents. Indeed, my working assumption is that the subsequent response to a critical incident by the media and other social institutions, the noisy cultural spectacle of nonstop images and voices cannot—in terms of public perception—be separated from the critical incident itself. The incident, and the spectacle of the incident, become one and the same in the public mind.

After a brief discussion of a recent example of this complex social and cultural context, the nature and specific role of media experts is addressed. What role do experts play in the overall cultural spectacle to which the public turns for facts and explanations? Might it be possible to develop a useful typology of media expertise that would help make useful distinctions between all these voices?

Finally, the impact of all this media scrutiny, cultural noise, and "expertise" on those who manage critical incidents is examined. What is it like to manage an ongoing challenge to public health and safety while simultaneously contending with pervasive media coverage? How can a responsible party deal with the sometimes contradictory voices that may be offering a running commentary on events that might still be unfolding?

## CRITICAL INCIDENTS IN THE AGE OF DIGITAL MEDIA

It is a tricky, if not impossible, task to select an illustrative example of how new media technologies have drastically changed the public experience of high-profile critical incidents. Social and cultural contexts differ, available media technologies constantly change, and every incident occurs in a particular context. While certain types of incidents do seem to recur according to predictable narratives—school and workplace shootings, environmental disasters, pandemic illness—to call any incident typical is to obscure its peculiarities and the unique insights that it might offer.

One recent and relatively brief critical incident experienced by the public in real time was the taking of three hostages at gunpoint by James Lee on September 1, 2010, at Discovery Channel's headquarters in Silver Spring, Maryland. Almost immediately, photographs of Lee carrying a weapon through the headquarters building were distributed globally through Twitpic, Twitter's photo-sharing service that quickly posts cell phone photos of users. An additional photo was distributed showing the arrival of Montgomery County's SWAT team with weapons and a bomb-detecting robot. As is now often the case, the whole concept of a first responder was confounded by the presence of citizens recording and distributing accounts of the event before the first responders even arrived. Many such accounts were distributed using Twitter. Those using Twitter to describe and comment on the events at the Discovery Channel even included the gunman himself, who distributed a tweet with his demands before he was killed (Farhi 2010).

This incident illustrates new aspects of the stage on which a critical incident is performed, a setting in which (1) the larger public can now often see aspects of a critical incident before the arrival of a first responder and (2) perpetrators are increasingly able to avoid official sources and communicate directly with the public. But the Discovery Channel incident involved relatively few people, was not related to some larger social upheaval, and hardly altered societal norms.

An even more dramatic example is the shooting death of Neda Agha-Soltan during the Iranian street demonstrations of June 2009. The shooting was an extraordinary example of how widely available video and cell phone technology can drastically expand the size of the global stage on which a critical incident unfolds.

Agha-Soltan had been viewing one of the sporadic street demonstrations in central Tehran, when, on June 20, she was shot by a member of the Basij government paramilitary. Several videos of the shooting—including one that appears to depict her on the ground, bleeding and losing consciousness—were posted on the Internet and received almost instantaneous international attention (Fathi 2009).

Of course, while the widely known YouTube platform provided a means for the speedy global distribution of the video, people still needed to learn that the video could be found on YouTube. Twitter helped build an audience when users began to send tweets that included a link to the video. Consider this hypothetical tweet:

> **stevegorelick** see the tragic killing of Neda Agha-Soltan at http://tiny.cc/x00qx #neda

This would have been a message from me directing readers to a link where the video of the shooting could be seen. Further, the addition of the keyword *#neda* (called a hashtag in Twitter parlance) would ensure that my message would be available to anyone searching for tweets including *#neda.*

This episode also represented a key, transitional moment in the role played by new media technologies in the depiction of critical incidents. For the first time, mainstream media outlets found that the flood of vivid information coming from these unconventional sources could no longer be dismissed as unreliable and lacking traditionally required sources. The *New York Times* and other mainstream media quickly incorporated these sources into their coverage of the Agha-Soltan killing and its tumultuous aftermath in the streets of Tehran. Tweets and anonymous video posted on YouTube now found their way onto the front page of those Web sites and became central to the coverage. This was a remarkable moment, given that it followed several years of heated arguments over what many in the mainstream press argued was the amateurishness of social media (Lemann 2006).

To be sure, these sources were accompanied by clear warnings about their reliability, but the Rubicon had been crossed. The stories of trained journalists, sometimes reporting directly from the scene of the incident, would now occasionally appear on the home pages of some of the world's most legitimate news sources alongside unsourced YouTube videos. In fact, the *New York Times* itself noted this transition and reported it in detail as a separate news story. The serious and honest argument about the reliability of such sources simply collapsed when the words and video coming directly from the incident became too compelling for audience-conscious publishers to ignore (Stelter 2009).

The Agha-Soltan incident also revealed the extent to which the Internet would become the terrain for countless video memorials. The impulse to create and post virtual tributes, so pervasive in the aftermath of 9/11, was on display in the hundreds of individually crafted videos posted on YouTube. Communities have always come together seeking both solace and answers in the aftermath of traumatic events and now YouTube would become a key site for such "gatherings" (Collins 2009).

As the case of Neda Agha-Soltan illustrates so vividly, the tools capable of communicating and amplifying critical incidents have changed, and continue to change, at a dizzying pace. The wide availability of cell phones with both still and video capability now allows witnesses to quickly capture and disseminate incidents that previously might never have come to public awareness. Further, each reduction in the size of those devices makes it more possible for them to be used surreptitiously. The text-messaging capability of cell phones, as well as Twitter, allows those witnessing events to transmit details that in the past might have either been censored or the sole province of mainstream news organizations.

But the availability of new technologies only partly describes the new cultural context in which critical incidents occur. Systems with enormous capacity—YouTube, Facebook, and other social networking sites—can store a virtually unlimited amount of text, audio, and video and make them available to a global audience. But interested audience members need the ability to mine all this data to find information about specific events and specific individuals. This is the point at which a variety of search engines, such as Google, provide the "maps" that direct an interested party through the flood of virtual traffic directly to the specific information being sought.

This searching is further facilitated by the wide use of "tags" that allow producers of content to provide keywords intended to lead potential viewers to their content. Even these tags have become more sophisticated, progressing from those for text to those for video to "geotagging," which embeds a photograph with the precise geographic coordinates at which the photograph was taken. Geotagging has made it even easier to find information about a particular incident at a specific location. It is no surprise, and no small challenge to established media and government institutions, that this entire process now often takes place before responsible parties reach the scene of an incident and before they are able to issue official statements.

This is the cultural context of critical incidents in the early twenty-first century, a din of simultaneous and contentious voices, constant repetition of the iconic catastrophic images, and limitless posts in cyberspace. Citizens receive instant images of, and commentary about, an event from the moment it happens. Groups of stunned witnesses, public safety officials rushing to the scene, and victims being rushed to hospitals are now part of visual vocabulary that begins the process of defining an incident for the public. And the public clamors for information to explain the sudden turn of events.

Enter the "expert" bearing answers.

## THE ROLE OF EXPERTS AND EXPERTISE IN A CRITICAL INCIDENT

Both media-anointed and self-proclaimed experts are recurring and central characters in most critical incidents. Because the role and impact of these experts are the focus of the remainder of this analysis, it is important to ask why their role deserves specific attention. Why select the media expert from the chorus when so many other voices can be heard? What is it about experts that make them such important participants in the unfolding of a critical incident?

The most important reason is that—with both positive and negative consequences—these voices actively seek, and often receive, the surplus symbolic value and legitimacy that comes with being labeled an expert. Their opinions and observations are heard above the din. To use a term borrowed from the constructionist perspective in sociology, they are "claims makers" asking for their claim to be privileged because of factors that might include their education, research, life experience, or social position. Pierre Bourdieu refers to these adornments as the "symbolic capital" adding to the perceived legitimacy, status, and value of an individual's actions and statements. *Dr.* Smith discussing a potential pandemic is not the same thing as *Mr.* Smith discussing it, regardless of how much someone really knows about a topic (Bourdieu 1984).

Indeed, a key challenge for anyone asserting expertise in a crowded media environment is constructing a persuasive personal narrative, adorned with adequate symbolic capital that bolsters the claim of special knowledge. Some institutions have actual rituals in which participants are given promotions, diplomas, or titles that recognize increasing levels of accomplishment. But the less-regulated public sphere is more like a competitive social problems marketplace, what Bourdieu referred to as "social space," in which many people are promoting versions of reality and arguing that their accumulated symbolic capital entitles their version, their collection of "facts," to be taken more seriously.

But the ability of experts to command attention with persuasive symbolic capital obscures the role that we all play as members of the audience. Reeling from a sudden trauma, we are often more than willing to grant a person the status of expert in our desperation to find the answers we think they can provide and the stability to which they can help us return.

Consider the sudden disorientation caused by a traumatic critical incident—episodes that present the greatest threat to existing forms of social organization and to widely accepted assumptions about how the world is supposed to operate. The most widely accepted beliefs, what Schutz and others have called "taken for granted reality," can quickly be rendered moot or irrelevant (Schutz 1972). Truth (or, more precisely, what had been taken for truth) can suddenly be rendered false. New facts jarringly present themselves and create a state of cognitive dissonance that must be resolved. The world has been turned upside down, and we need answers. This is why it is hard for us to avoid becoming active participants in the process of anointing and legitimizing experts who can provide the answers that we need (O'Connell and Mills 2003).

We are, after all, not simply curious about why something may have happened. In our trauma and dislocation, we want our world reassembled as we knew it, and we want to know how to do that. This means that authoritative, expert voices are a precious commodity at precisely those moments when answers seem scarce. This need to reassemble and feel whole, however, is also what makes us vulnerable to the answers and theories that come to us in the form of viral e-mails, urban legends, hucksters, or experts describing the most implausible conspiracy theories (DiFonzo and Bordia 2007).

Expert and authoritative opinion, regardless of the obvious problems in deciding who precisely is an expert, has been seen as a remedy for confusion since antiquity. After describing the shock felt by unshackled prisoners who discover that familiar shadows had actually always been concrete objects, Plato—in his "Allegory of the Cave"—proposes a select group of "philosophers" whose wisdom and special training will allow them to be guides through the darkness (380 B.C.E./1945).

Even the spiritual world has historically designated an individual of uncommon wisdom—the shaman—whose special position allows him or her to serve as an interpreter and mediator between the world of the sacred and the world of the profane. In the disciplines of sociology and anthropology, the borrowed concept of shamanism has been constituted as a discrete area of study, examining those individuals whose social status, special training, and even mystical qualities have conferred on them the special status of guide and expert (Harvey 2003).

Obviously, as long as people have confronted baffling problems, experts have been proposed—and have presented themselves—as those who could bring light to the darkness. It is hard to imagine, however, that any of them could have anticipated how dramatically the nature of experts and expertise would change in the crowded digital age.

Prior to the digital age, the mainstream media served as the primary arbiter of all these claims. Anyone could claim to be an expert, but being selected for introduction and presentation as an expert by legitimate media institutions was the sine qua non of expertise. The implication was that one had been selected because of an actual measure of expertise—that a powerful institution had decided that an expert claim should be taken seriously. A psychologist was given a talk show, a former military officer was asked to evaluate a tactic or strategy, or a doctor was asked about the importance of certain childhood vaccinations.

Of course, being chosen by a legitimate institution neither verified the person's claim to expertise nor proved the accuracy of the claims being made. But studies have revealed the extent to which the authority and perceived legitimacy of an expert is itself presented as the reason that we should take that experts' claim seriously. This may be an almost amusing tautology—that is, experts' claims are true because they are authoritative experts—but on close examination, many claims of accuracy and truth are buttressed only by the symbolic authority of the expert (Walton 1997).

In the age of social media, blogs, and Twitter, the role of the mainstream media as arbiter of expertise has diminished. An appearance on the *Today Show* or a brief quotation in a story on the evening news is still extraordinarily valuable to anyone seeking to bolster a claim of expertise. But these new digital tools have provided a number of alternate routes to public attention that, on occasion, have been enormously successful in publicizing an expert and his or her claims. Nate Silver, who first came to public prominence in 2008 with his blog FiveThirtyEight.com, a source of stunningly accurate and statistically based political predictions, had his blog acquired by the *New York Times* and renamed FiveThirtyEight: Nate Silver's Political Calculus. The very type of news source initially marginalized by mainstream journalists became part of the *New York Times,* conferring a level of legitimacy that few publications can match. The influence of mainstream media remains strong, but one way this is being accomplished is through their adoption and absorption of the very new media tools that they had originally seen as a threat.

The point is that, for good or bad, these selected voices are accorded a legitimacy and a chance at influencing the public agenda that other individuals might not have. Further, both the Internet and the advent of twenty-four-hour cable news have created a virtually bottomless news hole that must be filled with content. This has dramatically increased the amount of broadcast time and space that can be filled with pundits, commentators, satirists, and the most legitimate experts.

This makes them an important subject for discussion and analysis. What types of experts play what types of roles during critical incidents? What is their impact? Is there any useful typology that would help us understand the range of actors who are put forward, or who put themselves forward, as experts? How do we make sense of all the voices asserting expertise that can almost always be heard screeching over one another during the critical incident?

## A PROPOSED TYPOLOGY OF MEDIA EXPERTISE IN THE DIGITAL AGE

The size and complexity of the public sphere virtually guarantee a social space in which expert claims are constantly being publicized and suppressed, legitimized and debunked, adopted and rejected. Is there any useful way to classify these diverse claims-makers?

While I suggest such an exploratory typology, I put it forward with several important qualifications. Useful and revealing typologies cannot be imposed on data. At best, they emerge from careful data collection and data analysis, which—when searching for logical groupings in large data sets—often calls for either cluster or factor analysis. This, however, is an exploratory typology. A more thorough analysis of a large set of expert claims might either combine or eliminate some of these categories. These categories are intentionally tentative and inclusive. Also, this typology is minimally concerned with whether the expert claims being made are true. Obviously, the veracity of a claim is absolutely crucial in any specific incident, even given the fact that truth is not easily determined in discussions of nuanced and subtle issues. My guiding assumption, though, is that the chorus of expert voices plays a role that, to a large extent, is independent of the truth of the claims. In other words, this is about the categories of social actors who make claims and not about who might be more likely to be accurate. We should, of course, be deeply concerned about quick and thoughtless pronouncements made in the midst of horrible public catastrophes, claims that can cause unnecessary confusion and anxiety. But this exploratory typology simply assumes that many kinds of experts—self-appointed or media-anointed, marginal or mainstream—are central to shaping public perception.

Another caution about this analysis is that it isolates the variable of media expertise to the exclusion of other variables. In fact, critical incidents shape, and are shaped by, a wide variety of social institutions. Extracting the media expert as a special case is done with full awareness that no single factor in a complex social process operates independently of other factors. Experts are important voices, and they often have a privileged legitimacy, but they are only one factor in the complex process in which societies are destabilized by sudden incidents. Further, the brief examples chosen are not intended to be fully representative of the type, only to present one of what might have been many examples.

Finally, I caution that this analysis rejects what seems to be the popular belief in a clear distinction between the real experts and everyone else. In fact, given the complexity of the issues being debated, it is almost pointless to make quick and glib distinctions between truth-tellers and liars. Truths, lies, and all the nuance between them can come from any of the categories. Further, to make a stark claim in the public sphere that something is true is to misunderstand the extent to which truth is an evolving rather than static concept, especially with regard to expert opinion. Over time, supposedly uninformed lay expertise can come to be accepted as orthodox truth. And rigorous research findings published in respected, refereed journals can turn out to be wrong. This is not evidence that anyone is lying. It simply reveals the complex process in which truth and knowledge emerge and the public comes to see an assertion as true.

Who are these experts?

## THE LAY EXPERT

For several decades, the Gallup organization has conducted an annual poll measuring public confidence in major institutions. The poll released in August 2010 revealed that confidence in the accuracy and reliability of newspaper and television news was at a near-record low. With some variation by age and income, only 25 percent of Americans expressed a great deal of confidence in either medium. This is the lowest level of confidence since 1993, when the survey began to include media institutions. Although the data on confidence in institutions are complex, virtually every public institution, with the exception of the military, has either lost the confidence of the public or remained relatively stable in public confidence since 2000 (Saad 2010).

This means that, during roughly the same period that included the 9/11 attacks, the war in Afghanistan, and the war in Iraq, citizens lost confidence in the very sources and institutions to which they have historically turned for answers in traumatic times. One response to a loss of confidence in official institutions has been the rise of the lay expert whose asserted expertise arises from direct personal experience. Now, however, a distinctly anti-empirical and anti-scientific form of lay expertise has emerged that goes beyond simply asserting the primacy of personal experience. Some declare that the very fact that a claim is made by an expert from a legitimate social institution is precisely what makes that claim suspect. These forms of lay expertise have taken on the trappings of fundamentalism. One apt illustration of the ascendance of the lay expert is rise of the movement of parents who claim that certain vaccines cause autism, despite the paucity of anything but anecdotal evidence (Begley and Interlandi 2009).

A recent development in this long-simmering controversy was the decision in early 2010 by the British medical journal *Lancet* to retract an article by Dr. Andrew Wakefield that had previously claimed a connection between the measles, mumps, and rubella (MMR) vaccine and subsequent diagnoses of autism. The *Lancet* found that, on close review, his original findings could not be verified given the flawed data he had presented. Parent groups were livid, and the blogosphere crackled with criticisms and attacks on the journal. However, setting aside the scientific controversy, the reaction of one prominent anti-vaccine advocate illustrates the strategies of this new form of lay expertise.

Jim Moody, director of a parents' group that claims that vaccines cause autism, said that his group's willingness to attack the medical establishment would actually strengthen Dr. Wakefield's standing. "Attacking scientists and attacking doctors is dangerous," he said. "This is about suppressing research, and it will fuel the controversy by bringing it all up again" (Harris 2010).

Of course, lay expertise comes in many forms. Sometimes lay experts reject a widely accepted social practice. Others use personal experience to confirm a widely shared finding. In both cases, it is personal experience and not large-scale data that backs up the claim.

When it comes to the potential impact of lay experts on the nature of critical incidents, the most important point to make is that new media tools have provided them with the tools to make claims that do not require the imprimatur of the mainstream media. And, as Christopher Clarke and others have pointed out, the mainstream media—which strive to present balanced points of view—may give disproportionate attention to lay expertise that cannot be supported with scientific evidence (Clarke 2008).

For this typology, we only need note that, in an era in which confidence in legitimate mainstream institutions is in decline, lay experts are able to attract attention and vigorously participate in the debates that take place both during and after critical incidents.

## THE INTERESTED OR COMPENSATED EXPERT

The universe of those offering expert opinion has always included people whose economic or political interests are clearly served by the version of reality that they are promoting. Sometimes these interests are made explicit, yet often they have been hidden beneath the cultural camouflage of charitable foundations, quickly created front groups, or titles or claimed credentials designed to imply impartiality.

Of course, anyone asserting expertise has some interest in the data or opinions presented. For this discussion, however, the "Interested Expert" is limited to those who are purposely charged with entering the public sphere to promote a set of facts on behalf of someone who is their client, their employer, or someone with whom they share political or economic interests. Sometimes these experts are directly compensated, yet more often they promote facts that, if widely accepted, might increase the value of a commodity or enterprise with which they are associated.

A typical example of such camouflaged interests in the midst of a critical incident was the testimony offered by "Nayirah," a fifteen-year-old Kuwaiti girl, before a public hearing of the congressional Human Rights Caucus on October 10, 1990. "Nayirah" provided an eyewitness description of Iraqi soldiers who had entered a Kuwaiti hospital, thrown babies from incubators, and left them to die on the floor.

Eventually it was revealed that the group for which "Nayirah" spoke—Citizens for a Free Kuwait—had been formed by the public relations firm of Hill and Knowlton and had been the centerpiece of an $11.5 million government-funded public relations campaign. After the war, the group revealed that it was financed almost entirely by the Kuwaiti government. "Nayirah" was actually the daughter of Saud Nassir al-Sabah, Kuwait's ambassador to the United States, who was reported to have listened to her testimony while seated in the hearing room.

While it would be hard to prove that her riveting testimony was the key variable leading to public support for the war, a number of analysts see it as having greatly influenced public opinion in favor of the Kuwaitis (Al-Yasin and Dashti 2008; Rowse 1992).

If the case of Nayirah—daughter of an ambassador—is perhaps too facile an illustration of the interested expert, the subtle and complex relationship of financial interests and expert opinion was perhaps most fully revealed in a series of Pulitzer Prize–winning stories written in 2009 by *New York Times* reporter David Barstow.

Barstow's multipart report detailed how the majority of ostensibly objective military commentators on network television were, in fact, deeply connected to the leadership of the Defense Department and, in some cases, in the employ of various defense contractors. One military analyst for Fox News, retired Marine Col. John C. Garrett, served as a lobbyist at the influential Patton Boggs firm helping firms win defense contracts at the same time that he appeared on television offering his expertise on military matters. Numerous comparable examples were cited (Barstow 2008a, 2008b, 2008c).

As in every other category, even the revelation of extensive conflicts of interest does not necessarily disprove the expert advice that was offered. Opinions about the potential efficacy of military strategy and tactics are not the kind of clear assertions that can be disproven.

This category, however, does call our attention to the possibility that expert opinions about everything from law enforcement strategies to the actions of elected officials might be offered during critical incidents by those with personal or financial interests that may not have been fully disclosed.

## THE CREEPING EXPERT

This category may or may not deserve a place in a more fully developed and validated typology. It may be nothing more than a hazard confronting anyone who enters the public sphere to present expert opinion. The "creeping expert" is someone who, while having earned a significant amount of authority in the public sphere, finds himself slowly and incrementally offering observations in areas that "creep" away from his/her core specialty. The "creeping expert" seems to be unconstrained by any concern that someone else might be a more appropriate voice. What I call expertise creep is a phenomenon that I unexpectedly experienced myself and described in a satirical essay some years ago in the *Washington Post* (Gorelick 2001).

It is easy to see how the institutional pressures and commercial imperatives of each major medium conspire to create "creeping experts." Someone is called to offer an opinion on a subject with which they are well acquainted, and a producer might notice that the expert is able to provide pithy quotes and sound advice in an attractive package. The next time an expert opinion is needed, a producer might very well reason that the subject at hand really is no different from the previous subject. The almost-surreal result in my case was that initial comments I made about a traumatic shooting led to urgent requests for my observations on adoption. I chose to end the cycle, but the public sphere is replete with examples of those who eventually morph into experts about everything.

Of course, many people are capable of offering wise and perceptive comments about a wide variety of topics. Some people's expertise does cover an impressively broad range. I include this in the typology, though, primarily because of the speed and urgency with which experts are chosen during critical incidents. Informative voices are needed and needed quickly. The existence of the "creeping expert" should simply remind us that, in our desperation for answers, many informed voices might not, on closer examination, be the most appropriate ones to offer an expert claim.

Twenty-four-hour cable news, with its virtually unfillable news hole, both needs and produces its own creeping experts. These people are versatile, articulate, and loquacious. Whether an accomplished ophthalmologist should be offering advice about infectious diseases or a movie star should be offering advice about the efficacy of antidepressants are entirely different questions (Boniface 2006).

## THE PERVASIVE EXPERT

Some experts seem to fully "own" either a social issue or institution. Their presence as a voice in everything from routine news stories to coverage of traumatic events is ubiquitous when their specific terrain is at issue. Indeed, in several cases, the very fact of the expert's pervasiveness has itself become a news story.

Many journalists and media critics have expressed concern that the establishment of such permanent expert "brands" necessarily limits the range of perspectives reaching the public (Harris and Service 2006). It is possible that pervasive experts might emerge—becoming, to some extent, what I have called a "creeping expert"—as their ability to deliver compelling sound bites makes them increasingly attractive to news organizations.

For nearly three decades, Norman Ornstein, a political scientist and resident scholar at the American Enterprise Institute, and Larry Sabato, professor of political science at the University of Virginia, have been perhaps the most quoted experts on U.S. politics and government in the American mass media. Ornstein's expertise is Congress, while the body of Sabato's scholarship and public statements has focused on electoral politics. Both Sabato and Ornstein freely acknowledge their status as public experts, and each has been known to speak humorously about the frequency with which they are quoted. Indeed, the issue of their omnipresence as news sources has itself been a continuing issue of discussion in the mass media, with one major publication referring early on to Ornstein as "the King of Quotes" (Waldman 1986) and Sabato jokingly referring to himself as "the most quoted college professor in the land" (Cohen 2005).

Robert Thompson, Trustee Professor of Television and Popular Culture at the S.I. Newhouse School of Public Communications at Syracuse University, is a ubiquitous source on any topic related to American popular culture. As with Ornstein and Sabato, the fact of his pervasiveness has itself been analyzed in journalism reviews and other publications. In 2007, David Rubin, then dean of the Newhouse School, made the controversial assertion that several news organizations were making a concerted effort not to quote him as frequently. Thompson takes a similarly humorous perspective on his role, freely acknowledging the fact that he has been called a "quote whore" (Noveck 2007).

Again, my reference to critical stories about these pervasive experts is intended only to show the extent to which they themselves have become topics for discussion in the very culture on which they comment. The argument about whether the dominance of such voices might crowd out alternative points of view is important, but for now I simply include them as a category that belongs in any typology of expertise.

## THE LEGITIMIZING EXPERT

What I call the "legitimizing expert" might or might not be a legitimately credentialed expert. But this is less than crucial because that person's primary role is not actually to dispense expert advice. This individual plays the role of expert, whose symbolic capital is a function of his/her appearance, title, and claimed credentials. These are the experts that we virtually never see. They appear in the credits of documentaries, news specials, news broadcasts, and feature films, with countless titles from "consultant" to "cyberterrorism adviser." It is rare that their names lack academic honorifics, initials after their name that signify a claimed license or military rank. But the presence of the person's name is meant to imply that all of what came before was reviewed by someone with expert knowledge and directly relevant experience.

This is not to claim that legitimizing experts are necessarily phantoms. They may or may not actually provide useful advice, although there is no way of knowing what advice they have actually provided. But they are almost always compensated and frequently cited by producers of content when the accuracy or authenticity of their product has been questioned. In fact, the addition of a legitimizing expert as an adviser or consultant is sometimes a specific response to a controversy about the accuracy or inclusiveness of the content.

One might reasonably ask whether a nonspeaking expert used solely for his or her name belongs in such a typology. I suggest that the practice of listing advisers and consultants is, in fact, an extremely subtle and important way that the producers of content seek to establish their legitimacy and gravitas. They might very well provide some expertise, but the unstated implication is that the expert has carefully reviewed the content for accuracy. Of course, we have no idea whether that is actually the case.

## THE CELEBRITY EXPERT

The celebrity expert, whose standing to assert expert opinion is a function of being well known in an unrelated field—often sports or entertainment—is not completely new, but began to appear with great frequency in the last few decades of the twentieth century. Of course, public interest in the private lives and views of well-known personalities has been ubiquitous throughout human history (Turner 2004). Yet, while the lives of great people have always been fodder for public fascination, the well-documented erosion of privacy and the private sphere has also played a role. Some celebrities resisted this tendency and raged against alleged libels and invasions of privacy, while others seemed to realize that, in the age of new media, embracing publicness (or at least the illusion of publicness) could be a shrewd marketing strategy. By 2010, thousands of public personalities had pages on Facebook or contributed to Twitter feeds, discussing their families, their favorite movies, and even their illnesses. Regardless of whether the statements were true or whether the celebrity was even the one tweeting or updating Facebook, the idea that entire portions of private lives could be successfully hidden became outdated. And the idea that the details of private lives could become valuable commodities became widely accepted.

The standing of a celebrity and his or her ability to command attention in the midst of a critical incident is a function of celebrity status. That status might or might not include some substantive knowledge. The person's interest might be grounded in personal reasons. The talk show host Montel Williams, for example, has standing to talk about multiple sclerosis because of both his celebrity status and his own diagnosis.

Independent of their expertise, they do have one advantage: when they either speak or make an appearance, the entire game changes. Regardless of whether they have anything useful to offer, celebrities are drawn to high-profile critical incidents and their presence is widely covered by media regardless of whether it serves the interests of those affected by or responsible for dealing with the incident. Hurricane Katrina, in particular, attracted celebrities including Sean Penn and Brad Pitt, who in some cases had access to resources, the attention of the mass media, and freedom of movement that was infinitely greater than those of the residents attempting to return to their homes.

Celebrity experts belong in a typology for a simple reason. They will be there. They will be heard. They will often have leading roles in the critical incident cast and a prominent place on the public stage. And they will enter the arena with influence and public admiration that responsible parties simply cannot ignore.

## THE LEGITIMATE CREDENTIALED EXPERT

Why include such a category when it is clear that the legitimacy of an expert and the authenticity of his or her credentials are so subjective? If society has such a hard time reaching consensus on expert claims, and if most experts would place themselves in this category, how can someone decide who should be included?

These are fair questions. Evaluations of legitimacy and accuracy are subjective. The truth of expert claims evolve over time as new evidence and new truths are introduced. However, with full awareness of the problem of subjectivity, it is important to acknowledge the fact that—with all the emphasis on celebrity, sensationalism, and provocative quotations—some of those called on for expertise have such specific and widely acknowledged expertise about the issue being discussed that they should constitute a discrete category in a typology.

One can acknowledge the problem of subjectivity in the determination of expertise, but taking this problem to an extreme can lead to absurdity. The development of the polio vaccine, for example, might have been infinitely more complex and involved more lesser-known personalities than is widely known, but no one could reasonably argue that Jonas Salk was not an unquestionable expert on poliomyelitis and the vaccine that he developed to prevent it (Oshinsky 2005). Some people know what they are talking about and have information that we need to hear. Determining who those people are might quickly lead us to a tangled thicket of conflicting claims, but we ignore them at our peril.

The distinguished meteorologist warning of an impending hurricane or the Centers for Disease Control expert on infectious diseases suggesting the importance of a specific flu vaccine must be heard.

The legitimate credentialed expert actually faces his or her own unique challenge during a critical incident. Professional training renders many of them unable to offer opinions without the nuance and complexity that most topics deserve. A meteorologist might want to remind us that an oncoming hurricane could easily change directions. An infectious disease specialist would likely want us to know that a predicted influenza pandemic might be less severe than expected or might come in the form of a completely unexpected viral mutation. These qualifications are an important part of how scientists discuss their findings with appropriate caution.

The problem is that mass media institutions are organized such that nuance is often glossed over. Producers believe, and they may well be right, that audiences are drawn to clear, bold, and unequivocal assertions. The whole professional ideology of news-gathering is that news is telling people *what is known.* And this, of course, is hardly a controversial idea.

But consider the professional ideology of serious researchers in the social, behavioral, and life sciences. Research that ultimately reveals what is not known, or what was mistakenly known, is, in fact, very important "news." Such null findings can lead researchers to abandon experiments revealed as likely failures. They might even suggest that the use of a popular medication should, at least for the present, be suspended pending further research. The point is that news organizations are not organized to gather and value news about causal links not proven, medications that are only partially effective, or criminal investigations based on theories that ultimately prove inapplicable.

The aversion of the media to nuance (or their inability to describe it) is important in understanding the role of media expertise in a critical incident. The very people whose voices might be most needed—the people who might be less comforting than painfully honest, those who can carefully explain how anthrax can be both potentially fatal and extraordinarily difficult to spread—are not always welcomed as the bearers of riveting news. But if a well-informed public is to play any role in mitigating the trauma and dislocation of a critical incident, every possible effort must be made to ensure that the legitimate, credentialed expert is heard in the din of confusing and contradictory cultural noise.

## THE CHALLENGES OF CRITICAL INCIDENT MANAGEMENT IN THE AGE OF THE DIGITAL REVOLUTION

Why should all these voices—pervasive, ostensibly possessed of some special knowledge, and now able to quickly reach an enormous audience—matter in the midst of a pressing critical incident? A true crime author or self-styled profiler might indeed have acquired some legitimacy and social standing, might quickly join the chorus of "expert" voices urging investigators to take specific actions, and may confuse an anxious public with conflicting theories, but why pay so much attention? Isn't the immediate task at hand to deal with the critical incident, apprehend the perpetrator, or stop a series of violent crimes? Why shouldn't responsible parties simply ignore all this noise? After all, a critical incident becomes a cultural echo chamber only when these voices either gain, or are given, an audience.

I am sure I have already betrayed the disdain I have for so many of those who, without carefully considering the impact of their words, enter the public sphere at times of great trauma and social dislocation. I deeply believe that the truth is most likely to emerge in a free marketplace of ideas. But I also know that—despite the noise and confusion of that marketplace—some voices can be extraordinarily helpful in managing a period of social instability. It is hard, for example, to imagine any public-health catastrophe or epidemic without the assurance and advice of carefully nuanced thinkers like Dr. Anthony Fauci, the director of the National Institute of Allergy and Infectious Diseases at the National Institutes of Health. In other words, the noisy and often off-key cultural chorus has some valuable and brilliant soloists who seemed fully aware of the care with which any public statement must be invested.

But I have also seen how other voices—strident, opinionated, ax grinding—can present serious problems for the responsible officials, victims, perpetrators, and bystanders for whom the incident is not academic but a real threat to public safety. An invitation to serve a brief consultancy on a serial-murder investigation involving the Federal Bureau of Investigation, a statewide investigative task force, local government officials, and a medium-size local law enforcement agency allowed me a unique vantage point from which to see how and why these marginal voices are able to insinuate themselves into urgent investigations. At the time, I agreed not to discuss specific aspects of this incident, but a subsequent public report of the conference in which I participated provides ample evidence of the confusion that conflicting media voices can cause in a community already traumatized by the murders of several young people (FBI 2006).

In this case, the months-long investigation was an anxious and frustrating time for all involved parties. The failure to apprehend a suspect immediately created an information void that all sorts of self-appointed and media-anointed experts were willing to fill with conjecture, theories, and rage. Eventually, even a well-known statewide official expressed his anger with suggestions that it might be wise for citizens to arm themselves until the crimes were solved. True-crime novelists, self-styled profilers, and even a psychic walked onto this particular public stage and joined in the public clamor for a quick resolution. Families of victims visibly and understandably demanded a resolution to the case to ease their pain. Eventually a suspect was apprehended, charged, and convicted. I refer to this incident, though, for what it illustrates about the challenges of managing critical incidents in this new media age.

The responsible local investigators, aided by federal law enforcement agencies, spent thousands of hours working on the case and focusing on witnesses and possible physical evidence. In retrospect, however, those officials almost certainly did not realize the extent to which the new media reality had presented them with yet another challenge. All the peripheral and critical voices, with the entire chorus offering a running commentary, were now operating in an era in which attracting public attention was infinitely easier than in the past. Local television network affiliates, newspapers, and radio stations were no longer necessary for spreading public awareness. Not all the technologies to which I have referred above existed at the time, but even the wide use of e-mail and Web sites made it possible for diverse parties to promote a variety of expert claims.

Managing high-profile critical incidents has always required that close attention be paid to both the concrete evidence gathered by responsible parties and the soft evidence that forms the cultural context in which the event takes place. The effort to obtain concrete evidence, to carefully inventory a situation and respond accordingly, has to be the primary responsibility of critical incident managers.

However, in the crowded new media environment, which offers a seemingly unlimited amount of information from multiple sources quickly produced, those managing incidents must now contend with the evolving secondary responsibility of closely monitoring what might be called "folk evidence." These are the infinite fragments of chatter that reveal what people *think* may have happened and what they *say* happened. This includes rumors produced by anonymous Internet sources, welcome and unwelcome advice, callers venting their anxiety on talk radio, and even the sometimes vile and anonymous rants circulated in viral e-mails. This flood of information might deserve the scorn that it often receives for its unreliability and for the confusion it can cause, but in the digital age it must be respected for its power to help form public perceptions about an event. Managers of critical incidents must now take seriously their new responsibility to be monitors (and perhaps even managers) of public perception.

In some ways, this resembles the intelligence function that has long been part of the public safety enterprise. Any professional asking "what's the word on the street?" or "heard any talk lately?" has, to some extent, been struggling to get a feel for the cultural environment in which an event took place. All this noise might be far removed from the actual events, but it offers an invaluable insight into how people have constructed the event in their imagination.

Why should we worry about people's imaginations? Anyone with direct responsibility during a critical incident might justifiably ask whether jumping into a sea of confusion is worthwhile. Why pay any attention to rumors, falsehoods, or urban legends when an entire community infrastructure—physical and social—has been affected by real events? Why spend time pouring over social media like Facebook or Twitter while waters are rising, lives are being threatened, or emotional crowds are gathering? This is a fair question given the public and institutional pressure that incident managers are under to solve immediate, concrete problems.

I suggest that this is a nostalgic view that is no longer salient in an information environment characterized by speed and volume. The public, whose consent and cooperation are always important in dealing with a critical incident, is now enveloped in the same crowded media culture as those in positions of responsibility. They hear the same rumors, receive the same viral e-mails, and read the same blogs and Web sites as everyone else. Indeed, many have now become producers of content themselves, adding to the cacophony surrounding an incident. We live, in other words, at a time when even the most specious rumors can quickly and easily reach a large audience. If there ever was a time when rumors could be dissipated by ignoring them, that time has passed.

This means that many officials should begin to listen systematically to all the voices and noise and fragments of information that are actively forming public perceptions. Robert Merton's notion of a self-fulfilling prophecy helps explain why this is important and how easily perceptions can become reality.

> The self-fulfilling prophecy is, in the beginning, a *false* definition of the situation evoking a new behavior which makes the original false conception come "true." This specious validity of the self-fulfilling prophecy perpetuates a reign of error. For the prophet will cite the actual course of events as proof that he was right from the very beginning. (Merton 1968)

Consider how quickly what Merton calls "false definitions of the situation" can now reach the public. This is not to ignore the extent to which new media technologies allow information widely accepted as truth to reach the public. It is to suggest that the speed with which a simple rumor can morph into a "reign of error" has greatly increased. It thus becomes possible for a public official to discover quickly that what s/he considers an obvious assumption about the situation is not shared by the public. Stated differently, numerous eyewitness reports of two light-skinned males running away from the scene of an explosion cannot, in the current social and cultural media context, preclude wide public acceptance of a rumor that the suspects appeared to be of Middle Eastern origin. Public officials cannot avoid immersion in this admittedly topsy-turvy world of public perception. Any belief, however specious, can come to be seen as legitimate in the current media environment.

Finally, the speed and volume of information about critical incidents require an even more fundamental understanding by responsible officials of how completely cherished assumptions about the world have been overturned. Some extraordinary critical incidents do not simply alter public perception. They have the potential to create the conditions for a full-blown existential crisis in which nothing seems certain and nothing can be trusted. Depending on how directly one participated in an incident, some people might find themselves grasping for any reliable truth in the chaos.

Sometimes the first available "truths" are rumors, inaccurate news reports, or the glib pronouncements of a media expert who has burst onto the scene. But any official who hopes to respond to a critical incident effectively must fully understand the complex beliefs—however marginal or outlandish—that might be common in the community that they are serving. Moreover, officials who wish to influence or alter those beliefs must fully appreciate the fear and desperation that can fuel public acceptance of even the most bizarre assertions. A media consumer experiencing a critical incident might be searching for meaning and for answers that do not exist and might be listening carefully to experts who are unprepared to provide those answers. We must, however, acknowledge how much that desperation is a quintessentially human reaction and be prepared to do battle on the perilous terrain of media and culture that creates the context for many critical incidents in the early twenty-first century.

## REFERENCES


Al-Yasin, Y., and A. A. Dashti. 2008. "Foreign Countries and U.S. Public Relations Firms: The Case of Three Persian Gulf States." *Journal of Promotion Management* 14 (3/4): 355–374. doi: 10.1080/10496490802637713.

Barstow, D. 2008a. "Behind TV Analysts, Pentagon's Hidden Hand." *New York Times,* April 20, 1.

———. 2008b. "One Man's Military-Industrial-Media Complex." *New York Times,* November 30, 1.

———. 2008c. "Pentagon Suspends Briefings for Analysts." *New York Times,* April 26, 15.

Begley, S., and W. J. Interlandi. 2009. "Anatomy of a Scare; When One Study Linked Childhood Vaccines to Autism, It Set Off a Panic. The Research Didn't Hold Up, But Some Wounded Families Can't Move On." *Newsweek,* March 2, 43.

Boniface, S. 2006. "Cruise Sorry Now; Tom's Apology for Anti-Depressant Rant at Brooke." *Sunday Mirror,* September 3, 10.

Bourdieu, P. 1984. *Distinction: A Social Critique of the Judgement of Taste,* trans. Richard Nice. Cambridge: Harvard University Press.

Cantril, H. 1940. *The Invasion from Mars: A Study in the Psychology of Panic.* Princeton, NJ: Princeton University Press.

Clarke, C. E. 2008. "A Question of Balance: The Autism-Vaccine Controversy in the British and American Elite Press." *Science Communication* 30 (1): 77–107.

Cohen, M. F. 2005. "The Quote Machines: They're Everywhere, a Handful of Scholars and Pundits with an Opinion for Every Reporter's Phone Call. Is There Anything Wrong with Turning Again and Again to the Usual Suspects, or Should Journalists Try Harder to Diversify the Expert Pool?" *American Journalism Review* 27 (2): 54–56.

Collins, D. 2009. "Neda the Martyr; Iranian Killed by Militia Becomes Opposition Icon on YouTube Footage." *Mirror,* June 23, 16.

DiFonzo, N., and P. Bordia. 2007. "Rumor, Gossip and Urban Legends." *Diogenes* 54 (1): 19–35. doi: 10.1177/0392192107073433.

Drabble, M. 2001. "A Beastly Century." *American Scholar* 70 (1): 160.

Farhi, P. 2010. "Twitter Gets First Jump on Discovery Channel Gunman; Tweet Stream Alerts News Media to the Unfolding Crisis." *Washington Post,* September 2, C02.

Fathi, N. 2009. "In a Death Seen Around the World, a Symbol of Iranian Protests." *New York Times,* June 23, 1.

Federal Bureau of Investigation (FBI). 2006. "Serial Murder: Multi-Disciplinary Perspectives for Investigators." Behavioral Analysis Unit, Washington, DC.

Gorelick, S. 2001. "Don't Quote Me; Hi. My Name Is Steven, and I'm a Recovering Talking Head." *Washington Post,* August 27.

Greenberg, B. S. 1964. "Diffusion of News About the Kennedy Assassination." *Public Opinion Quarterly* 28 (2): 225–232. doi: 10.1086/267239.

Harris, G. 2010. "Journal Retracts 1998 Paper Linking Autism to Vaccines." *New York Times,* February 3, 9.

Harris, M., and C. N. Service. 2006. "The Professor of Pop Culture Punditry." *Kingston Whig-Standard* (Ontario), July 11, 20.

Harvey, G. 2003. *Shamanism: A Reader.* New York: Routledge.

Herring, S. D. 1989. *From the Titanic to the Challenger: An Annotated Bibliography on Technological Failures of the Twentieth Century.* New York: Garland.

Lang, K., and G. E. Lang. 1953. "The Unique Perspective of Television and Its Effect: A Pilot Study." *American Sociological Review* 18 (1): 3–12.

Lemann, N. 2006. "Amateur Hour; Journalism Without Journalists." *New Yorker,* August 7, 44.

McLuhan, M. 1964. *Understanding Media: The Extensions of Man*, 2d ed. New York: New American Library.

Merton, R. K. 1968. *Social Theory and Social Structure* (enlg. ed.). New York: Free Press.

Murrow, E. R. 1958. Speech delivered to the Radio and Television News Directors Association. October 15. www.rtnda.org/pages/media_items/edward-r.-murrow-speech998.php.

*New York Times.* 1949a. "Nation Saddened by Rescue Failure." April 11.

———. 1949b. "One Little Girl" [Editorial]. April 11.

Noveck, J. 2007. "Syracuse Prof. Is Pop Culture Ambassador." *Washington Post,* May 14.

O'Connell, C. J., and A. J. Mills. 2003. "Making Sense of Bad News: The Media, Sensemaking, and Organizational Crisis." *Canadian Journal of Communication* 28 (3): 323–339.

Ochberg, F.; S. Cinti; E. Goodman; E. Kirby; H. Melia; S. Prior; and R. Taggart. 2007. "The Critical Incident Model." PowerPoint presentation delivered at the meeting for Critical Incident Analysis at John Jay College of Criminal Justice (CUNY), New York, January 4.

Oshinsky, D. M. 2005. *Polio: An American Story.* New York: Oxford University Press.

Plato. 380 B.C.E./1945. *The Republic of Plato,* trans. F. M. Cornford. New York: Oxford University Press.

Rosenberg, B., and D. M. White. 1957. *Mass Culture: The Popular Arts in America.* Glencoe, IL: Free Press.

Rossino, A. B. 1999. "Eastern Europe Through German Eyes." *History of Photography* 23 (4): 313–321.

Rowse, A. E. 1992. "How to Build Support for War." *Columbia Journalism Review* 31 (3): 28–29.

Saad, Lydia. 2010. "Congress Ranks Last in Confidence in Institutions." Gallup, July 22. www.gallup.com/poll/141512/congress-ranks-last-confidence-institutions.aspx.

Schutz, A. 1972. *The Phenomenology of the Social World.* London: Heinemann Educational.

Stelter, B. 2009. "Journalism Rules Are Bent in News Coverage from Iran." *New York Times,* June 29, 1.

Turner, G. 2004. *Understanding Celebrity.* Thousand Oaks, CA: Sage.

United Press International. 1949. "Dog Takes Up a Vigil on Girl's Front Lawn." *New York Times,* April 11.

Waldman, S. 1986. "The King of Quotes; Why the Press Is Addicted to Norman Ornstein." *Washington Monthly* 18 (December).

Walton, D. N. 1997. *Appeal to Expert Opinion: Arguments from Authority.* University Park: Pennsylvania State University Press.

CHAPTER 19

# Ethical Challenges Inherent in Critical Incident Dynamics

NED BENTON

In May 1889, 2,209 people were killed when a dam burst in the Appalachian Mountains above Johnstown, Pennsylvania.[1] The dam was owned by a sports club with a membership primarily of wealthy financiers such as Andrew Carnegie, Andrew Mellon, and Henry Clay Frick. No club member ever accepted any moral responsibility for the disaster, and no lawsuit against the club was ever successful. Studies of the incident have documented many instances in which people and organizations failed to take responsibility for the dam, failed to disclose defects in the design and maintenance of the dam, failed to plan for the foreseeable risks of disaster, shirked their professional responsibilities during the disaster, made hard choices that resulted in the rescue of some victims and death of others, and generally failed to acknowledge personal and organizational responsibility for the consequences of the disaster.

By any definition, the Johnstown tragedy was a disaster. As Erikson explains, "a disaster is an 'event' with a distinct beginning and a distinct end, and is by definition extraordinary—a freak of nature, a perversion of the natural process of life" (1976, 253). Disasters cause extensive trauma. Some disasters, such as being struck by an asteroid, might qualify as genuine "acts of God," but such "morally innocent" disasters can present moral and ethical challenges to responders and victims. Most incidents and disasters, however, have more complex causation and moral responsibility, and the processes of emergency response and aftermath management raise additional moral and ethical challenges.

After the Johnstown disaster, the editor of the *Johnstown Tribune* opined: "We think we know what struck us, and it was not the work of Providence. It was the work of man." If people and organizations acted wrongly before, during, and after the Johnstown disaster, what were the wrongs involved?

This chapter examines what people and organizations might or might not do before, during, and after an incident or disaster. Based on a review of what emergency management and rescue organizations assert to be their values about decisions and conduct related to incidents and disasters, the chapter concludes that there are four ethical imperatives, which are listed below with examples from the Johnstown incident.

- *The duty of disclosure and transparency:* When the engineers evaluated the dam above Johnstown, they were wrong not to disclose their findings widely.
- *The duty to plan and prepare:* When the sports club engineer designed "improvements" to the dam, he was wrong not to prepare a plan for rising waters in the reservoir.
- *The duty to prioritize access to services, care, or rescue:* When Clara Barton led a team of nurses and volunteers to treat the sick and injured, she was right to make hard choices in deploying scarce treatment resources to the injured.
- *The challenge of exceptional circumstances as justification for exceptional actions:* Survivors were right to take over Alma Hall and to form an ad hoc government inside because the exceptional circumstances warranted a set of actions that would normally be illegal.

While the validity of the imperatives and the examples might seem to be a matter of common sense, it is important to examine the imperatives in greater depth to assess what the precise moral determinations are, how and when they arise in incidents and disasters, and what duties and expectations are created for the victims, responders, managers, and official leaders when confronted with incidents and disasters.

## MAPPING THE MORAL TERRITORY

In his article "Ethics and Disasters: Mapping the Moral Territory," which followed the Katrina disasters in New Orleans, Msgr. Charles J. Fahey set out to map the moral and ethical territory involved in the many decisions that are made during disasters. He observed, "responsibility and accountability are in play before, during and subsequent to potentially or actually catastrophic events. Imbedded ethical issues must be identified and analyzed" (Fahey 2008, 65). Responding to his challenge, this paper seeks to explore the following questions.

- What ethical obligations arise as critical incidents take place?
- Can ethical principles be applied to reconcile competing claims?
- Are there moral values and ethical standards that should apply to all incidents?

Fahey's concept of "mapping" the moral and ethical[2] territory suggests that challenging situations, forcing hard choices between competing ethical aspirations, can emerge in distinct conceptual areas of any incident. Kirby (2012) presents a conceptual model for the analysis of incidents that is helpful in defining where ethical challenges can arise (see Figure 19.1).

Incidents can be viewed as interactions between three spheres of action: The event (blob), the political arena (box), and the bureaucracy and adhocracy (triangle). Beginning

Figure 19.1 **Critical Incident Analysis Model**

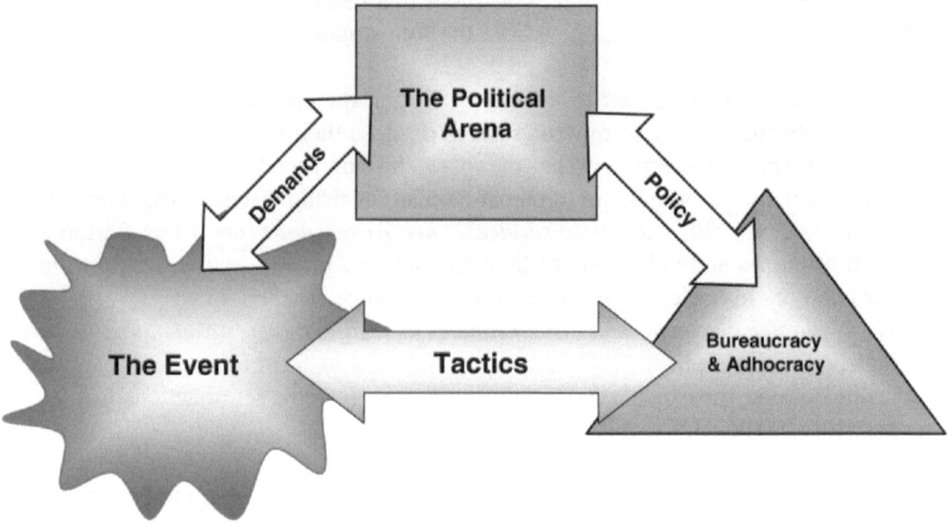


*Source:* Ochberg et al., 2007.

on the left and continuing in a clockwise flow, the event or critical incident (blob figure) creates a demand eventually received in the political arena (box figure). Regardless of whether a political leader is involved, through existing or ad hoc rules, there is policy and a delegation of authority from this zone to the next. The triangle figure represents bureaucracy and adhocracy and symbolizes the tools and the tactics that people deploy to manage the event. Directional arrows between the three components demonstrate the ebb and flow of the model. For example, the demands of participants and targets in the event zone flow back and forth from those in top levels of governance to those engulfed at the scene. The interveners in the management zone, the bodies and agencies through which elected government official work, subsequently have their own demands generally associated with policy needs. The tactics utilized by the interveners are also dynamic in nature. Tactics deployed to the event zone cause a reaction, and, based on that reaction, new tactics may be applied to modulate various aspects of the situation (Kirby 2012).

The Critical Incident Model includes a fourth domain—the media (see Figure 19.2). Kirby observes:

> The media is powerful and can influence the outcome of an incident in positive or negative ways. For example, media can be a positive force criticizing the government, exposing corruption, and educating the public. On the other hand, the media can leak information at the wrong time, push the timeframe of the incident too fast, sensationalize the incident, and distort reality. A vital part of critical incident analysis is in understanding how the media operates and how it affects an incident. (2010, 14)

Figure 19.2 **A More Detailed Version of the Critical Incident Analysis Model**

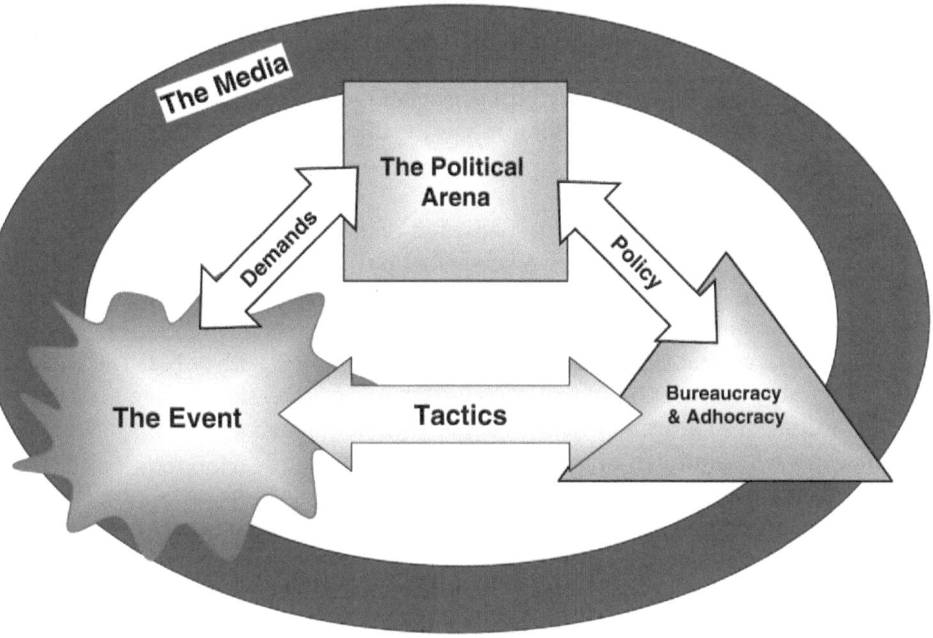


*Source:* Ochberg et al., 2007.

Kirby's model suggests that a "map" of conflicting ethical aspirations that might arise in incidents should consider what takes place inside the event itself, what the incident managers do to respond to and attempt to control the event, the decisions that political leaders and policymakers might make before, during, and after the incident, and the role of the media.

## COMMON VALUES IN INCIDENT MANAGEMENT

Ethics are prescriptions for decisions and behaviors, and they are necessarily grounded in values of right and wrong that are sometimes explicitly stated and sometimes only implied. In order to examine ethical challenges, it is first necessary to examine the values—some shared and some not—that the ethical prescriptions are intended to reinforce.

The appendix summarizes assertions of values in policy statements of national and international aid and emergency management organizations. In most cases, the values asserted by one organization are shared by the other organizations, but framed and phrased differently.

## TRANSLATING VALUES INTO CODES OF ETHICS

Is it possible to assimilate these diverse values into sets of codes of ethics for policymakers and professionals involved in emergency management and response? Has this been done already?

Because of the different purposes and circumstances of the various professions and official roles involved in emergency management and response, universal ethical standards are insufficient, and profession-specific codes of ethics are necessary and appropriate.

For example, the American Medical Association (AMA) Code of Medical Ethics Opinion 9.067, "Physician Obligation in Disaster Preparedness and Response," states the following:

> National, regional, and local responses to epidemics, terrorist attacks, and other disasters require extensive involvement of physicians. Because of their commitment to care for the sick and injured, individual physicians have an obligation to provide urgent medical care during disasters. This ethical obligation holds even in the face of greater than usual risks to their own safety, health or life. The physician workforce, however, is not an unlimited resource; therefore, when participating in disaster responses, physicians should balance immediate benefits to individual patients with ability to care for patients in the future.
>
> In preparing for epidemics, terrorist attacks, and other disasters, physicians as a profession must provide medical expertise and work with others to develop public health policies that are designed to improve the effectiveness and availability of medical care during such events. These policies must be based on sound science and respect for patients. Physicians also must advocate for and, when appropriate, participate in the conduct of ethically sound biomedical research to inform these policy decisions. Moreover, individual physicians should take appropriate advance measures to ensure their ability to provide medical services at the time of disasters, including the acquisition and maintenance of relevant knowledge.

The statement codifies expectations with respect to some of the values listed in Table 19.1, but the emphasis is on the duty to offer medical care and assistance and the factors limiting that duty, as well as the commitment to science as the basis for policy and practice.

The International Association of Fire Fighters (IAFF) has adopted a Code of Ethics (2009) that reads as follows:

> As a firefighter and member of the International Association of Fire Fighters, my fundamental duty is to serve humanity; to safeguard and preserve life and property against the elements of fire and disaster; and maintain a proficiency in the art and science of fire engineering.
>
> I will uphold the standards of my profession, continually search for new and improved methods and share my knowledge and skills with my contemporaries and descendants.
>
> I will never allow personal feelings, nor danger to self, deter me from my responsibilities as a firefighter.

Table 19.1

**Ethical Hazards: Disclosure and Transparency**

| | |
|---|---|
| Ethical Hazard | Disclosure of information can facilitate risk management and evaluation, but can violate rules and organizational expectations, and might have adverse secondary consequences. |
| Example in the Incident Scene (Blob) | Should a first responder tell an injured victim that rescue will not be attempted due to the inherent dangers associated with the rescue operation? |
| Example in Emergency Management (Triangle) | Would an unnecessary evacuation impose costs on the people and organizations involved and impair response to future evacuations? |
| | Would alerting potential victims in a large-scale emergency result in panic and impair evacuation? |
| Example in Public Policy (Box) | Should officials with awareness of potential flaws in critical infrastructure reveal the problems when official organization expectations require non-disclosure? |
| Example in Journalism | Can the reporting of an incident impair the management of the incident? |
| Possible Ethical Expectations | Disclosure and transparency might be presented as presumptive expectations. Decisions not to fully disclose relevant information or decision rationales would be appropriate only when all of the following conditions are met: (a) there is an officially documented decision not to disclose; (b) there is clear and present danger associated with disclosure; and (c) the adverse consequences are minimal. |

> I will at all times, respect the property and rights of all men and women, the laws of my community and my country, and the chosen way of life of my fellow citizens.
>
> I recognize the badge of my office as a symbol of public faith, and I accept it as a public trust to be held so long as I am true to the ethics of the fire service. I will constantly strive to achieve the objectives and ideals, dedicating myself to my chosen profession—saving of life, fire prevention and fire suppression.
>
> As a member of the International Association of Fire Fighters, I accept this self-imposed and self-enforced obligation as my responsibility.

Unlike the AMA code, the IAFF statement asserts a virtually unlimited duty to preserve life and property, without referencing the safety limitations that are commonly recognized in incident management. For example, Occupational Safety and Health Administration (OSHA) standard 1910.134(g)(4) requires that, in interior structural fires involving dangerous conditions, at least two firefighters enter and remain in visual or voice contact with each other at all times while at least two firefighters are located outside. The standard also provides that "Nothing in this section is meant to preclude firefighters from performing emergency rescue activities before an entire team has assembled." However, this language does not define a duty or responsibility to take risks regardless of the conditions.

The International Association of Emergency Managers (IACM n.d.) has adopted a code of ethics that reads as follows:

> **Preamble**
>
> Maintenance of public trust and confidence is central to the effectiveness of the emergency management profession. The members of the International Association of Emergency Managers (IAEM) adhere to the highest standards of ethical and professional conduct. This Code of Ethics and Professional Conduct for IAEM members reflects the spirit and proper conduct dictated by the conscience of society and commitment to the well-being of all. The members abide by the association's core values of respect, commitment, and professionalism.
>
> **Values**
>
> Respect: Respect for supervising officials, colleagues, associates, and most importantly, for the people we serve is the standard for IAEM members. We comply with all laws and regulations applicable to our purpose and position, and responsibly and impartially apply them to all concerned. We respect fiscal resources by evaluating organizational decisions to provide the best service or product at a minimal cost without sacrificing quality.
>
> Commitment: IAEM members commit themselves to promoting decisions that engender trust and those we serve. We commit to continuous improvement by fairly administering the affairs of our positions, by fostering honest and trustworthy relationships, and by striving for impeccable accuracy and clarity in what we say or write. We commit to enhancing stewardship of resources and the caliber of service we deliver while striving to improve the quality of life in the community we serve.
>
> Professionalism: IAEM is an organization that actively promotes professionalism to ensure public confidence in emergency management. Our reputations are built on the faithful discharge of our duties. Our professionalism is founded on education, safety, and protection of life and property.

The code emphasizes the values of professionalism, efficiency, effectiveness, and compliance with laws and regulations, but does not address hard questions implied by many of the values presented in Table 19.1. For example, while the AMA and others struggle with the duty to provide service in settings where resources are insufficient to meet all urgent needs, the IAEM code simply asserts: "We respect fiscal resources by evaluating organizational decisions to provide the best service or product at a minimal cost without sacrificing quality." The IAEM code might also be criticized as constituting more a statement of values rather than prescriptive ethics; and as drifting between aspirational statements (striving for impeccable accuracy) and flattering but unsupported conclusions (compliance with all laws and regulations) about the merits of the organization and its membership

The Federal Emergency Management Administration (FEMA) published a report setting out "Principles of Emergency Management"; it recognized that "While no single code of ethics has yet been agreed upon for the profession, the Code of Ethics of the International Association of Emergency Managers, with its emphasis on respect, commitment and professionalism, is generally accepted as the standard for emergency managers" (2007, 8).

The Society of Professional Journalists has developed, through a broadly consultative process with its membership, a Code of Ethics (Society of Professional Journalists 1996). The code set out values and then lists a set of expectations for journalists. Each expectation completes a sentence beginning "Journalists should . . ."

## MAPPING THE ETHICAL TERRITORY

If there is no agreed-upon code of ethics for emergency management, how would it be possible to map potential areas where official responsibilities and ethical aspirations might come into conflict? In a field with a more developed set of professional expectations, it would be possible to juxtapose common expectations with common situations and then comment on the conflicts that might arise.

However, since there are no officially recognized common ethical expectations, an alternative approach is to juxtapose common situations with commonly asserted values. The analytical process can define areas where ethical expectations might need to be clarified or codified.

At this point in a scholarly paper in applied ethics, it would be appropriate to delve into the methods of such an inquiry.[3] However, the commonly asserted values in the Appendix almost universally limit the selection of approaches to those that are based on objective evidence and articulated in transparent and equitable rules. This almost universal aspiration limits us to a form of rule utilitarianism.

> According to rule-utilitarianism, a behavioral code or rule is morally right if the consequences of adopting that rule are more favorable than unfavorable to everyone. Unlike act utilitarianism, which weighs the consequences of each particular action, rule-utilitarianism offers a litmus test only for the morality of moral rules, such as "stealing is wrong." Adopting a rule against theft clearly has more favorable consequences than unfavorable consequences for everyone. The same is true for moral rules against lying or murdering. Rule-utilitarianism, then, offers a three-tiered method for judging conduct. A particular action, such as stealing my neighbor's car, is judged wrong since it violates a moral rule against theft. In turn, the rule against theft is morally binding because adopting this rule produces favorable consequences for everyone. (Feiser 2009)

Another approach that might apply is casuistry or "case-based" analysis. Ethical analysis begins with a thorough analysis of the facts of the case involved, without reference to theoretical principles. Analysts with different moral or ethical perspectives might reach agreement about ethical decisions and actions in a particular case because they are persuaded by compelling facts even though their ethical reasoning might differ. An advantage of this approach for the analysis of incidents and emergencies is that disagreements about the ethics of decisions and actions are often based on disagreements about the facts of the case involved, and after the facts are clarified, consensus develops about ethical decisions and actions. However, the approach has limited utility for the development of codes of ethics because the method rejects the idea of abstract codifications of expectations.

In a report about ethical considerations in responding to pandemics, the World Health Organization summarized the complex relationship between ethical obligations and official actions during emergencies.

> In balancing competing interests and values, policy makers can draw on ethical principles as tools for weighing conflicting claims and for reaching appropriate decisions. Ethics does not provide a prescribed set of policies; rather, ethical considerations will be shaped by the local context and cultural values. The principles of equity, utility/efficiency, liberty, reciprocity, and solidarity are especially helpful in the context of influenza pandemic preparedness planning. Although these principles often give rise to competing claims, they provide a framework for policymakers to assess and balance the range of interests that follow from them. All ethical deliberations must take place within the context of the principles of human rights, and all policies must be consistent with applicable human rights laws. (2007, 3)

The International Federation of Red Cross and Red Crescent Societies identified five "moral hazards" in providing humanitarian aid during disasters:

> complicity in abuses (feeding refugees may help armed factions regroup); legitimizing violations (prioritizing aid over investigating rights violations may encourage a climate of impunity); aid's negative effect (too much aid may undermine local markets or depopulate areas); targeting and triage (the most needy may be left to die if others can be more effectively helped); advocacy or access (condemning abuses can mean agencies are expelled). (IFRC 2003)

Because incidents usually involve government and organizational responses, it is necessary to consider the ethical issues from more than a personal perspective. The ethical challenges that arise involve decisions and actions of people who must consider their own personal values and morals while also considering the ethical standards and policies of the organizations with which or for which they work. Burke (1986) has set out a framework for the understanding of an analysis of ethical obligations in democratic societies. Ethical responsibilities can be based on external sources such as laws, religion-based rules, professional codes of ethics, or agency regulations. Ethical responsibilities can also be based on personal values such as dedication to the common good, religious norms and values, internalized norms of a profession, or values such as service to humanity or respect for the environment. Burke finds flaws in both frameworks as guides to ethical decisions by public officials at all levels of public service. He proposes that externally imposed and personally derived values be reconciled with a commitment to the ideals of democratic governance and the goals of legitimately adopted public policies. The proper actions to be taken when an official is faced with potentially conflicting duties and ethical obligations are discovered in the process of reconciliation. Burke defines the steps that an official should take when official duties result in unethical or immoral actions or consequences.

## APPLYING THE CRITICAL INCIDENT ANALYSIS MODEL TO MAP ETHICAL TERRITORY

There are six relatively distinct areas of potential conflict between the duties of victims, responders, incident managers, and public officials and their ethical obligations. These can be examined in each of the four domains of incident dynamics.

## THE DUTY OF DISCLOSURE AND TRANSPARENCY

Transparency is a value asserted in several of the statements in Table 19.1.

> The reasoning behind choices made is fully articulated (in language appropriate to particular audiences) and the values and principles justifying those decisions are clearly identified and open for examination. This commitment to clarity and openness, which is based on a deep respect for all individuals and communities involved, exists in balance with the understanding that those with the authority and responsibility of making decisions must often make decisions in a timely manner. (Kinlaw and Levine 2007, 3)

> Public engagement and involvement of relevant stakeholders should be part of all aspects of planning. Policy decisions and their justifications should be publicized and open to public scrutiny. In order for public engagement in preparedness planning to be meaningful, effective modes of communicating with and educating the public about the issues involved are essential. The principles of outbreak communication are: trust; transparency; communicating to the public early, dialogue with the public; and planning. Advance planning will allow the development of strategies that will reach the entire population and that are linguistically and culturally appropriate. (World Health Organization 2007, 4)

However, aspirations for transparency and full disclosure are not always achieved in emergencies and incidents. Table 19.1 sets out some examples of potential conflicts between official duties and ethical aspirations.

During an incident, managers might withhold information from victims about dire circumstances when informing them might make the situation worse and rescue less possible. These choices include situations in which:

- Panic might impair evacuation or rescue.
- Fully informed victims might be less cooperative than partially informed victims.
- The victim is perceived to have impaired judgment.
- Information is incomplete, and the manager is concerned about appearing to provide inaccurate information.

Helsloot and Ruitenberg (2004) wrote a comprehensive review of citizen response during disasters, and they generally concluded that victims do not panic during disasters and that providing information has a positive influence on behavior during emergencies. Thus, while there might be a theoretical situation in which withholding information might avoid panic, there is little evidence that these situations arise with any frequency.

Before a disaster takes place, public officials, contractors, and technical experts can be aware of flaws in critical infrastructure or critical procedures and policies—flaws that could have catastrophic consequences. These choices include a range of technical problems:

- failures in design of systems
- failures in manufacture or installation of systems
- failures to maintain systems in reliable working order
- failures to develop appropriate emergency plans
- failures to realistically assess risks.

Ethical issues involve conflicting perceptions of duty and responsibility with respect to supervisors, organizations, government, society at large, individual and collective rights, and moral obligations.

A well-known example involves the failure of the levees during the flooding after Hurricane Katrina hit New Orleans (Seed et al. 2005).

Burke extensively explores the ethical issues involved in the unauthorized disclosure of private, confidential, restricted, or classified information (1986, 189–196). His analysis involved the balancing of the positive value of information to policy formulation and implementation, against the potential harm such as a risk to national security or impairment to organization or official reputation.

Kremer (2006) examined the ethical obligations of architects and engineers to reveal design flaws before or after an incident. He pointed out that while there is little official guidance, there is legal and practical benefit in timely and truthful disclosure, and the architect or engineer rarely has a legal or professional obligation to assist a client in concealing a dangerous or illegal situation.

Codes of ethics might address disclose and transparency using the following standard:

- Disclosure and transparency should be the expected policy and practice.
- Decisions not to fully disclose relevant information or decision rationales would be appropriate only when all of the following conditions are met: (a) the decision not

to disclose is officially documented; (b) there is clear and present danger associated with disclosure; and (c) the adverse consequences are minimal.

## THE DUTY TO PLAN AND PREPARE

Individuals, organizations, and communities should evaluate, prepare for, and mitigate foreseeable risks. The following are examples of statements reflecting this value:

> Public health officials have a responsibility to maximize preparedness in order to minimize the need to make allocation decisions later. Preparedness also includes determining and articulating what rules will govern public health decision making in advance of the time that decision making must commence. Though every specific choice or contingency cannot be foreseen, comprehensible foundational guidelines and procedural action plans provide coherence and direction and build trust. (Kinlaw and Levine 2007, 3)

> Emergency managers anticipate future disasters and take preventive and preparatory measures to build disaster-resistant and disaster-resilient communities. (FEMA 2007)

However, aspirations for comprehensive planning and preparation are not always achieved in emergencies and incidents. Table 19.2 sets out some examples of potential conflicts between official duties and ethical aspirations.

Individuals, organizations, and communities should evaluate, prepare for, and mitigate foreseeable risks. The failure to assess and manage risk has consequences for the individuals and entities who fail to undertake risk management, but the consequences can also affect third parties. If a homeowner is reckless with fire and fire protection, the homeowner's house might burn down, but the fire might also spread to adjacent homes. If a community fails to provide adequate firefighting services, is it right for the community to benefit from firefighting provided through mutual aid by other adjacent communities? If a citizen fails to heed an evacuation warning before a hurricane, do first responders have an obligation to risk their lives to rescue the person after the hurricane arrives?

All these examples illustrate the concept of moral hazard, which arises when a person, organization, or community fails to take responsibility for foreseeable risks, thereby creating additional risks for third parties. The concept of moral hazard provides the rationale for the ethical duty of people, organizations, and communities to assess and mitigate risk through planning and preparation.

The same concept also informs the analysis of the ethical duties of emergency managers and emergency service providers when the people needing rescue bear significant responsibility for the circumstances. Public officials and incident managers might make choices to withhold services based on moral hazard considerations. These choices might include:

- situations in which victims ignored warnings to evacuate, in particular, warnings notifying the victims that services would not be provided.

Table 19.2

**Ethical Hazards: Planning and Preparation**

| | |
|---|---|
| Ethical Hazard | The assessment and management of risk has costs and consequences for the individuals and entities involved, but the consequences can also adversely affect third parties. |
| Example in the Incident Scene (Blob) | Should emergency managers attempt to rescue people who made deliberate decisions to ignore instructions to evacuate in advance of a disaster? Should resources be expended and lives risked? |
| Example in Emergency Management (Triangle) | Should emergency managers limit the rescue of victims who ignored planning instructions? |
| Example in Public Policy (Box) | How do public officials balance the costs of planning and preparedness against other pressing needs? |
| | Can risk management encourage potential victims to inappropriately depend on public protection? |
| Example in Journalism | Should journalists take risks in covering incidents that might require dangerous rescue operations? |
| Possible Ethical Expectations | Risk assessment and management should be a fundamental ethical obligation for public officials and emergency managers. |

- situations in which an incident is framed with a narrative that assigns responsibility for the consequences of an incident to a set of victims, so that they are viewed as undeserving of services.
- situations in which, in planning to mitigate an emergency, services like flood insurance are withheld or priced expensively to deter individuals or organizations from making decisions that create risk.

In an article about humanitarian intervention, Kuperman (2008) examines government's responsibilities and moral hazard. He posits several cases in which local uprisings took place in revolt against government oppression, in which the people engaged in the uprising might have expected international intervention to support and protect them. In the absence of the belief that intervention would take place, they might not have initiated the uprising. He concludes that governments should consider whether claimed intentions to intervene to protect oppressed minorities cause more harm than good, in cases when the expectations of interventions motivate actions that would otherwise be reckless, dangerous, and counterproductive.

## THE DUTY TO PRIORITIZE ACCESS TO SERVICES, CARE, OR RESCUE

During disasters and large-scale emergencies, public officials, emergency managers, first responders, health-care providers, and victims are frequently faced with limited resources and competing needs. Therefore, they must make decisions to prioritize care or rescue for one person or set of persons over others. Situations can include rescues, medical triage, evacuation, and, subsequent to an incident, the prioritization of recovery and reconstruction. In the Katrina incident, the hard choices even included decisions about the rescue of pets.

However, aspirations for comprehensive planning and preparation are not always achieved in emergencies and incidents. Table 19.3 sets out some examples of potential conflicts between official duties and ethical aspirations.

A well-known example involves treatment and rescue decisions during the Katrina disaster at St. Rita's Hospital in New Orleans (Fink 2009). When an evacuation of the flooded hospital was arranged after some delays, medical personnel were faced with decisions about whether and how to treat or rescue patients on a long-term-care floor who were extremely ill and were assessed as being unlikely to survive evacuation. Should the medical personnel have arranged for evacuation anyway? If evacuation of some of the patients was not possible, and ongoing care was also not possible due to the evacuation of the staff, what should be done?

The World Health Organization summarized the following principles for the allocation of scarce medical resources during a pandemic.

1. During the first phase, everyone who needs treatment will receive it. This phase will continue until the number of those requiring treatment exceeds the capacity of the enhanced treatment facilities. In this phase treatment will be administered to individuals on a "first come, first served" basis or to those who are already being treated for another illness.
2. The second phase begins when it is no longer possible to treat everyone because therapeutic capacity is exhausted and some have to be turned away. In this phase, scarce therapeutic resources will be reserved for those whose conditions are most threatening.
3. Finally, the third phase corresponds to the triage used in war or disaster situations. From the outset of this phase scarce resources should be reserved for patients with life-threatening conditions. When all those with life-threatening conditions can no longer be treated, priority will be given to those who are expected to have the best chance of survival as a result of treatment. Conversely, treatment in this phase will, if possible, be withheld only from those who are unlikely to benefit from it. Individuals with a poor prognosis will be given palliative treatment only in this phase. (2007, 6)

While these principles apply to emergency health care, the principles could also be applied to other types of emergency services by other providers. Two basic rationales underlie the principles: utility and equity.

Eastman, Philips, and Rhodes examine triage models and justifications. They observe: "There is no 'ethically right' method of triaging. However, it is important that any method adopted is not only explicit and transparent, both in its technical and ethical underpin-

Table 19.3

**Ethical Hazards: Prioritization of Access to Services, Care, or Rescue**

| | |
|---|---|
| Ethical Hazard | The neediest persons might be left to die if others can be helped more effectively. |
| Example in the Incident Scene (Blob) | How should first responders make decisions to prioritize care or rescue for one person or set of persons over others? |
| Example in Emergency Management (Triangle) | How should emergency managers allocate resources to prioritize care or rescue for one person or set of persons over others? |
| Example in Public Policy (Box) | How do public policymakers balance the need to respond to assessments of future risk with immediate needs? |
| Example in Journalism | How do journalists reconcile their duty to report the incident, and the humanitarian ethic to alleviate the human suffering that is being reported? |
| Possible Ethical Expectations | Risk assessment and management should be a fundamental ethical obligation for public officials and emergency managers. |

nings, but is also capable of ethical justification" (2010, 1077). They note the potential ethical distinct between withholding and withdrawing treatment:

> The legal basis for withdrawal of treatment is that of "futility," such that it can be adjudged to be in the patient's "best interest" that withdrawal should occur; the test is therefore applied solely in relation to the individual patient under consideration. As things stand, the law has not sanctioned "rationing" of health care resources beyond its relevance to the application of treatment; that is, the English courts have not applied the principle to withdrawal of care from one patient in the interest of another. (2010, 1080)

Thus, in a conventional health-care context, treatment might be withdrawn because further treatment would be futile based on the progression of the patient's condition. But in an emergency or disaster situation, treatment might be withheld or withdrawn because, as the WHO policy states, "therapeutic capacity is exhausted and some have to be turned away" (2007, 6). If treatment is not provided because of a lack of resources, responsibility shifts, in some degree, from the patient's condition and the cause of that condition to the public officials and emergency managers whose actions or inactions may have resulted in the shortage of resources.

## EXCEPTIONAL CIRCUMSTANCES AS JUSTIFICATION FOR EXCEPTIONAL ACTIONS

During incidents, officials, managers, and victims in critical incidents might decide to take and use property that does not belong to them or take actions that would otherwise be illegal or unethical because they view the incident as creating an exceptional situation. The following are some examples from the appendix illustrating the concept of exceptional circumstances:

> During the course of a pandemic, the functioning of society may be threatened. Our moral tradition embodies an understanding that it may be ethically acceptable (or perhaps even ethically mandatory) to suspend some (but not all) ordinary moral rules in such circumstances. For example, limits on individual freedom or choice may be necessary to protect individuals as well as entire communities during pandemic influenza. Yet, individual liberty should be restricted with great care and only when alternative approaches to realizing the goal of weathering the pandemic are not likely to be effective. Suspensions of ordinary moral rules should be anticipated and the conditions calling for such suspensions should be specified. (Kinlaw and Levine 2007, 4)

> Preparedness planning for an influenza pandemic involves balancing potentially conflicting individual and community interests. In emergency situations, the enjoyment of individual human rights and civil liberties may have to be limited in the public interest. However, efforts to protect individual rights should be part of any policy. Measures that limit individual rights and civil liberties must be necessary, reasonable, proportional, equitable, non-discriminatory, and in full compliance with national and international laws. (World Health Organization 2007, 3)

However, reconciling ordinary laws and policies with the duty to save lives and preserve property cannot always be achieved in emergencies and incidents. Table 19.4 sets out some examples of potential conflicts when exceptions to ordinary laws and policies have to be considered.

With respect to extraordinary measure to be taken by governments and official actors, Sorell observed:

> The application of exacting moral standards to public emergencies is in some ways as inappropriate as the application of such standards to private emergencies. Moralizing tends to assume normality, or abnormality that can be correlated with the normal by a standard of proportionality. Neither normality nor an ability to proportion powers to an abnormal situation need be present when decent governments confront emergencies. (2002, 32)

Sorell recognized that emergencies may require exceptional responses, but that they need to be proportional to the situation.

Greenawalt examined the jurisprudence of justifications and excuses in the formulation of penal codes. While penal codes assume wrongdoing is to be sanctioned or punished, and exceptional interventions in emergencies are not inherently wrong, the analysis is useful in assessing the "rightness" of exceptional interventions. He offers the following definitions:

Table 19.4

**Ethical Hazards: Exceptional Circumstances**

| | |
|---|---|
| Ethical Hazard | Suspension of legal and moral expectations may be justified or excused by higher moral imperatives such as saving lives. |
| Example in the Incident Scene (Blob) | Should victims steal from stores in order to feed their families? |
| | Should first responders use force or restraint without following the usual processes of law and policy? |
| Example in Emergency Management (Triangle) | Should emergency managers authorize the confiscation of private goods or issue "shoot to kill" orders? |
| Example in Public Policy (Box) | Should elected leaders suspend civil liberties to preserve life and property? |
| Example in Journalism | How do journalists report on looting, when they might encourage more people to engage in looting? |
| Possible Ethical Expectations | Official measures that limit individual rights and civil liberties must be necessary, reasonable, proportional, equitable, nondiscriminatory, and consistent with legal authority. |

- A claim of justification is a claim that one's act was warranted, that what one did was right, or was within a legally permissible range of behavior. Thus, one may break the speed limit if one is driving a seriously injured person to the hospital. As far as the law is concerned, someone who offers a successful justification has acted appropriately.
- Someone who claims an excuse does not deny that he acted wrongfully. Rather, he urges that special conditions relieve him of the responsibility or blame that most people committing such acts would carry. A claim that someone killed because he was insane is an excuse. (1998, 24)

A generalization is that a claim of justification would apply to everyone faced with the same circumstances, but a claim of excuse would be based on the specific attributes of the person involved. The concept of "duress"—the immediate and inescapable threat of serious harm—is also legally important, particularly to a claim of justification when the law implicitly or explicitly recognizes it.

In an incident, individuals who decide to break the law are entitled to a defense based on concepts of justification, excuse, and duress. Authorities should recognize when emergency circumstances constitute justification for exceptional conduct for everyone involved.

Official measures that limit individual rights and civil liberties must be necessary, reasonable, proportional, equitable, nondiscriminatory, and consistent with legal authority.

## VALUES, ETHICS, AND CRITICAL INCIDENTS

Critical incidents are a subset of all emergencies, incidents, and disasters. Kirby observes:

> Unlike individual incidents, critical incidents in a social context are large-scale events that shock communities or nations and may draw global attention as they unfold. They are usually traumatic and tragic, causing death or destruction. Because they are unexpected and may involve new and unique methods or targets, they shake individual's assumptions and can generate fear or wonder. (2010, 5)

Ochberg et al. offered the following definition of critical incidents:

> A critical incident is a relatively brief occurrence involving injury, loss, conflict, discovery or change of significant proportion, usually unscripted and unanticipated, with the potential to alter existing societal norms. Critical incidents are usually traumatic, threatening the bonds of trust that bind communities, but may be positive, initiating historic consequents. (Schwester et al, Chapter 3)

When an incident evolves in a way that "threatens the bonds of trust that bind communities," sometimes the threat to trust is based on an evil act like a shooting of innocent people. Such acts are inherently immoral, and they violate common ethical principles.

However, the threat to trust is sometimes the result of unethical acts by public officials, emergency managers, or participants in the incident. Public trust is violated when public officials are revealed to have misled or lied to victims or when officials are revealed to have been reckless in fulfilling their duties to protect people from known risks. Public trust is violated when emergency managers are perceived to prioritize rescue or treatment for the wealthy or privileged over the poor and vulnerable. Public trust is violated when victims of incidents, taking exceptional actions in attempting to protect or rescue themselves or others, are inaccurately presented in the media as looters, murderers, or criminals.

The ethics of decisions and actions by everyone involved in incidents can be central to the escalation of an ordinary incident into a critical incident. Floods, for example, are typically not considered critical incidents, because flooding is unfortunately part of the natural order of many environments. But a flood like the Johnstown flood of 1889 becomes a critical incident precisely because of the unethical conduct of the members and managers of the sports club and the heroically ethical conduct of many of the victims and first responders as the disaster unfolded.

The example illustrates how concepts of ethics and values are central to making the distinction between a routine disaster or crime and a critical incident and are therefore central to critical incident analysis.

# APPENDIX
# VALUES IN EMERGENCY MANAGEMENT

| Value | Definition | Source |
|---|---|---|
| Humanitarianism | To prevent and alleviate human suffering wherever it may be found . . . to protect life and health and to ensure respect for the human being. | International Federation of Red Cross and Red Crescent Societies (IFRC), 2003 |
| Neutrality and Independence | Through disinterest in the politics of war or disaster, humanitarians hope to gain access to all those in need. | International Federation of Red Cross and Red Crescent Societies (IFRC), 2003 |
| Goal Identification | Individuals most at risk of experiencing the serious negative health consequences of hospitalization or death if infected are given priority. | Kinlaw and Levine, Centers for Disease Control, 2007, 2 |
| Preservation of Societal Functioning | Those individuals who are essential to the provision of health care, public safety and the functioning of key aspects of society should receive priority in the distribution of vaccine, antivirals and other scarce resources. | Kinlaw and Levine, Centers for Disease Control, 2007, 3 |
| Transparency | The reasoning behind choices made is fully articulated (in language appropriate to particular audiences), and the values and principles justifying those decisions are clearly identified and open for examination. This commitment to clarity and openness, which is based on a deep respect for all individuals and communities involved, exists in balance with the understanding that those with the authority and responsibility of making decisions must often make decisions in a timely manner. | Kinlaw and Levine, Centers for Disease Control, 2007, 3 |
| Public Engagement and Involvement | Public engagement and involvement are essential to build public will and trust and should be evidenced throughout the planning and response process. The public is seen as a partner with other experts, with particular attention to vulnerable or historically marginalized members of society. Clear mechanisms must be created for public involvement in planning and for feedback throughout the process. | Kinlaw and Levine, Centers for Disease Control, 2007, 3 |
| Maximize Preparedness | Public health officials have a responsibility to maximize preparedness in order to minimize the need to make allocation decisions later. Preparedness also includes determining and articulating what rules will govern public health decision making in advance of the time that decision making must commence. Though every specific choice or contingency cannot be foreseen, comprehensible foundational guidelines and procedural action plans provide coherence and direction and build trust. | Kinlaw and Levine, Centers for Disease Control, 2007, 3 |
| Scientific Evidence | The current knowledge basis should serve as a foundation for ethical guidelines, and a commitment to ongoing scientific and ethical evaluation of interventions should be made. | Kinlaw and Levine, Centers for Disease Control, 2007, 4 |

*(continued)*

Appendix: Values in Emergency Management *(continued)*

| Value | Definition | Source |
|---|---|---|
| Global Partnership | This recognition is not based simply on the potential of global involvement to benefit U.S. citizens (an "instrumental" reason), but on a deep recognition of the common good and our interdependence globally. | Kinlaw and Levine, Centers for Disease Control, 2007, 4 |
| Balancing of Individual Liberty and Community Interests | During the course of a pandemic, the functioning of society may be threatened. Our moral tradition embodies an understanding that it may be ethically acceptable (or perhaps even ethically mandatory) to suspend some (but not all) ordinary moral rules in such circumstances. For example, limits on individual freedom or choice may be necessary to protect individuals as well as entire communities during pandemic influenza. Yet, individual liberty should be restricted with great care and only when alternative approaches to realizing the goal of weathering the pandemic are not likely to be effective. Suspensions of ordinary moral rules should be anticipated and the conditions calling for such suspensions should be specified. | Kinlaw and Levine, Centers for Disease Control, 2007, 4 |
| Diversity | Given numerous historical examples of abuse of individuals, particularly those who are considered vulnerable, in the name of the public good . . . public health officials must adequately acknowledge and respond to strong currents of suspicion and distrust of the healthcare system. This acknowledgement is, of course, a part of a much larger healthcare dialogue. Addressing this distrust should be a strong and enduring commitment and not viewed as merely instrumental to inducing individuals to comply with recommendations. Diverse public voices should be involved in determining the need for restrictions and in articulating the ethical justification for these restrictions. | Kinlaw and Levine, Centers for Disease Control, 2007, 5 |
| Procedural Fairness and Justice | We recommend an approach to justice that focuses on the procedures to be followed with the hope that good procedures will lead to fair outcomes. Following are the elements of an ideal procedural justice approach:<br>• Consistency in applying standards across people and time (treating like cases alike).<br>• Decision makers who are impartial and neutral.<br>• Ensuring that those affected by the decisions have a voice in decision making and agree in advance to the proposed process.<br>• Treating those affected with dignity and respect.<br>• Ensuring that decisions are adequately reasoned and based on accurate information.<br>• Communications and processes that are clear, transparent and without hidden agendas.<br>Inclusion of processes to revise or correct approaches to address new information, including a process for appeals and procedures that are sustainable and enforceable. | Kinlaw and Levine, Centers for Disease Control, 2007, 5 |

| Value | Definition | Source |
|---|---|---|
| Balancing Rights, Interests and Values | Preparedness planning for an influenza pandemic involves balancing potentially conflicting individual and community interests. In emergency situations, the enjoyment of individual human rights and civil liberties may have to be limited in the public interest. However, efforts to protect individual rights should be part of any policy. Measures that limit individual rights and civil liberties must be necessary, reasonable, proportional, equitable, non-discriminatory, and in full compliance with national and international laws. | World Health Organization (2007, 3) |
| The Evidence Base for Public Health Measures | Public health measures that involve significant costs and/or burdens should be reserved for situations where they can be reasonably expected to make a difference to the consequences of a pandemic. Preparedness plans should be flexible enough to allow timely adaptation as new evidence about the nature of the disease arises. | World Health Organization (2007, 3) |
| Transparency, Public Engagement and Social Mobilization | Public engagement and involvement of relevant stakeholders should be part of all aspects of planning. Policy decisions and their justifications should be publicized and open to public scrutiny. | World Health Organization (2007, 3) |
| Information, Education, and Communication | In order for public engagement in preparedness planning to be meaningful, effective modes of communicating with and educating the public about the issues involved are essential. The principles of outbreak communication are: trust; transparency; communicating to the public early, dialogue with the public; and planning. Advance planning will allow the development of strategies that will reach the entire population and that are linguistically and culturally appropriate. | World Health Organization (2007, 4) |
| Resource Constraints | While all countries must make reasonable efforts to prepare for an influenza pandemic, differences in access to resources mean that what is reasonable for one country may not be reasonable for another. In developing countries, limited resources and immediate health-care needs may make it difficult to develop and implement comprehensive plan. . . . However, at some point countries will have to make difficult decisions about the relative weight to be given to pandemic preparedness compared to other important public health priorities, such as HIV and malaria. These decisions about resource allocation should be informed by a process of public engagement, and their rationale should be clearly communicated to the public. | World Health Organization (2007, 4) |
| Seek Truth and Report It | Journalists should be honest, fair and courageous in gathering, reporting and interpreting information. | (Society of Professional Journalists, 1996) |
| Minimize Harm | Ethical journalists treat sources, subjects and colleagues as human beings deserving of respect. | (Society of Professional Journalists, 1996) |
| Act Independently | Journalists should be free of obligation to any interest other than the public's right to know. | (Society of Professional Journalists, 1996) |
| Be Accountable | Journalists are accountable to their readers, listeners, viewers and each other. | (Society of Professional Journalists, 1996) |

*(continued)*

Appendix: Values in Emergency Management *(continued)*

| Value | Definition | Source |
| --- | --- | --- |
| Respect | Respect for supervising officials, colleagues, associates, and most importantly, for the people we serve is the standard for IAEM members. We comply with all laws and regulations applicable to our purpose and position, and responsibly and impartially apply them to all concerned. We respect fiscal resources by evaluating organizational decisions to provide the best service or product at a minimal cost without sacrificing quality. | International Association of Emergency Managers (2010) |
| Commitment | IAEM members commit themselves to promoting decisions that engender trust and those we serve. We commit to continuous improvement by fairly administering the affairs of our positions, by fostering honest and trustworthy relationships, and by striving for impeccable accuracy and clarity in what we say or write. We commit to enhancing stewardship of resources and the caliber of service we deliver while striving to improve the quality of life in the community we serve. | International Association of Emergency Managers (2010) |
| Professionalism | IAEM is an organization that actively promotes professionalism to ensure public confidence in emergency management. Our reputations are built on the faithful discharge of our duties. Our professionalism is founded on education, safety, and protection of life and property. | International Association of Emergency Managers (2010) |
| Comprehensive | Emergency managers consider and take into account all hazards, all phases, all stakeholders and all impacts relevant to disasters. | FEMA (2007) |
| Progressive | Emergency managers anticipate future disasters and take preventive and preparatory measures to build disaster-resistant and disaster-resilient communities. | FEMA (2007) |
| Risk-driven | Emergency managers use sound risk management principles (hazard identification, risk analysis, and impact analysis) in assigning priorities and resources. | FEMA (2007) |
| Integrated | Emergency managers ensure unity of effort among all levels of government and all elements of a community. | FEMA (2007) |
| Collaborative | Emergency managers create and sustain broad and sincere relationships among individuals and organizations to encourage trust, advocate a team atmosphere, build consensus, and facilitate communication. | FEMA (2007) |
| Coordinated | Emergency managers synchronize the activities of all relevant stakeholders to achieve a common purpose. | FEMA (2007) |
| Flexible | Emergency managers use creative and innovative approaches in solving disaster challenges. | FEMA (2007) |
| Professional | Emergency managers value a science and knowledge-based approach based on education, training, experience, ethical practice, public stewardship and continuous improvement. | FEMA (2007) |

## NOTES

1. The incident is the subject of many reports and books, including *The Johnstown Flood* (McCullough, 1968).

2. In this paper, the terms "values, morals, and ethics" are frequently used. Values are core ideals that motivate decisions and behaviors. Morals are individual and social beliefs about right and wrong. Ethics are prescriptive expectations, often codified, about what should and should not be done, and professional ethics define such expectations and duties for members of a profession.

3. See Zack (2006) for a concise academic article about the role of philosophical inquiry in emergency management. *The Internet Encyclopedia of Philosophy,* which is a peer-reviewed academic resource, presents a concise summary of the conceptual tools involved in such an inquiry (Feiser 2009).

## REFERENCES

American Medical Association (AMA) Code of Ethics Opinion 9.067. 2002. "Physician Obligation in Diaster Preparedness and Response." December.

Burke, J. P. 1986. *Bureaucratic Responsibility.* Baltimore: John Hopkins University Press.

Eastman, N.; B. Philips; and A. Rhodes. 2010. "Triaging for Adult Critical Care in the Event of Overwhelming Need." *Intensive Care Medicine* 36 (6): 1076–82.

Erikson, K. T. 1976. *Everything in Its Path: Destruction of Community in the Buffalo Creek Flood.* New York: Simon & Schuster.

Fahey, C. J. 2008. "Ethics and Disasters: Mapping the Moral Territory." *Generations* 31 (4): 61–65.

Federal Emergency Management Administration (FEMA). 2007. "Principles of Emergency Management Supplement." Report, September 11. Washington, DC.

Feiser, J. 2009. "Ethics." In *Internet Encyclopedia of Philosophy,* May 10. www.iep.utm.edu/ethics/ (accessed June 14, 2010).

Fink, S. 2009. "The Deadly Choices at Memorial." *New York Times Sunday Magazine,* August 30, MM28.

Greenawalt, K. 1998. "Justifications, Excuses, and a Model Penal Code for Democratic Societies." *Criminal Justice Ethics* 17 (1): 14–29.

Helsloot, I., and A. Ruitenberg. 2004. "Citizen Response to Disasters: A Survey of Literature and Some Practical Implications." *Journal of Contingencies and Crisis Management* 12 (3): 98–111.

International Association of Emergency Managers (IAEM). n.d. "IAEM Code of Ethics and Professional Conduct." www.iaem.com/about/IAEMCodeofEthics.htm (accessed June 14, 2010).

International Association of Fire Fighters. 2009. "Firefighter Code of Ethics." Associated Fire Fighters of Illinois, January 30. www.affi-iaff.org/?zone=/unionactive/view_article.cfm&HomeID=64489/ (accessed June 14, 2010).

International Federation of Red Cross and Red Crescent Societies (IFRC). 2003. "World Disasters Report 2003: Focuses on Ethics in Aid." http://www1.reliefweb.int/rw/lib.nsf/db900SID/LGEL-5PKCUR.

Kinlaw, K., and R. Levine. 2007. "Ethical guidelines in pandemic influenza." Centers for Disease Control and Prevention, Washington, DC.

Kirby, E. A. 2010. "A Conceptual Model for Critical Incident Analysis." *Journal of Critical Incident Analysis,* 1(1), 3–16.

Kremer, E. 2006. "When Buildings Fail: Ethics for the Worst-Case Scenario." *Architectural Record* 194 (10): 1–3.

Kuperman, A. J. 2008. "The Moral Hazard of Humanitarian Intervention: Lessons from the Balkans." *International Studies Quarterly* 52: 49–80.

McCullough, D. G. 1968. *The Johnstown Flood.* New York: Simon & Schuster.

Ochberg, F.; S. Cinti; E. Goodman; E. Kirby; H. Melia; S. Prior; and R. Taggart. 2007. "The critical incident model." PowerPoint presentation at the meeting for Critical Incident Analysis at John Jay College of Criminal Justice (CUNY), New York, January 4.

Seed, R. B., et al. 2005. "Preliminary Report on the Performance of the New Orleans Levee Systems in Hurricane Katrina on August 29, 2005." Report No. UCB/CITRIS-05/01, University of California, Berkeley, November 2.

Society of Professional Journalists (SPJ). 1996. "SPJ Code of Ethics." www.spj.org/ethicscode.asp (accessed June 15, 2010).

Sorell, Tom. 2002. "Morality and Emergency." *Proceedings of the Aristotelian Society* 103(1): 21–37.

World Health Organization. 2007. "Ethical Considerations in Developing a Public Health Response to Pandemic Influenza." WHO/CDS/EPR/GIP/2007.2. Geneva.

Zack, N. 2006. "Philosophy and Disaster." *Homeland Security Affairs* 2: 1–13.

# About the Editor and Contributors

**Richard W. Schwester,** PhD, is associate professor of Public Management at John Jay College of Criminal Justice (CUNY), where he holds a BA from Johns Hopkins University and an MA and PhD from Rutgers University. His research centers on the use of technology in government, inter-local shared services, and police and emergency service delivery models. Some of Dr. Schwester's most recent work appears in *Public Budgeting and Finance, Public Performance and Management Review, Public Administration Quarterly, International Journal of Public Administration, International Review of Public Administration,* and *the International Journal of Organization Theory and Behavior.* He is also the co-author of *Public Administration: An Introduction.* Dr. Schwester serves as associate director of the Academy for Critical Incident Analysis (ACIA) at John Jay.

**Ned Benton,** PhD, is chair of the Department of Public Management at John Jay College of Criminal Justice. His academic discipline is psychology, and he completed his clinical internship at Decatur Memorial Hospital in Decatur, IL. He then joined the National Clearinghouse for Criminal Justice Planning and Architecture in the Architecture Department of the University of Illinois and became its director of planning, directing consulting and master planning studies for prison systems. In 1975, he was appointed director of the Oklahoma Department of Corrections. In 1979, he accepted an appointment as a professor at John Jay College. In the 1980s, Dr. Benton wrote *Execucomp: Maximum Management with the New Computers,* which won a Library Journal "Best Business Books of 1983" Award. He currently serves as the monitor in *United States of America v. Commonwealth of Puerto Rico.*

**Dr. Divya Chandrasekhar** is an assistant professor in the Department of Urban Planning and Environmental Policy of the School of Public Affairs at Texas Southern University. She earned her PhD in regional planning from the University of Illinois at Urbana-Champaign in 2010. Her research focuses on the developmental aspects of long-term postdisaster recovery and other forms of environmental change.

**Sandro Cinti,** M.D., is an associate professor of internal medicine and infectious diseases at the University of Michigan (UM) Medical School and the Ann Arbor Veteran's Affairs Health System. He specializes in HIV/AIDS and biopreparedness. Dr. Cinti coordinates biopreparedness efforts at the UM Health System and the Ann Arbor VA. He is also a member of the national VA Influenza Advisory Board and serves on several state of Michigan public health committees.

**Dr. Jeanne-Marie Col** is associate professor of emergency management and public administration at John Jay College of the City University of New York. Before joining John Jay College in 2003, she served as United Nations senior interregional adviser in governance and public administration. While at the UN, she founded a global program for the integration of public administration and the science of disaster. Within the framework of disaster management, this program contributed best practice case study research, national pilot projects, and international conferences to the continuing issues of science policy and public administration action on science issues. Dr. Col has published articles in *Public Administration Review* and *Chinese Public Administration Review.* She continues her research on international and comparative disaster management.

**Roddrick Colvin,** PhD, is an associate professor in the Department of Public Management at John Jay College of Criminal Justice. His research interests include employment policy, gay rights, hate crimes, and international human rights policies. He currently teaches courses in public administration, human resources management, and policy analysis.

**Glenn P. Corbett,** P.E., is associate professor in the Department of Protection Management at John Jay College of Criminal Justice, and he serves as a member of the National Institute of Standards and Technology Investigation. Professor Corbett has extensive experience in different facets of fire protection including the fields of firefighting and fire prevention, as well as in-depth knowledge of the development and trends of the American fire protection profession.

**Meredith Dank,** PhD, is a research associate at the Urban Institute (UI) in Washington, DC. She is currently working on a number of federally funded projects to combat human trafficking. Dr. Dank received her PhD in criminology from the City University of New York Graduate Center in 2009. Her dissertation was on the commercial sexual exploitation of children in New York City and the role social capital plays within this population.

**Matthew Durak** is an undergraduate working toward his bachelor of science degree in computer engineering at Michigan State University. He expects to graduate in May 2012 and plans to attain a master's degree in computer science. Currently, he is working on research in agent-based modeling and evolutionary computation with his mentoring professor, Dr. Erik Goodman.

**Erik D. Goodman,** PhD, is professor of electrical and computer engineering and professor of mechanical engineering at Michigan State University. He received a B.S. (1966, mathematics) and an MS (1968, systems science) from MSU, and his PhD (1971, computer and communication sciences) from the University of Michigan. He has been working in evolutionary computation since 1970 and has also published extensively in ecosystem modeling and simulation and in manufacturing process simulation and verification. His current research includes automated design of dynamic systems using genetic programming and use of agent-based modeling to allow optimization of regulation of telecommunications and planning of emergency response/evacuation.

**Steven M. Gorelick** is a professor of media studies at Hunter College of the City University of New York. His PhD in sociology from the Graduate Center of New York included advanced specializations in both criminology and media studies. Dr. Gorelick is a senior fellow at Hunter College's Center on Health, Media, and Policy, and his current research focuses on the depiction of disease and catastrophic incidents in media and culture. He is currently at work on a book examining the changing role of self-appointed and media-anointed experts in the construction of social problems in media and culture.

**Norman Groner** is an associate professor in the Department of Protection Management at the John Jay College of Criminal Justice, the City University of New York. He has worked in the human factors field for twenty-five years, much of it in the area of cognitive factors related to fire safety, emergency planning, and security management. Dr. Groner has an MS and PhD in general psychology from the University of Washington.

**James Hawdon,** PhD, is a professor of sociology at Virginia Tech. He recently completed two National Science Foundation projects focused on the relationship between critical incidents and community solidarity. He has published dozens of articles and two books in the areas of crime, the sociology of drugs, and policing. His book, *Drug and Alcohol Use as Functions of Social Structure,* won the Adele Mellen Prize for Contributions to Scholarship.

**Alexander C. Henderson** is an assistant professor of health care and public administration at the C.W. Post Campus of Long Island University. His current research examines administrative discretion and influence among street-level bureaucrats in emergency medical services organizations. He received his PhD from the School of Public Affairs and Administration at Rutgers University-Newark in 2011. Prior to his doctoral studies he served as a chief administrative officer, operational officer, board member, and volunteer with several emergency services organizations in suburban Philadelphia.

**Amber M. Horning** is a PhD student in the Department of Criminal Justice at the Graduate Center, CUNY/ John Jay College of Criminal Justice. She holds an MA in forensic psychology from John Jay. She was the Archive Graduate Scholar for the Academy of Critical Incident Analysis (ACIA) from 2008–2009. Her current research interests include classification systems, narratives, and cross-national/cultural comparisons.

**Elizabeth A. Kirby** currently teaches at Central Michigan University's College of Education and Human Services in the Educational Leadership Department. She received an Ed.D. in educational leadership from Central Michigan University. Kirby teaches courses in organizational theory, culture, and change and serves as the director of three departmental degree programs for graduate students. Critical incident analysis, school leadership, and teen victimization provide the context for much of her research.

**Dr. Patricia Levy** teaches at Fort Hays State University. She has had extensive background in multicultural practice with Native American, African American, and a wide range of Middle Eastern and European populations both in Israel and the United States. Dr. Levy has also published in the United States and internationally in England, Australia, and Greece on mass trauma response to disaster and terrorism, and on social work and health with rural and aged populations.

**Ruth A. Maurer,** PhD from the Colorado School of Mines, is Professor Emerita of Mathematical and Computer Sciences at the Colorado School of Mines. Dr. Maurer has forty years of experience teaching mathematics, statistics, and operations research in the United States and internationally. She has served as Visiting Professor of Engineering, United States Military Academy (West Point); Visiting Professor of Mathematics, Luleå Technical University (Luleå, Sweden); Chair, Applied Management and Decision Sciences, Walden University. Dr. Maurer has further served as consultant on mining projects in the United States and Sweden, and as consulting energy economist to a large bank.

**Dr. Marian E. Mosser** currently works for the Department of Justice, Bureau of Alcohol, Tobacco, Firearms, and Explosives, as an intelligence research specialist for the Terrorist Explosives Device Analytical Center. She teaches part time at several major universities. Dr. Mosser received a master's degree in criminal justice from Youngstown State University; a master's degree in military arts and history from the Command & General Staff College, Leavenworth, KS; and a PhD in applied management and decision sciences from Walden University, Minneapolis, MN. In 2007, while working as the Region 4 Coordinator for the State of Florida Division of Emergency Management, she lectured at the Florida Emergency Management Academy and Florida Emergency Preparedness Academy. In 2009, she spoke at the DHS Higher Ed Summit at George Washington University on Law Enforcement Preparedness Against Terrorism and Information Sharing to Facilitate Situational Awareness and Improve Multi-Jurisdictional Response to a Crisis/Disaster.

**Johanna Nurmi** is preparing her PhD dissertation from the University of Turku, Finland, on the community level consequences of mass violence. She holds a master of social sciences degree from the University of Turku, Finland. She has carried out interviews of local residents and crisis workers in two Finnish communities that have experienced school shootings, Jokela and Kauhajoki. The focus of her research is on the collective identity and the dynamics of solidarity and conflict in the recovery process of the targeted communities.

**Frank Ochberg,** M.D., is a founding board member of the International Society for Traumatic Stress Studies and recipient of their highest honor, the Lifetime Achievement Award.

He edited the first text on treatment of post-traumatic stress disorder (PTSD), and served on the committee that defined PTSD. Ochberg founded and secured the funding for the Dart Center for Journalism and Trauma, served as its first chairman, and now is chairman emeritus of the center. He helps journalists understand traumatic stress, and he helps traumatic stress experts understand journalists. He was associate director of the National Institute of Mental Health and director of the Michigan Mental Health Department. At Michigan State University, he is clinical professor of psychiatry, formerly adjunct professor of criminal justice, and adjunct professor of journalism. Ochberg developed, with colleagues, the National Center for Critical Incident Analysis, Global Youth Connect (a young persons' human-rights organization), Gift from Within (a charity for persons with PTSD), and the Committee for Community Awareness and Protection (responding to serial-killer threats). For the latter activity, he is the first physician to receive the Law Enforcement Medal of the Sons of the American Revolution. As a Red Cross volunteer, Ochberg has helped families at sites of earthquakes, floods, fires, and aircraft disasters. He represents the Dart Foundation and directs their support of victimization programs around the world.

**Atte Oksanen,** Dr. Soc. Sci., is an adjunct professor of social psychology and works as a researcher in the University of Tampere, Finland. He is co-leading the research project Everyday Life and Insecurity: Social Relations after the Jokela and Kauhajoki School Shootings in Finland (Aaltonen Foundation 2009–2012) with Professor Pekka Räsänen.

**Pekka Räsänen** is a professor of economic sociology at the University of Turku, Finland. He has studied a variety of topics connecting with mass violence, culture, and consumer behavior. His research focuses on the ways that contemporary social life is affected by the new information and communication technologies.

**John Ryan** is a professor of sociology and chair of the Sociology Department at Virginia Tech, where he teaches the Sociology of Law. He received his PhD in sociology from Vanderbilt University. His research interests include the study of violence and crime control within communities. One line of research has been to examine the relative effectiveness of violence intervention programs that focus on structural issues (for example, enforcing housing codes) versus those that focus on cultural issues (for example, parenting or self-esteem classes). He has also researched dimensions of community solidarity and recovery from critical incidents, the effectiveness of community policing, and the relationship among grass-roots neighborhood anti-crime organizations, community policing, social disorganization and crime.

**Vincenzo A. Sainato** is an assistant professor of criminal justice at Loyola University, New Orleans. He earned his PhD in criminal justice from John Jay College of Criminal Justice at the CUNY Graduate Center in 2009. His research crosses a wide spectrum of subject areas, including economic development and crime, technology and criminal justice, police administration and governance, critical incident analysis, crime prevention, and cybercrime.

**Jeffrey D. Stern** leads the Policy Analysis Division at the Homeland Security Studies and Analysis Institute. A former battalion fire chief and emergency management professional in local government, he helped lead incident management teams to Hurricanes Charley (2004) and Katrina (2005) and supported medical operations in Haiti and the Dominican Republic in the aftermath of the Haitian earthquake (2010). He previously served in the federal government as Executive Director of the Homeland Security Advisory Council at the U.S. Department of Homeland Security, and as a White House Fellow. Stern holds an MPA (with honors) from the American University School of Public Affairs, has a B.A. in government from the College of William and Mary, and is currently a PhD candidate at Virginia Tech's Center for Public Administration and Policy.

**Robert C. Till,** PhD, P.E., is an associate professor in the Department of Protection Management at John Jay College. He is involved in teaching and research in fire science. His research interests include the use of Discrete Event Simulation, Agent Based Modeling, and the application of eye-tracking to simulate how humans may behave during emergency situations. Till is a Licensed Professional Engineer (PE) in the state of Connecticut and a special expert member of the National Fire Protection Association Standard for Fixed Guideway Transit and Passenger Rail Systems (NFPA 130) Committee.

# Name Index

*Italic* page references indicate illustrations or charts.

# Subject Index

*Italic* page references indicate illustrations and charts.